ECOLOGY

1926 Photo IDs

1 Anna M. Starr, Mount Holyoke College, South Hadley MA

2 Wladysaw Szafer, Botanical Institute, University of Krakow, Poland

3 Karel Domin, Charles University, Prague, Czechoslovakia

4 Arthur G. Tansley, Cambridge University, Cambridge MA

5 Homer Leroy Shantz, University of Arizona, Tucson AZ

6 Eduard August Rubel, Technical University, Zurich, Switzerland

7 Carl Johan Fredrik Skottsberg, Botanical Garden, Gothenburg, Sweden

8 Thomas William Woodhead, Technical College, Huddersfield, UK

9 Gustav Einar DuRitz, Plant Biological Institution, Uppsala, Sweden

10 Greta DuRitz, Plant Biological Institution, Uppsala, Sweden

11 Alexandre Borza, University of Cluj, Romania

12 Henry Chandler Cowles, University of Chicago, Chicago IL

13 George Damon Fuller, University of Chicago, Chicago IL

14 William L. Bray, Syracuse University, Syracuse NY

15 Eleonora E. Sagerstrom, Syracuse University, Syracuse NY

16 Mrs. Karel Domin, Prague, Czechoslovakia

17 James Willilam Toumey, Yale University, New Haven CT

18 Barrington Moore, Washington DC

19 Harriet George [Barclay], University of Chicago, Chicago IL

20 Frank Caleb Gates, Kansas Agricultural College, Manhattan KS

21 John B. Egerton, Calvert Hall College, Baltimore MD

22 Charles Christian Plitt, University of Maryland, Baltimore MD

23 John H. Ehlers, University of Michigan, Ann Arbor MI

24 Lena Bondurant Henderson, Rockford College, Rockford IL

25 Aravilla M. Taylor, Lake Erie College, Painesville OH

26 Norman Taylor, Brooklyn Botanic Garden, Brooklyn NY

27 Adolf E. Waller, Ohio State University, Columbus OH

28 Ethel M. Mygrant, University of Minnesota, Minneapolis MN

29 Helen J. Brown, Ohio State University, Columbus OH

30 Henry Teuscher, Morton Arboretum, Lisle IL

31 Marie McElreath, University of Chicago, Chicago IL

32 Mathildc H. Krcnz, Univcrsity of Illinois, Urbana IL

33 Carol Y. Mason, University of Illinois, Urbana IL

34 Laetitia Morris Snow, Wellesley College, Wellesley MA

35 Mrs. Theodore MacFarlane (Nellie Malura Cross) Knappen, George Washington University, Washington DC

36 Edgar Theodore Wherry, U.S. Department of Agriculture, Washington DC

37 George Elwood Nicholas, Yale University, New Haven CT

38 Henry Allan Gleason, New York Botanical Garden, New York NY

39 Elizabeth H. Cowles, University of Chicago, Chicago IL

40 William Skinner Cooper, University of Minnesota, Minneapolis MN

41 Meyer Halushka, John Marshall High School, Chicago IL

42 Clarence Ferdinand Korstian, U.S. Forest Service, Asheville NC

43 Arthur P. Kelley, Rutgers University, New Brunswick NJ

44 Herbert T. Vogt, Western State Normal College, Kalamazoo MI

45 Alvar Palmgren, Helsinki University, Helsinki, Finland

46 Rafael Zon, U.S. Forest Service, Washington DC

47 Millard S. Markle, Earlham College, Richmond IN

48 Samuel T. Dana, Northeastern Forest Experiment Station, Amherst MA

49 Charles Clemon Deam, State Forester, Bluffton IN

50 Selden Richarad Warner, Sam Houston State Teachers College, Huntsville TX

A Centennial History of the Ecological Society of America

Past Presidents at the 75th Anniversary Meeting, 1990. (*Standing*: Jane Lubchenco, Forest Stearns, Bruce Menge, Ron Pulliam, Larry Bliss, Francis Evans, Gwen Bliss, Bob Paine, possibly Mrs. Francis Evans (?), Dick Root, Mrs. Gene Likens, Gene Likens, Carole Levin, Simon Levin, Paul Risser, Margaret Davis, Frank Golley, Dwight Billings, Dennis Knight, Judy Knight. *Seated*: Mrs. Forest Stearns, Eugene Odum, Jean Langenheim, Hal Mooney, Rexford Daubenmire, possibly Mrs. Art Hasler (?), Art Hasler.)

A Centennial History of the Ecological Society of America

Frank N. Egerton

CRC Press
Taylor & Francis Group
Boca Raton London New York

CRC Press is an imprint of the
Taylor & Francis Group, an **informa** business

CRC Press
Taylor & Francis Group
6000 Broken Sound Parkway NW, Suite 300
Boca Raton, FL 33487-2742

First issued in paperback 2019

ISBN-13: 978-4987-0069-6 (hbk)
ISBN-13: 978-0-367-37763-2 (pbk)

Visit the Taylor & Francis Web site at
http://www.taylorandfrancis.com

and the CRC Press Web site at
http://www.crcpress.com

Dedication

In recognition of ESA ecologist–historians, particularly

Allee, Barrett, Beidleman, Billings, Blair, Bowers, Burgess, Clements, Coleman, Croker, F. Davis, Dexter, Fretwell, Frey, Golley, Gorham, Hasler, Hutchinson, Kaspari, Kendeigh, Kingsland, Lussenhop, McGinnnies, McIntosh, Mills, E. Odum, T. Park, P. Price, Sagarin, Simberloff, Slack, Wilson, Zedler,

and ESA ecologists who have published autobiographies, particularly

Beebe, Bormann, E. Clements, Colinvaux, Ehrlich, Gorham, Harshberger, Hutchinson, Keever, Lovejoy, Norse, Patrick, Pearse, Slobodkin, Sumner, Whittaker, Wilson, and Woodwell

Contents

List of Figures

ECOSKETCHES

Acknowledgments

I have discussed this history with more Ecological Society of America (ESA) members than I can name here, and I appreciate their comments even if not thanked individually. My deepest gratitude is to the readers of my entire manuscript, named in alphabetical order: David C. Coleman, Eville Gorham, Edward A. Johnson, and Eric A. Mills. Additionally, I thank R. P. (Mac) McIntosh, age 92, who critiqued Chapters 1 and 2, and for his other contributions to the history of ecology. Officers and staff of ESA have been helpful when I have had questions or requests, though most of my requests went to Katherine S. McCarter, executive director of ESA's headquarters in Washington, DC, who always came up with helpful responses. Robert L. Burgess deserves my posthumous acknowledgment for all his efforts on behalf of the history of ESA.

For the use of the ESA *Bulletin*, I thank my University of Wisconsin–Parkside colleague Greg Mayer for lending me his collection of paper copies, and I thank JSTOR for permission to use and copy from online issues before 1970. I also have locally available for my use a complete set of *Ecology*.

These living subjects of ecoSketches—Art Cooper, Paul Ehrlich, Gene Likens (who also checked the one for deceased Herbert Bormann), Jane Lubchenco, Robert Paine, Ed Wilson—kindly checked the sketch of himself or herself for accuracy. Living subjects who did not respond presumably found their ecoSketches to be accurate. Charles Hall commented on my ecoSketch of H. T. Odum.

Don Lintner, University of Wisconsin–Parkside photographer, has once again (as in all my histories) provided copy for the illustrations. Librarians from the University of Wisconsin–Parkside, especially Dina Kaye and David Gehring, have once again provided invaluable assistance. I thank Bill Mitsch for sending me a copy of *Ecological Modelling*, October 15, 2004, devoted to H. T. Odum's career. My good friend, Jerry Hershberger, has been very helpful in solving problems concerning my computer and computer files. My wife, Andrea, has again provided unfailing assistance and support.

Ecological Society of America History: Introduction

What should be the scope of an ecological society's centennial volume? In writing his book-length 75-year history of the British Ecological Society (BES), John Sheail (1987) described the general history of British ecology—not limited to BES history. Robert McIntosh (1976) described the history of American ecology since 1900 in 20 pages, and Robert Burgess provided "The Ecological Society of America: Historical Data and Some Preliminary Analyses" (1977). Sharon Kingsland (2005) discussed aspects of the history of American ecology, 1890–2000, in a brief book. Collectively, American limnologists wrote their own history in a spectacular volume (Frey 1963a) consisting of chapters on different geographic regions or on themes. George Lauff's chapter in that volume is "A History of the American Society of Limnology and Oceanography" (1963), which covers about three decades in 16 pages. Historian Joseph Kiger wrote a history of the American Society of Parasitologists in 10 pages (1962), and Norman Benson surveyed "The American Fisheries Society, 1920–1970" (1970b) in a dozen pages. None of these efforts provide a satisfactory example for me to follow.

There could be different kinds of ESA histories. My purpose has been to write an even-handed, accurate, and interesting history. There may be enough data available for a business–bureaucratic history, but I do not know how to write it and make it interesting. The ESA's geographical chapters and ecological sections (animal behavior, applied ecology, aquatic ecology, international affairs, paleo-ecology, physiological ecology, and others) are very meaningful for some members. A history of chapters and sections perhaps could be written by a committee of participants, but it begins after World War II. By the 1980s, there were regular columns in the *Bulletin* on technological tools, mostly computer software; perhaps an ecotech could write a history of that endeavor, but I suspect it would mainly interest other ecotechies. I am a historian, and for me satisfactory history tells stories. Others may lament stories I did not tell, but choices had to be made. For some periods of time the focus is on society activities, and for other periods the focus is on scientific achievements. There is no objective formula for these shifts, but each focus seems to work better for different periods of time. I began writing a history of ecological sciences in the *Bulletin* in January 2001, but for obvious reasons I do not discuss that series in this history.

Ecologists studying biotic communities or ecosystems are analyzing the most complex aspects of nature, and their science can, presumably, come to reflect that degree of complexity. However, only mathematicians and philosophers study complexity itself; it is too abstract for science. Like all sciences, ecology began with simple questions that led to fairly simple answers. As ecological sciences progressed, complexity emerged into both questions and answers. The ESA has documented that process in its publications since 1917.

Think of this as a conversation with earlier colleagues. If we do not know their stories, we cannot easily appreciate their work and so as a result may take their achievements for granted. Ernst Haeckel had named and defined "Oecologie" in 1866, and four ecological sciences began to become organized in the later 1800s: plant ecology, animal ecology, limnology, and marine biology. During the 1910s, British and American ecologists independently decided to organize, so that members could more easily interact with colleagues. Their efforts were quite successful, because their groups were initially small, and they understood from other scientific societies what to do and expect.

This history is organized into 6 chapters of 15 years each and a seventh chapter of 10 (or 11) years. The *Bulletin* and *Ecology* are the major sources for the first five chapters, the *Bulletin* is the main source for Chapter 6, and *Frontiers* is the main source for Chapter 7. Biographies of ecologists who are active in ESA are also important, and biographical sources published before 1996 were listed by Robert L. Burgess in "American Ecologists: A Biographical Bibliography" (1996). Burgess' bibliography assisted me in selecting 37 ecoSketches of prominent ESA members, which are included here, five per chapter (excepting six in Chapter 7), arranged chronologically by birth dates. Since there are obviously more than 36 "stars" in ESA history, this collection is a sample of the stars. EcoSketches tilt this history toward the stars, but the chapters include many less well-known members who helped run the society and meetings and published in the journals. This history of a scientific society, in that respect, differs from histories of particular sciences, which focus on a few scientists who made the most noteworthy achievements.

Some ecologists, reading this, may wonder: Why did he include *that*, which is less important than my study on *this*, which he omitted? The answer is that the author relied upon his limited knowledge and fallible judgment.

1 1914 to 1929

Origins

The earliest ecological institutions were aquatic research stations, mostly marine, which began to flourish in the 1870s in Europe and North America (Juday 1910; Kofoid 1910; Jack 1945; Hiatt 1954, 1963; Croker 2001:143–161; Egerton 2014a,b). Four ecological sciences began to emerge during the 1890s: plant ecology, animal ecology, limnology, and marine ecology. In 1899 three research stations were founded in America's Rocky Mountains by faculty members of three universities, and four other Rocky Mountain biological stations were founded by the year 1914 (Vetter 2011:111–119, 2012). The Carnegie Institution of Washington, DC, established a desert field station in 1903 on the (then) outskirts of Tucson, Arizona, and in 1908 a seashore Carmel Laboratory in California (McGinnies 1981; McIntosh 1983; Bowers 1988; Craig 2005; Egerton 2013:364). In 1908, a botanist at Lenox College, Iowa, Thomas H. Macbride, persuaded the University of Iowa Alumni Association to buy land for a biology field station on the shore of the West Branch of Lake Okoboji (Lannoo 2012:14–19), which opened in 1909. Macbride commented: "the factors of ecology and distribution are all here." It was and is a center for both summer courses and research, and during its first decade, almost 50 biological papers were published on research performed there (Lannoo 2012:30).

Next, ecological schools began to emerge at universities; three informal ones appeared in America by the early 1900s. At the University of Chicago, there was a terrestrial plant and animal school (Engel 1983:137–153; Mitman 1992:3–5; Cittadino 1993a; Cassidy 2007:25–39); at the University of Wisconsin, a limnological school gradually emerged (Frey 1963b; Beckel 1987; Egerton 1987); and at the University of Nebraska, a prairie school of plant ecology was developed (Tobey 1977, 1981:9–121)—all in the Midwest, where universities were still expanding, in contrast to the relative stasis in eastern universities. Early ecologists also tended to either come from the Midwest or study there, apparently indicating that they were recruited as undergraduates. Notable botany professors Charles Bessey (1845–1915) at Nebraska (Shor 1999), and John Coulter (1851–1928) at Chicago (Smocovitis 1999), were not ecologists, yet they encouraged students to study it.

In Britain, plant ecologists organized much earlier than the animal, marine, and limnological ecologists. In 1911, Arthur G. Tansley organized a four-week International Phytogeographical Excursion for August (Sheail 1987:32–35; Schulte-Fischdick & Shinn 1993:110–112; Ayres 2012). In 1913, British plant ecologists led by Tansley, organized the British Ecological Society, and consequently, its *Journal of Ecology* was virtually a plant ecology journal, though all ecologists were welcomed into the British Ecological Society (BES) (Sheail 1987:97).

Frederic Clements and Henry Cowles, with their wives, attended the British Excursion in 1911, and while there, Cowles agreed to host a similar one in the United

States and did so in the summer of 1913 (Tansley 1913–1914; Nichols 1914; Sears 1969:130–131; Beidleman 2009; Egerton 2013b:365). Both the British and American excursions were so successful that a third was planned for 1915 in Switzerland but was postponed until 1923 because of World War I (Harshberger 1924). In 1915, American ecologists organized the Ecological Society of America. Did the British example stimulate American action? Without contrary evidence, that is a reasonable assumption. However, there is contrary evidence. The initial suggestion came from professor of zoology Robert Wolcott (1868–1934) at the University of Nebraska, in Lincoln, whose interests ranged from insects to birds (Cattell & Brimhall 1920:520–521; Ward 1934; Burgess 1996:115). On March 27, 1914, he wrote to a colleague, Victor Shelford: "What do you think of such a thing as a society of ecologists to include both botanists and zoologists and to be a society for field work rather than a society for the reading of papers" (Croker 1991:120)? Shelford preserved Wolcott's letter and quoted the relevant part in *Ecology* (Shelford 1938:164), along with his response. Wolcott got the idea from reading Shelford's *Animal Communities in Temperate America as Illustrated in the Chicago Region* (1913); he then wanted to visit sites Shelford described. Wolcott envisioned spring or summer field trips to Midwestern places of ecological interest. At the time, Shelford was an instructor of zoology at the University of Chicago, and he took Wolcott's idea to the university's plant ecologist, Cowles, under whom he had taken a course. Cowles liked the general idea but thought a national organization would be better than a Midwestern one (Cassidy 2007:77–78). Zoologists Wolcott and Shelford may not have known about the founding of the British Ecological Society, but presumably Cowles did.

Cowles, a good organizer with an engaging personality, decided that the time for ecologists to gather and discuss Wolcott's proposal would be at the 1914 meeting of the American Association for the Advancement of Science (AAAS) in Philadelphia, Pennsylvania. On the evening of December 30th, Cowles gathered 22 ecologists in the lobby of the Hotel Walton, whose names Shelford recorded, and later repeated (1917, 1938:165): C. C. Adams, H. H. Bartlett, F. H. Blodgett, W. L. Bray, C. T. Brues, W. A. Cannon, Cowles, A. P. Dachnowski-Stokes, R. F. Griggs, J. W. Harshberger, A. F. Hill, O. E. Jennings, D. T. MacDougal, Z. P. Metcalf, G. E. Nichols, R. C. Osburn, A. S. Pearse, H. L. Shantz, Shelford, F. Shreve, N. Taylor, and Wolcott. All except Bartlett and Brues became charter members, and nine of them later served as ESA presidents (Burgess 1977:2). The assembled ecologists appointed an organizing committee that included Cowles, Shelford, and Wolcott, but also botanists John William Harshberger, who served as chairman, and Forrest Shreve. A year later, at the AAAS meeting in Columbus, Ohio, on December 28, 1915, at the Hotel Hartman, about 50 ecologists convened and organized ESA. Harshberger had about 50 letters from ecologists who were not present but who supported the plan. Dues were set at $1 a year. The ESA was later incorporated in the State of Wisconsin, supposedly in 1915, but perhaps later (Burgess 1977:2). Shelford—who became the "George Washington" of ESA—was elected as the first president (from the Midwest); Harvard Professor of Entomology William Morton Wheeler as vice president (from the East); and Shreve at the Carnegie Institution's Desert Laboratory, Tucson (from the West) as secretary–treasurer. In 1938, the secretary and treasurer would become separate offices.

Henry C(handler) Cowles (1869–1939)

(From *Ecology* 16, 281, 1935. With permission.)

Cowles, from Connecticut, earned his B.A. degree (1893) at Oberlin College and Ph.D. at the University of Chicago (1998). His doctoral dissertation was on plant succession in the Indiana Dunes along the south shore of Lake Michigan. It was indebted to Eugenius Warming's *Plantesamfund* (1895); he learned Danish in order to read it. From 1899 to 1901, he published his dissertation in three important articles (1901a,b,c), which are reprinted in Victor Cassidy's *Henry Chandler Cowles: Pioneer Ecologist* (2007:100–220). Cowles continued writing minor articles on plant ecology until 1933 (Cassidy 2007:221–264, 297–341). The July 1935 issue of *Ecology* was dedicated to Cowles in recognition of his contributions to the science (Cooper 1935). However, Cassidy judged (2007:52):

> The IPE [International Phytogeographic Excursion] in America was the pinnacle of Cowles' academic career. Soon after the scientists returned home, World War I broke out and made further excursions impossible. After this time, he produced no important research. While he always had abundant ideas and insights, he shared them with classes instead of publishing.

Cassidy lamented that Clements, not Cowles, wrote the great synthesis on plant ecology for the first generation of American ecologists. However, Cowles' special talent lay in personal interactions, and his training of a new generation of plant ecologists was a great contribution (Sears 1958; Cook 1980; Engel 1983:137–159; Hagen 1992:16–20; Mitman 1992:16–19; Cittadino 1993a; Burgess 1996:31–32; Erickson 1999a).

Although officers tended to be prominent ecologists, not all prominent ecologists became officers. For example, Clements dominated American plant ecology for half a century, and his magnum opus, *Plant Succession*, appeared in 1916. Yet, he was never an ESA officer, and at the 1916 ESA Annual Meeting in New York, the ESA appointed a chairman for a committee on succession, who was zoologist Charles Adams; he was instructed to find a botanical co-chair. This probably indicates that Clements was disinterested in co-chairing that committee, though he gave two talks at the meeting. Evelyn Hutchinson, at Yale, was the leading ecologist of the 1900s, more gregarious than Clements, yet he was never an officer either.

Shelford later remembered (1938:165):

> In March [1916] a special circular announced a field trip to Dismal Swamp, conducted by Professor Harshberger; to the Chicago dune area by Cowles, Shelford, and Fuller; a land and a sea trip near San Diego; and an excursion in the vicinity of Albuquerque, N. M., by A. O. Weese. All but the last were carried out.

The San Diego meeting was with the Western Society of Naturalists on August 10–14, 1916, and Forrest Shreve (1916) reported on it in *Science*. His report contains the abstracts for 10 of the speakers and titles of other speakers who did not provide abstracts. The ESA Archives in Georgia has a copy of the actual program provided to attendees. The meeting included two field trips, one to the Scripps Institution (as it was then called) and the other to the Cuyamaca Mountains and the edge of the Colorado Desert.

The December 1916 ESA meeting was at Barnard College, New York City, held jointly with the Botanical Society of America; there were two symposia, one on soil temperature and the other on climatic conditions. The ESA already had committees to study both topics. Shreve was not only writing the *Bulletin*, he was also chairman of the Soil Temperature Committee and reported that cooperation had been arranged among a number of men, not all in ESA, to measure soil temperature, and the committee sought standardization of measurements. Burton Livingston, chairman of the committee on climatic conditions, reported that progress was being made organizing the committee and more time was needed to formulate a report. His report in the February *Bulletin* was similar to his January report. Shreve reported that the American Phytopathological Society had initiated the soil temperature project, and members of both societies had joined to establish a series of 30 stations that installed thermographs. Some committee members had already installed their instruments and others would do so in the spring. Their goal was to collect records for at least two years.

Shreve had sent abstracts for the papers to be presented to members before the annual meeting; therefore, he did not publish abstracts in the *Bulletin*. The ESA Archives has a copy of the program. Furthermore, Shreve (or someone) sent the abstracts and news about the founding of ESA to Tansley, the editor of the BES *Journal of Ecology*, who printed the news and abstracts in that journal (ESA 1917). There were 36 abstracts, and there may have been other talks given without abstracts provided. The talks covered a wide range of topics. Four speakers—Clements, Gates, Lowe, Pearse—gave two talks and provided two abstracts. Speakers came mainly from the Midwest and Northeast; the West was represented by speakers from Texas, New Mexico, and Arizona; the South had no speaker; and Canada's only speaker was zoologist Archibald G. Huntsman, University of Toronto. The talks were about 10 to 15 minutes each, and the abstracts indicated the time needed.

Frederic E(dward) Clements (1874–1945)

(Courtesy of the Hunt Institute for Botanical Documentation.)

Clements was born in Lincoln, Nebraska, and he earned B.Sc., M.A., and Ph.D. degrees (1894–1898) at the University of Nebraska, where he taught from 1897 to 1907. He collaborated with Roscoe Pound on *The Phytogeography of Nebraska* (1898, edition 2, 1900, reprinted 1977), which served as their Ph.D. dissertations. In 1900, Clements and his wife, botanist Edith, established an Alpine Laboratory on Pikes Peak (Colorado), where they and others spent many summers doing research. Clements published *Research Methods in Ecology* (1905), the first such book ever published. During the decade 1907 to 1917, he served as chairman of the Department of Botany at the University of Minnesota. His successes received a setback when his *Plant Physiology and Ecology* (1907) was unfavorably reviewed by a plant physiologist. Henry Cowles developed ideas on plant succession for his doctoral dissertation, which he published in three articles (1899–1901). Clements enthusiastically accepted this concept and wrote his *magnum opus* on *Plant Succession* (1916). It strongly influenced plant ecology for four decades, though it was criticized for its over-reaching theoretical claims. This monograph also included a discussion of paleo-ecology, in which he coined that term as "paleo-ecology" (Clements 1916:279; Clements & Chaney 1924). In 1917, he became a research associate for the Carnegie Institution and remained one until his retirement in 1942. In American terrestrial ecology, the community concept was initially the domain of plant ecologists, with Clements taking the lead (Whittaker 1962:51–52). He provided a formal classification, with the formation being the basic unit and other designations within that. However, as with succession theory, some found his system too elaborate and rigid. Animals were initially placed in zoologist C. Hart Merriam's life zones (1890) based upon the temperature in the Western mountains (Grinnell 1935; Daubenmire 1938; Sterling 1977:212–236). Several animal ecologists considered the community problem (Whittaker 1962:62–63) before Clements and animal ecologist Victor Shelford collaborated on *Bio-Ecology* (1939). They received praise for expanding the community concept to include animals but were also attacked in Evelyn Hutchinson's book review (1940) for not going beyond descriptive ecology to include nascent studies on community dynamics. Clements is remembered in recent times for two related dogmatic sins: his concept of a biotic community as a superorganism and his claim that biotic communities always move in a given climate toward a single climatic climax. This is a case of the evil one goes on living while the good is interred with one's remains, and reinforced by memory of his rigid personality. Ed Wilson refers to an ant colony as a superorganism (Hölldobler & Wilson 2009), and I have offered some praise of Clements as a historian of ideas on the succession of plant communities (Egerton 2009). With a fresh look, others may also rediscover his other forgotten virtues (Schantz 1945; Tansley 1945; Pool 1954; Pound 1954; Clements 1960; Humphrey 1961:51–53; Ewan 1971; Sears 1973; Tobey 1977, 1981; Anon. 1994a; Burgess 1996:26–27; Hagen 1988, 1992: see index, 1999; Cittadino 1997). The April 1954 issue of *Ecology* became the Frederic E. Clements Memorial Issue, which included his bibliography and nine invited papers in his honor by Raymond Pool, Roscoe Pound, John Phillips, Vernon Young, John Weaver and W. E. Bruner, Victor Shelford, Homer Shantz, Ralph Chaney, and Charles Kendeigh.

Clements, in a talk entitled "The Development and Structure of Biotic Communities" (1917:120–121), coined the terms "biotic community" and "biome." These terms have persisted, but many others that he also coined, have not. He was even accused of naming a spade a "geotome," but that myth has been laid to rest (Taub 1995). Clements has also been credited with founding the first paradigm in ecology (Simberloff 1980a:13; McIntosh 1985:42–43) of the plant community as a superorganism, which he discussed in "America's first ecology book," *Research Methods in Ecology* (Clements 1905:199), in *Plant Succession* (1916:3), and later works. Paleo-ecology began in Europe, probably with Japetus Steenstrup's publication (1842) of his analysis of peat bogs in the Netherlands and Denmark (Egerton 2009:49). Fritz Rehbock (1983) thought paleo-ecology arose with Edward Forbes' articles around the same time, though Rehbock did not specify a particular article or date. Clements (1916:279) coined the term: "it is proposed to designate the study of past vegetation as paleo-ecology, or *palecology* in the contracted form." Possibly the earliest introduction of paleo-ecology to American ecologists was in the publications of Andrew Douglass. He called attention to the correlation of tree rings with climate in "Weather Cycles in the Growth of Big Trees" (*Monthly Weather Review*, June 1909, cited from Clements 1916:478). However, most ecologists learned of his investigations from his *Climatic Cycles and Tree Growth* (3 vols., 1919–1936) and "Evidence of Climatic Effects in the Annual Rings of Trees" in the first issue of *Ecology* (1920). Douglass believed that 11-year sunspot cycles influenced rainfall, and that this pattern could be detected in the annual rings of trees, which are broader in rainy than in dry seasons. By aligning overlapping tree-ring patterns in trees of different ages, he traced climate back 3,200 years in the arid Southwest. His goal was to determine climatic cycles, not past vegetation, but others could use his findings or technique as a means for correlating past vegetative composition with climate.

Earlier, Hart Merriam had led a small expedition from August 12 to September 27, 1889, to the San Francisco Mountain and adjacent areas at the Grand Canyon, Arizona, to explore the influence of temperature on the distribution of plant and animal species (Sterling 1977:212–219). In a published report on that survey (1890), Merriam developed his very influential life-zone theory, in which he identified seven different "life zones" of plants and animals, extending from Alpine (above 3500 m or 11,500 ft) at mountaintop down to Desert (1200–1800 m or 4000–6000 ft elevation), and he provided lists of plants and animals found in each zone. He then classified these Life Zones into larger Regions, and the Regions into larger Provinces, so that he could hypothetically extend his system to all of the United States. With thermometers, he could measure the temperature at different elevations, and with barometers he could determine elevations. However, during that arid time of year he could not measure, and so did not consider, possible precipitation influence upon plant and animal species distributions. His life-zone system achieved some popularity, but also criticism (Kendeigh 1932; Daubenmire 1938). The scope of Clements' biome and Merriam's life-zone projects differed, but both attempted to integrate plants and animals living in the same environment into a classification system.

John Harshberger (1869–1929) was a lifelong resident of Philadelphia. He earned his B.S. and Ph.D. degrees (1892, 1893) at the University of Pennsylvania, and then taught there for his entire career. He was one of the very first American ecologists to use the term "ecology" in the title of his articles (1900, 1902, 1903). He was active in ESA from the first organizing meeting in 1914 until his death and served as ESA vice president in 1917 and president in 1926. His interests were broad, and the bibliography in his autobiography (1928) contains more than 300 titles. His most notable publication was *Phytogeographic Survey of North America* (1911, lxiii +790 pages, 1 map, 18 plates). Harvard botanist Merritt Fernald (1873–1950) reviewed it in *Rhodora*, which he edited (on Fernald: Humphrey 1961:85–88). Reading Fernald's 11.5-page review must have been the most humiliating experience in Harshberger's life: "Had the author been more familiar with the painstaking work of these energetic students [mentioned above] and had he realized the need of emulating their accuracy and critical judgment the unfortunate volume which is now before us might never have been written" (Fernald 1911:214). (Eville Gorham [personal comm.]: Fernald was known for contemptuous reviews.) For the sake of his self-esteem, Harshberger could be thankful that not many ecologists read *Rhodora*. Norman Taylor's 10-page review in *Torreya* (1911) lacked Fernald's contempt but still had a long parade of errors and omissions. Harshberger's publishing career continued, and he even published a phytogeographical article in the first volume of the *Brooklyn Botanic Garden Memoirs* (1918). Since this was Taylor's institutional publication, he may even have solicited Harshberger's contribution to clear the air between them after his unfavorable review. Harshberger's obituary in *Ecology* was by Yale plant ecologist George Nichols (1930), who called his "Phytogeographic Survey of North America" article "monumental." (Since Nichols in his obituary mistakenly dated it to 1913, he perhaps relied upon memory when writing his obituary.) Publisher Hafner's decision to reprint Harshberger's *Phytogeographic Survey* in 1958 was based upon Nichols' evaluation of it (Burgess 1996:52; Hunt 1997; Erickson 1999b). No other botanist published a phytogeography of North America that might have met Fernald's approval (nor did he).

Three ESA committee chairmen were appointed at the December 1916 meeting but were not mentioned in print until the following February *Bulletin*: Arthur S. Pearse as chair of the Committee on Freshwater Fish and Fisheries with Special Reference to Food and Life Histories; Charles C. Adams as first co-chair of the Committee on Succession and the Interaction of Organisms in Communities Created; and Stephen A. Forbes as chair of the Committee to Formulate Ecological Needs of Economic Entomology (Croker 2001).

Forrest Shreve (1878–1950)

(From Billings et al., *ESAB* 32, 5, 1951. With permission.)

Shreve was a field ecologist who served as ESA's first secretary–treasurer (1916–1919) and its sixth president (1922). He was also the first editor of the

Bulletin (1916–1919), and editor of *Plant World* (1910–1919) before it was given to ESA to become *Ecology* in 1920 (Billings et al. 1951). Other than early studies of temperate plant communities in his native Maryland and then tropical plant communities in Jamaica, he spent his career at the Carnegie Institution's Desert Laboratory at Tucson, Arizona, where he was among the earliest ecologists to study arid plant communities (Burgess 1996:99–100). Shreve married plant physiologist Edith Bellamy (1909) whose scientific judgment he highly respected (Shantz 1951:365). He and A. E. Cameron ran the ESA Committee on Soil Temperature for a decade but never felt their data were worth publishing. Shreve also assisted in running *Botanical Abstracts* from its origin in 1917 until it merged with *Abstracts of Bacteriology* in 1926 to become *Biological Abstracts*. His bibliography covers 4 pages in Janice Bowers' *A Sense of Place: The Life and Work of Forrest Shreve* (1988). The ESA's Forrest Shreve Student Research Award is presented annually to one or two students for desert ecology research.

The choice of Shreve as ESA secretary–treasurer was a fortunate one. He had been editing the monthly journal *Plant World* since 1910 (Bowers 1988:31–32) and was able to apply that experience toward producing ESA's *Bulletin*, which he began as a very modest 3 pages each for the January and February issues in 1917. A *Handbook of the Ecological Society of America* was the third issue, in March 1917, an ambitious project with 56 pages (ESA 1917). It was primarily an annotated directory of ecologists, with an introduction (3 pages) by Shelford, who was president in 1916, ESA's Constitution (1 page), and a summary of three field trips sponsored by ESA in 1916 (2 pages). The directory begins with statistical information, in which the 307 members were grouped according to their major interests and locations, as well as the number of ESA members who were also in related professional societies, such as the Botanical Society of America, listed in a paragraph. (In comparison, the British Ecological Society reached only 250 members in 1932 [Figure 1.1] [McIntosh 1985:66].)

This annotated directory, compiled by or under Shreve, was slightly abridged from a questionnaire that each new member completed upon being elected a member. The questions were as follows (p. 8):

1. On what ecological subjects have you published papers?
2. In what ecological subjects have you a special interest or works in progress?
3. In what localities have you carried on ecological work?
4. With what regions are you slightly familiar?
5. In what taxonomic groups are you particularly interested (italics indicates willingness to identify material in that group)?
6. With what experimental methods have you had the most experience?
7. With what field instruments have you worked?

This directory was the first significant document that ESA published and the first thematic issue of the *Bulletin*. (The April 1918 *Bulletin* added the names of 40 members

MEMBERSHIP OF THE ECOLOGICAL SOCIETY GROUPED ACCORDING TO SUBJECTS OF MAJOR INTEREST

Plant ecology	88
Animal ecology	86
Forestry	43
Entomology	39
Marine ecology	14
Agriculture	12
Plant physiology	7
Plant pathology	4
Climatology	4
Geology	4
Animal parasitology	3
Soil physics	3
	307

Approximately 50 members of the society are members of the American Society of Zoologists, and 45 are members of the Botanical Society of America. About 70 members not included in the above groups are Fellows of the American Association, and nearly 100 others are members of the Entomological Society of America, the American Association of Economic Entomologists, the Society of American Foresters, the Association of American Geographers, the American Phytopathological Society, the American Society of Naturalists, and other scientific organizations.

GEOGRAPHICAL DISTRIBUTION OF THE MEMBERS OF THE ECOLOGICAL SOCIETY

Illinois	32	Montana	4
District of Columbia	30	New Jersey	4
New York	30	British Columbia	3
California	20	Indiana	3
Massachusetts	14	Nebraska	3
Minnesota	14	Quebec	3
Michigan	10	Texas	3
Pennsylvania	10	Hawaii	2
Colorado	9	North Dakota	2
Ohio	9	Philippine Islands	2
Ontario	9	South Carolina	2
Wisconsin	9	Vermont	2
Maryland	8	Alberta	1
Iowa	7	British Guiana	1
Kansas	7	Canal Zone	1
New Mexico	7	Florida	1
Arizona	6	Louisiana	1
Connecticut	6	Maine	1
Missouri	6	North Carolina	1
Washington	6	New Hampshire	1
Oregon	5	Sweden	1
Utah	5	Tennessee	1
Idaho	4	Wyoming	1
Total			307

7

FIGURE 1.1 Statistical summaries on ESA members. (From the ESA *Bulletin* 1(3), 7. With permission.)

who joined after the directory was printed.) Any ecologist who wanted to collaborate with another for an investigation could browse the directory and find colleagues whose interests matched those which the investigator sought. One could also check the background of ecologists whose work one had encountered in print or in lectures.

A conspicuous ESA member, who was not very active in ESA activities, was New Yorker (Charles) William Beebe (1877–1962), the best known and most widely read American naturalist of the first half of the 1900s (Welker 1975:75; Sterling 1997b). He wrote 24 books and 821 articles (Berra 1977). As a teenager, he enjoyed collecting and studying invertebrate animals along the shore of the Bay of Fundy, where his family had a summer home, but year-round, his first love was birds (Gould 2004:85–240). He studied zoology at Columbia University for three years and then obtained employment at the new Bronx Zoological Park, where he spent his career (Bridges 1974:63; Sterling 1997a, 1999; Gould 2004:54–55). Director of the Bronx Zoo, William T. Hornaday, founded the journal, *Zoologica*, in which Beebe published his findings. Its first article was Beebe's "Geographical Variation in Birds with Especial Reference to the Effects of Humidity" (1907). On an expedition to Pará, Brazil, he studied the "Fauna of Four Square Feet of Jungle Debris" (1916), in which he found: flatworms (Platyhelminthes) 2%, roundworms (Nemathelminthes) 2%, true worms (Vermes) 3%, and most abundant were ants, which included two new genera (identified by Wheeler). Back home, he compared that tropical fauna with similar sized plots in the New York Zoological Park and in Labrador. Yet, his entry in the first ESA *Handbook* (directory) only lists his interests in the ecology of birds (ESA 1917:11). In 1923, attracted by Darwin's *Beagle* account of the animals, he led an expedition to the Galapágos Islands. "But once in the archipelago the sea and its tenants forced themselves upon our attention and interest" (Beebe 1924:126). The land animals had somber coloring and often moved slowly, but marine species were often brightly colored and moved quickly. However, he also saw that "three-fifths of the birds of this island [Indefatigable] were wholly dependent on water life for food, while a number of others included fish and crustaceans as a part of their diet" (Beebe 1924:127). He became a pioneer in undersea exploration, as seen in his *Beneath Tropical Seas: A Record of Diving among the Coral Reefs of Haiti* (1928).

The *Bulletin* announced in February 1917 that there would be a meeting in April of the Pacific Division of the American Association of Western Societies of Naturalists, which would include an ecological session for which ESA members could submit abstracts to organizer Leonas Burlingame, a botanist at Stanford University. This second ESA summer meeting in the West with another biological society established a tradition, with Western meetings usually held in June. One California member, who may have been at that meeting, was zoologist Joseph Grinnell, who famously used the term "niche" in a paper on the California thrasher (1917), though he had previously used it in another publication (Grinnell & Swarth 1913; Cox 1979:99–100, 1980).

The most consequential early committee was on the Preservation of Natural Conditions (Kinchy 2006). In 1917, president Ellsworth Huntington appointed Shelford as its chairman, after Shelford's term as ESA president ended (Croker 1991:122–123). Shelford had soon recruited a committee with 25 members (and by 1921, 50 members). The May *Bulletin* urged all members to send information to Shelford on valuable undisturbed areas in danger. The March 1918 *Bulletin* announced that he had sent cards to all members on which to report areas they wanted preserved. All members were concerned about preservation of natural areas, so they responded. He also listed in the March 1918 *Bulletin* the names and regions of 19 committee members who had assumed responsibility for collecting information in specified states or Canadian providences. The committee identified some 600 areas preserved or deserving preservation in North America, which they wanted to ensure would remain natural for ecological studies and other purposes (Croker 1991:122–125). The committee initiated ESA's first lobbying effort with government officers, legislators, and the public. Foremost among the committee's priorities was preservation of natural areas in every national forest and national park. The January 1918 *Bulletin* announced that the committee was influential in securing preservation of Skokie Marsh, near Chicago, Illinois. Shelford had moved in 1914 from the University of Chicago to the University of Illinois, but he remembered that marsh and was undoubtedly active in securing it. In the March 1919 *Bulletin*, Shelford announced that citizens of Georgia had organized an Okefinokee Society to save that swamp from logging and drainage, and that those who wanted to help that effort should contact the Society's secretary, Dr. J. F. Wilson, Waycross, Georgia. The April 1920 *Ecology* (p. 137) issue carried discussions of the Okefenokee Society and Save the Redwood League. More generally, the October 1920 issue of *Ecology* (pp. 310–311) announced ESA's opposition to building dams, irrigation ditches, and making other commercial changes in national parks and monuments, and urged other professional societies to join this lobbying effort.

Victor Ernest Shelford (1877–1968)

(From *ESAB* 36, 116, 1955. With permission.)

Why do we consider Shelford the George Washington of the ESA? Not just because he was the intermediary among the three who initiated the founding of ESA, not only because he was its first president. He also wrote the book that prompted Robert Wolcott's suggestion for a society: *Animal Communities in Temperate America as Illustrated in the Chicago Region: A Study in Animal Ecology* (1913). His bibliography runs to 4 pages in Robert A. Croker's *Pioneer Ecologist: The Life and Work of Victor Ernest Shelford, 1877–1968*

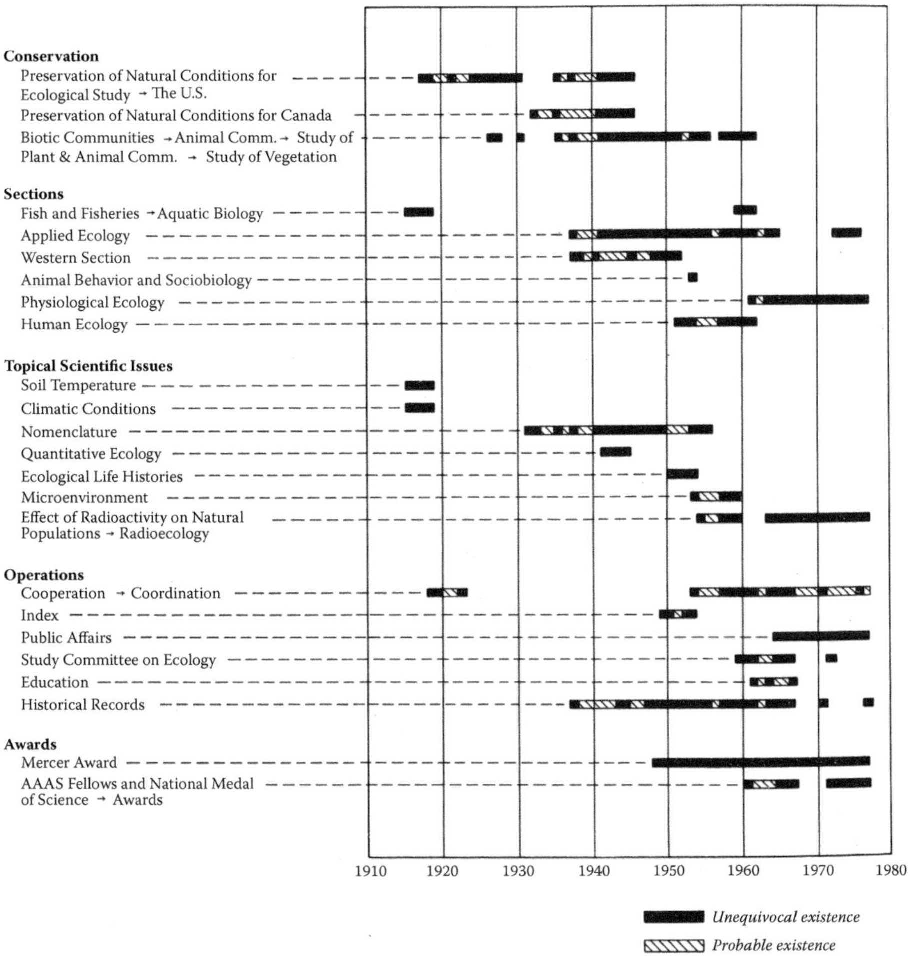

FIGURE 1.2 Selected committees of the Ecological Society of America, grouped by general categories and time spans. (From Burgess 1977:12.)

(1991). Shelford also suggested establishing a Committee on the Preservation of Natural Conditions and was its first chairman. That committee was among the most important ones ever established (Figure 1.2), and sponsored an important *Naturalist's Guide to the Americas* (Shelford & Shreve 1926). When ESA decided in 1945 to end that committee and not advocate for nature preservation, he wrote a personal check to establish a separate Ecologist Union, which later changed its name to the Nature Conservancy, which became the largest preservation organization in the world (Kimler 1990; Mitman 1992:38–45, 1997; Burgess 1996:99; Egerton 1999b; Birchard 2005).

At the December 1918 meeting, president Cowles appointed a Committee on Cooperation (*Bulletin* 3, no. 1, Jan. 1919, p. 2), with a general goal of listing problems on which work was needed among different phases of the work of ESA (whatever that means); and a specific goal to study the factors that were limiting the distribution of species on the mountains of the New England states. In retrospect, there seems to be an odd disconnect between the general and specific goals. Two members were appointed from plant ecology, Homer L. Shantz and Norman Taylor; two from forestry, George P. Burns and Barrington Moore; with representatives from animal ecology to be chosen later. The objective of the appointments was that committee members live close enough to be able to get together. Shantz did not remain on the committee (but is mentioned below); he was at the U.S. Department of Agriculture in Washington, perhaps too far away from northeastern states. Since the foresters had botanical training, it may not have seemed necessary to replace Shantz with another botanist. Two animal ecologists later became members, Charles Adams and Thomas Hankinson, though Hankinson in Illinois was no closer to New England than was Shantz.

This committee cooperated to produce an impressive three-part study on the "Plants and Animals of Mount Marcy, New York," published in the April, July, and October issues of *Ecology*, volume 1 (Adams et al. 1920). The committee set a good example on cooperation among ecologists. It was a pioneering study, for plant and animal ecologists studied together the ecology of a specific location. When Clements and Shelford collaborated to write *Bio-Ecology* (1939), they began with a historical introduction of 14 pages on previous ecological studies concerning both plants and animals of a community, 1877–1937, yet they overlooked this three-part Mount Marcy study—possibly because the discussion of animals was much briefer than of plants.

Mount Marcy, which rises 5344 feet in elevation and is the third highest peak east of the Mississippi River, is in the Adirondacks about 25 miles west of Lake Champlain. This team decided to focus upon the ecology of species at timberline. Part 1 was on the environment at timberline: topography, geology, climate, soil, and general aspects of plants and animals. It was illustrated with nine photographs of landscapes, vegetation, a bog lake, and three stations where data were collected. The data included four tables and six graphs, indicating general weather, evaporation, and solar radiation, and maximum and minimum temperatures during August. Parts 2 and 3 covered plants and animals. Part 2 had six vegetation–landscape photographs, and Part 3 featured a map, a graph, and three tables, all on tree growth. There was a discussion on animals in different environments, but it was less dominant and conspicuous than the vegetation discussion, and received proportionately less emphasis with no illustrations, graphs, or tables. This committee set a good example for other committees, and this three-part article was considered to be the second important document produced by ESA.

Illinoian Charles Adams (1873–1955) earned a B.S. degree from Illinois Wesleyan (1895), an M.S. degree from Harvard (1899), and a Ph.D. from Chicago (1908) in zoology and animal ecology, and served as both a faculty member and museum director at several places, before finally settling at Albany (Palmer 1956; Mitman 1992:36–37; Collins 1994a; Burgess 1996:10; Harmond 1997a). In addition to ecology, his early biological interests included evolution and biogeography (Ilerbaig 1999:446–452). His *Guide to the Study of Animal Ecology* (1913, 1977) was the earliest published organization of animal ecology, which also featured helpful bibliographies. He initiated the establishment of the Roosevelt Wild Life Forest Preserve and its Experiment Station, where the joint Mount Marcy study was conducted (1919–1920), in which he participated and coauthored the report (Adams et al. 1920). He was the seventh ESA president, in 1923.

Norman Taylor recalled (1938) that in June 1919, when he and other committee members ran their ecological survey on Mount Marcy, they agreed that ESA needed a journal. A journal requires financial backing, and the other surveyors suggested that Taylor appeal to the Brooklyn Botanic Garden, where he worked, for assistance. The Garden agreed and *Ecology* was launched in 1920, with Barrington Moore (1883–1966) as editor (Dana 1967; Burgess 1996:78). (Moore was also ESA president in 1919 and 1920—the only president who served two terms.)

Taylor's note triggered a recollection from Moore (1938), which was that the owners of the journal *Plant World* had given it to ESA so it could become ESA's *Ecology*, using *Plant World*'s subscribers list and printing arrangements. Moore described this as an "extremely generous action of the owners of the *Plant World*, those who put their hard cash in to the *Plant World* and their devotion to building it up" since "they relinquished the opportunity of ever getting back the money they had invested in that publication…" This warm memory does not accord with the facts. *Plant World* had been founded in 1897, and its editor before Shreve had been Professor Volney Morgan Spalding (1849–1918), who had retired from the University of Michigan to the Desert Laboratory, where he researched the distributions of several plant species while also editing *Plant World* (Burgess 1996:100–101). When his health declined in 1909, he relinquished his research and editorship and moved into a sanitarium in Loma Linda, California (Bowers 1988:31). *Plant World* then struggled along until its editorship fell into Shreve's lap. Its owners, the Plant World Association, consisted of 15 men, 9 of whom were charter members of ESA (Burgess 1977:17). In late 1919, Daniel Trembly MacDougal, director of the Desert Laboratory, offered *Plant World* to ESA. The assistance by the Brooklyn Botanical Garden, which Taylor had arranged, was for the business management of *Ecology* for three years, which began with the first issue in 1920, but this was not explained until the April 1921 issue (p. 157) by Committee on Publication chairman Moore. Dr. Stuart Gager (1873–1943), who was the director of the Brooklyn Botanic Garden, served as the business manager of *Ecology* from its first issue until his death in August 1943 (Svenson 1944). The Brooklyn Botanic Garden continued publishing *Ecology* through volume 29, number 1, January 1948, after which Duke University Press began publishing it.

The first issue of *Ecology* appeared in January 1920, and its first article was Moore's presidential address, delivered December 31, 1919. In 1919, Shreve had resigned from both the secretary–treasury position and editor of the *Bulletin*, and his editorship of *Plant World* ended because of its transfer to ESA. To reward him for his service, the ESA members elected him president for 1922. While adjusting to the changes, ESA was unable to publish the *Bulletin* during 1920, 1921, and most of 1922. During that period, Society news appeared in *Ecology*. From its first issue, *Ecology* was a fine, diverse, professional journal. It included animal, plant, and human ecology. Coverage of human ecology gradually disappeared from *Ecology*, probably from a lack of articles about it submitted for publication (Cittadino 1993b). The subject either migrated to anthropology or was revived by anthropologists, as Bernard Campbell's *Human Ecology* (1983, edition 2, 1995) attests.

The first death notice of a member was for Charles A. Hart (1859–1918), in the *Bulletin* (volume 2, number 4, p. 8), without an obituary, though three obituaries for him appeared elsewhere (Burgess 1996:52). The first ESA obituary was for marine zoologist Ellis L. Michael (1881–1920), written by the director of the Scripps Institution for Biological Research (as then named), William Ritter (1921), in *Ecology*, since the *Bulletin* was not published in 1921. The next issue of *Ecology* (April 1921) published an article that Michael had been asked to write on "Problems of Marine Ecology." He died before its completion, so it was finished by Winfred E. Allen. The problems they outlined concerned environmental conditions and organisms. The environmental conditions they discussed related to any oceanic inhabitants, but the problems they discussed were limited to plankton.

A persistent challenge in the history of ecology is the classification of vegetation. Connecticut native George Nichols (1882–1939), a Yale man in education (B.A. 1904, Ph.D. 1909) and career, was a founding member of ESA and served as both ESA vice president (1920) and president (1932); he offered "The Interpretation and Application of Certain Terms and Concepts in the Ecological Classification of Plant Communities" (1917). He was optimistic because he was building upon the work of Cowles, and he thought his system had the virtue of simplicity. Yet, if it was so simple, why did it take a two-part article of 28 pages? The units to be classified were associations, within formations. He based his system of formations upon succession, and he identified three causes of succession—climatic, edaphic, and biotic—but then these causes could be modified by topography. At the end, he did not provide a comprehensive outline for all the associations within all the formations, but he did provide an example of how his system worked, in an outline on "The Vegetation of Northern Cape Breton" (Nichols 1917:350–351). That outline was based upon his pending monograph, "The Vegetation of Northern Cape Breton Island, Nova Scotia" (1918). He was also an authority on the ecology of bryophytes (Olmsted 1941; Clark 1994; Burgess 1996:81).

Some plant ecologists thought what Nichols had attempted was a step in the right direction, but maybe his system was simple because it was simplistic. In 1923 he tried again, in another two-part article: "A Working Basis for the Ecological Classification of Plant Communities." Eight other ecologists critiqued his manuscript before he submitted it to *Ecology*. This 39-page article (versus 28 pages in 1917) reflected the critiques of his manuscript by his readers, and was undoubtedly a stronger system. His closing example still included Cape Breton vegetation, and his outline had expanded from 31 lines to 35 lines (Nichols 1923:177).

Other plant ecologists might have, on practical grounds, used Nichols' classification. However, in 1917 Henry Gleason had also published "The Structure and Development of the Plant Association," which contained a different view of the plant association. Gleason was one of the ecologists who had critiqued Nichols' 1923 article before publication, but obviously Nichols' 1923 paper did not satisfy Gleason, because three years later he published his now-famous paper, "The Individualistic Concept of the Plant Association" (1926), which was an alternative to Nichols' classification. Gleason had five ecologists critique his paper prior to its publication (1926:26), but Nichols was not one of them.

H(enry) A(llan) Gleason (1882–1975)

(From *ESAB* 34(2), 1953, cover. With permission.)

Illinoisan Gleason earned B.S. and M.S. degrees at the University of Illinois (1901, 1904) and a Ph.D. at Columbia University (1906). Before leaving Illinois, he had found evidence that challenged Clements' claim that succession always goes in one direction to one climax in a given climate (Gleason 1953:40). Gleason then taught at the universities of Illinois (1906–1910) and Michigan (1910–1919) before spending the rest of his career at the New York Botanical Garden. For much of his ecological career, he was a prophet crying in the wilderness, but he lived long enough to receive ESA special recognition as a distinguished ecologist (1953) and became Eminent Ecologist (1959) (Cain 1959; McIntosh 1975:253). The difficulty was that he challenged the dogma of Clements' dominant plant ecology theory of climatic climaxes. Gleason failed to find a botanical journal that would publish his "Vegetational History of the Middle West," so he had it published (1922) by the Association of American Geographers. However, the Torrey Botanical Club did publish both his "Some Applications of the

Quadrat Method" (1920), and his seminal "Individualistic Concept of the Plant Association" (1926). Clements had collected enormous amounts of evidence to support his theory, and Gleason's own accumulation of detailed evidence was in plant taxonomy, not ecology (Gleason 1953, 1975; McGuire 1975; Stuckey 1981; Nicolson 1990; Hagen 1992:28–31; Cohn 1994; Nicolson & McIntosh 2002; Kingsland 2005). He was one of 50 botanists awarded the Certificate of Merit by a jubilee committee of the Botanical Society of America "for his work on tropical and temperate floras of America and for the ideas and inspiration which he has supplied to the field of systematic botany" (Meyer 1958:17). John Curtis and his students developed Gleason's ideas in their studies on the vegetation of Wisconsin.

In the April 1918 *Bulletin*, Preservation Committee member Professor John Davidson, University of British Columbia, suggested that committee members write descriptive articles on areas proposed for preservation. The September 1919 *Bulletin* announced that the committee during two years had compiled a list of natural areas to save and wanted to publish it with information on three topics for each area: (1) plant communities with map, (2) relation of preserves to agriculture, and (3) relation of preserves to forestry—what methods of logging cause the least disturbance?

A concern related to saving areas was saving species living in those areas. W. G. Van Name in the January 1921 issue of *Ecology* (pp. 76–78) warned about the "Threatened Extinction of the Bald Eagle," from hunting. He reminded readers that the bald eagle is our national emblem. Despite that symbolism, the Alaska Legislature in 1917 had voted a bounty on eagles. Van Name cited protests already published in several popular natural history journals and urged ESA members to actively support the call for a federal law to protect bald eagles. However, it was not until 1940 that Congress finally enacted a $500 penalty for shooting our national symbol (Hogner 1958:127–128).

An early success of the Committee for the Preservation of Natural Conditions was the lobbying for Glacier Bay to become a national monument. In 1914, plant ecologist William S. Cooper went to Glacier Bay to judge its suitability for studying glacial influences upon plant succession, following up on a suggestion from Dr. Lawrence Martin of the U.S. State Department. Cooper subsequently read John Muir's *Travels in Alaska* (1915) and studies of the glaciers by the U.S. Geological Survey. He began his research in the summer of 1916, and in 1923 he published his findings in a three-part study in *Ecology*, "The Recent Ecological History of Glacier Bay, Alaska" (Cooper 1923). The ESA Committee's ability to persuade the federal government to declare the Glacier Bay region a National Monument in 1925 (and a National Park and Preserve in 1980) was dependent upon Cooper's study that provided evidence of the ecological importance of the Glacier Bay region. His study was actually the first of four, the others were published in 1929, 1935, and 1939 (Cooper 1931; Bohn 1967:80–83; Rumore 2009, 2012). The success of the Committee's lobbying in 1925 was not, however, the start of a string of successes, because it was unable to point to such impressive evidence for other valued sites.

William Skinner Cooper (1884–1978)

(From Buell, *ESAB* 44, 108, 1963. With permission.)

As a child in Detroit, Cooper fell in love with flowers, so his mother bought him a copy of Asa Gray's *Manual of Botany*. Before college, he had a herbarium of 3,000 specimens and traded with collectors nationwide. At Alma College in Michigan he studied under Cowles' student, Edgar Transeau (1875–1960; Burgess 1996:106), who "influenced my scientific development more than any other person" (Cooper 1994:130). In graduate school at Johns Hopkins University, he became interested in the relationship between glacial geology and vegetation. This interest arose during two summers of study in Colorado's Rocky Mountains. Cooper's first publication was "Alpine Vegetation in the Vicinity of Long's Peak, Colorado" (1908). He transferred to the University of Chicago, where he discovered that he had already learned Cowles' teachings from Transeau. Instead, "The greatest event at the University of Chicago was contact with Rollin D. Salisbury. Besides his influence upon my geological thinking, he was the greatest teacher I have ever studied under." Cooper's doctoral dissertation was on "The Climax Forest of Isle Royale and Its Development" (1913), much of which he published in six articles, 1912–1914. He spent two years teaching ecology at Stanford University (1914–1915) before settling at the University of Minnesota. He supervised the Ph.D. research of 13 graduate students. In 1916, he made the first of five trips to Glacier Bay, Alaska (also 1921, 1929, 1935), where his dual interests in ecology and geology merged in his studies on the vegetation of glaciated soils. In 1926, he revisited Isle Royale and found several locations where he had photographed vegetation during 1909–1910. He re-photographed the vegetation from the same locations, and related the changes that had occurred during

17 years in *Ecology* (Cooper 1928). From 1923 to 1942, many of his published papers described his research at Glacier Bay. His research results were the main evidence for the ecological significance of the region, which led to its becoming a national monument in 1925 and a national park in 1980 (Bohn 1967:80–83; Ramsay 1984). He served as ESA vice president in 1927 and president in 1936. He was one of 50 botanists awarded a Certificate of Merit by a jubilee committee of the Botanical Society of America, for "being one of the creators of an American tradition in ecology. His deep feeling for the relatedness and parallel developments of geology, physiology, taxonomy, and vegetation science has been a guiding light to a whole generation" (Meyer 1958:16; McCormick 1994a; Burgess 1996:29–30; Pickett 2007). He was ESA's Eminent Ecologist for 1963 (Buell 1963).

Shelford offered his committee a challenge to describe all the natural areas of North America, from Canada and Alaska south to northern South America and including the Caribbean Islands and some Pacific islands. The Preservation Committee published *Preservation of Natural Conditions* (32 pages), a guide to focus its activities, and the third important document from ESA, with 7 pages of photographs and 2 pages of diagrams, listing the names of 27 committee members, but edited by Shelford. (The cover is dated 1921, but the page with the printer's name is dated 1922.) Barrington Moore spread the word about preservation of natural areas for ecological research to upper-class hunters in a Boone and Crockett Club book, *Hunting and Conservation* (1925).

Shelford found about 75 ESA members who were willing to undertake a descriptive project for natural areas. It was quite an achievement for both Shelford (senior editor) and ESA to complete a fine *Naturalist's Guide to the Americas* (xv + 761 pages, 1926), with Shreve as publication editor and seven associate editors of selected topics. It included useful maps and references and was the fourth important ESA document. It would have been helpful if the table of contents had listed the 16 maps, because their appearance depended on finding someone to provide one, but since the maps were few, one could not know until one went looking whether a particular account included a map. For example, for Canada there were only maps for British Columbia and Ontario, and for the United States, only for Minnesota, Ohio, and West Virginia. The only photograph included was a frontispiece of Long's Peak, Rocky Mountain National Park, Colorado. The *Guide*'s price was $10, but $6 for contributors.

After this *Guide* appeared, whenever ecologists published studies on species in a particular area, readers could place the species within a clear context of ecological information. Yet, the Committee on the Preservation of Natural Areas was by no means ready to disband. At the time, its most recent chairman, Warren Waterman, published a 3-page report in the April 1926 issue of *Ecology* that was as ambitious as any others. He thought that Canada should have its own subcommittee and subcommittee chairman. Shelford and some other ecologists aspired to a second volume. For that volume, the June 1926 *Bulletin* reported that $169 was contributed or pledged. A second volume never appeared, perhaps a casualty of the Depression. Or maybe it did appear, in the modified form of Shelford's last book, *The Ecology of North*

America (1963), which covered mostly the same ground as the *Naturalist's Guide*, though with new maps, photographs, and references—a climax of his life's work (Croker 1991:156).

Calls to action for preservation of natural areas were not limited to Preservation Committee members, and in 1920 some calls appeared in *Ecology*. An unsigned commentary on "The National Parks" (pp. 310–311) protested a Federal Water Power Bill that Congress had passed, which empowered a commission to grant licenses to build reservoirs, irrigation ditches, and power lines and plants within national parks and monuments. There was no mention of the Hetch Hetchy Dam for Yosemite National Park that had been approved in 1913, after a long political fight (Nash 1982:161–181; Worster 2008:418–439, 448–453), though it had made little progress by 1920. Utilitarian projects for national parks and monuments were difficult problems politically. For example, when Rocky Mountain National Park was created in 1915, it already had 18 reservoirs and irrigation ditches; and in 1933 when a tunnel was proposed to carry Grand Lake water from the western side of the park 13 miles through a tunnel to the eastern side, both the pump station and its exit were just outside the park (Buchholtz 1983:189–192 and map, p. 2). However, in 1955, when a dam was proposed for Echo Park in Dinosaur National Monument, the preservationist lobby prevailed (Stegner 1955; Nash 1982:209–219).

A service performed by *Ecology* for five years was to carry zoological abstracts of ecologically relevant articles. Zoology Professor Lee R. Dice, University of Michigan, organized the project. These abstracts began in the July 1921 issue, occupying 2 pages (230–231). The abstracts were written by various animal ecologists, though not abstracts of articles they had themselves written. In later issues of *Ecology*, the number of pages of abstracts gradually increased. The reason botanical abstracts were not published was that they were already being published in *Botanical Abstracts*. The last "Zoological Abstracts" was 9 pages in the last issue of *Ecology*, volume 6 (1925, 460–468). It ended because *Biological Abstracts* began appearing in 1926.

The Fourth International Congress of Plant Sciences met at Cornell University, August 16 to 23, 1926 (Figure 1.3). One of the official photographs from the Congress is of 50 ecologists, 39 of whom were Americans. At that Congress, Nichols attacked Gleason's individualistic concept (Nichols 1929). He allowed Gleason to read his lecture before he delivered it, and afterward Gleason spoke on "Plant Associations and Their Classification: A Reply to Dr. Nichols" (Gleason 1929). Gleason's biographer has a detailed discussion of Nichols and Gleason's arguments at that meeting (Nicolson 1990:139–143). Two volumes of *Proceedings* were edited by Benjamin Duggar, at the University of Wisconsin, and boxes of papers concerning the *Proceedings* are now at the University of Wisconsin Herbarium (Burgess 1981b:203).

When Robert Whittaker published his comprehensive "Classification of Natural Communities," his modest goal was an "inquiry into the theory, the general meaning and underlying problems, of the classification of natural communities" (1962:2). His monumental study began with European classifications, which began earlier than in America, and continued simultaneously with American attempts. Whittaker described both the Nichols and the Gleason systems in neutral terms, in the same paragraph (1962:53). After his vast survey, however, he acknowledged: "Diverse

FIGURE 1.3 Ecologists at the Fourth International Congress of Plant Sciences, 1926. (From *ESAB* 62, 204. With permission.) Their identities and locations are provided in the front of this volume.

approaches to classification may be equally justified, and may to some extent complement one another" (Whittaker 1962:159).

Scientific societies encourage the teaching of their subjects. Courses in ecology very gradually moved from the fringes of offerings to mainstream. Robert Burgess (1990:15–19) surveyed the history of ESA efforts to address education, and his account is summarized here. He found that ESA efforts were sporadic but increasing in frequency as time went by. In 1917, Stephen Forbes was appointed chairman of a Committee on Economic Entomology, which was to encourage an ecological perspective in teaching and research. At the Cambridge ESA meeting, 1922, plant ecologist Lawrence Palmer of Cornell spoke on "Ecology through Nature Study in the Elementary Schools," about the Cornell Rural School Leaflets, 120,000 of which were distributed free in over 15,000 rural schools. By April 1924, ESA's *Bulletin* listed summer courses taught at field stations across America. At the 1935 ESA Annual Meeting, a resolution was passed to urge educational institutions that had not already done so to offer courses in plant and animal ecology in order to equip students with knowledge for administering natural resources. In addition to Nebraska and Chicago before World War I, Duke, Georgia, and Wisconsin universities began training ecologists before World War II.

An important function that professional society journals provide is the reviewing of books. Such reviews influence what members read and what academically employed members select as textbooks. (Reviews also provide a window into ESA history.) The first review published in *Ecology* was in the second issue of volume 2 (April 1921, pp. 145–149), by Adolph E(dward) Waller (1892–1975), on two monographs by Clements' former student, John E(rnst) Weaver (1884–1966) on root development. (Weaver had been the plant ecologist at the University of Nebraska since 1915: Dale 1967; Laycock 1994; Burgess 1996:109.) Many textbooks were reviewed in *Ecology*. Zoologist Arthur Pearse (1877–1956) wrote an early textbook, *Animal Ecology* (1926, x + 417 pages), and botanist Walter McDougall (1883–1980) published *Plant Ecology* (1927, vi + 326 pages). Although a reviewer referred to Pearse's book as the first textbook on animal ecology, it had been preceded by American Charles Adams' bibliographical *Guide to the Study of Animal Ecology* (1913) and Englishman Lancelot A. Borradaile's *The Animal and Its Environment* (1923). Borradaile's book was very well illustrated but did not contain the word "ecology." Thus reflecting the fact that terrestrial animal ecology was at the time little studied in Britain—somewhat at Rothamsted Experimental Station—and Charles Elton was beginning at Oxford—and so Borradaile found no critical mass of animal ecologist colleagues with whom to interact.

George T. Jones of Oberlin College wrote *Ecology*'s review of Pearse's text, and although praising it as the first such book, nevertheless complained that "it will not serve as a text, because the organization, the classification of communities, shows almost no influence of the recent advances in ecology" (1926:503). Jones then provided details:

> Pearse's fundamental errors in his classification are three in number: first, an emphasis on habitat without reference to the facts of actual distribution; second, a shifting point of view; and third, the entire omission of the idea of succession from consideration.... After a long and unsatisfactory attempt to classify animal communities on the basis of physiological reaction, it has become clear that the only successful way to determine the limits of animal communities is by the use of data on the distribution of the animals themselves.

Jones cited Shelford and E. D. Towler's study on "Animal Communities on San Juan Channel and Adjacent Areas" (1925) to illustrate his point. Pearse might have responded, if only to himself, that his book was already in press before he knew of their study. Pearse was a founding member of ESA and was president in 1925. By the time he published his autobiography, it seems that long before, he ceased attending ESA meetings; his only mention of ESA was in a list of societies of which he had been president. However, his text survived Jones' criticisms, for he reported that it "has been used for twenty-five years" (1952:67)—this despite Englishman Charles Elton's rival, the very popular *Animal Ecology* (1927).

Thomas Park, in his history of animal ecology, 1900–1940 (1949:57), compared Pearse and Elton's textbooks. Pearse "laid a general background of physical and biotic factors and then classified animals ecologically according to their major habitats. The treatment was primarily descriptive." Elton, on the other hand,

> was concerned more with organizing ecology around principles, and most of his principles centered around the animal community and the natural population. Unlike Pearse, he was interested, not so much in whether an animal was found in a desert or a lake, but rather in the environmental factors limiting the distribution of such a form. Elton stressed also the quantitative aspects, particularly in connection with the number

> of animals that occupy any community and the impact that these numbers make on their total environment. He viewed food chains as the most important integrating factor of the community, and his treatment of this subject is outstanding.

In other words, Elton's book was much more theoretical than Pearse's book. For that time, it seems fair to see these differences as a reflection of differences in educational focus of the two countries, with America's focus being more empirical and descriptive and England's being more theoretical. A stock example has been the niche: Joseph Grinnell's concept (1917) emphasized place, and Elton's concept (1927:63–68) emphasized function. However, Thomas Schoener (1989:79–80) has spoiled this simplistic contrast by reviewing five Grinnell publications concerning niche, 1913–1928, and he judged the differences between Grinnell's and Elton's concepts are rather slight, and that the significant contrast is rather between Grinnell and Elton's versus Hutchinson's (1957).

George Fuller from the University of Chicago, reviewed McDougall's text in a half-page review, which was full of praise. However, Fuller recommended it for "junior colleges, normal schools, and possibly in some high school classes" (1928:99). McDougall might have wondered why, with such praise, Fuller did not recommend it for undergraduate college courses. Be that as it may, two years later plant ecologists had a choice for their courses between McDougall's text and another, 200 pages longer, by Weaver and Clements, using the same title, *Plant Ecology* (1929, xx + 520 pages). When McDougall saw Arthur Vestal's 7-page review of Weaver and Clements' book, he must have felt slighted when comparing it to Fuller's review of his own *Plant Ecology*. Retrospectively, Ronald Tobey (1981:143) called Weaver and Clements's book "the leading textbook for the specialty."

Since Weaver and Clements had both published previous books separately, Vestal's lengthy review seems justified. Vestal cited five reviews of it already published (including one by Fuller of over 2 pages) and justified his lengthy review by "the year which has elapsed since its publication has afforded the opportunity of very detailed study of its content" (Vestal 1931:232). Vestal goes through all the chapters, explaining their pros and cons, sometimes with alternative suggestions. For Weaver and Clements' bibliography with 611 publications, Vestal had only praise, and his conclusion on the book was that it "is indispensable to botanists and all ecologists, and will prove useful to geographers, foresters, agronomists, grazing workers, and landscape architects."

The ESA journals contained many discussions of instruments and methods of research. This enabled members to stay abreast of innovations and also encouraged standardization, so that results of different ecologists could be compared. This effort began with the first two committees established by the ESA, on soil temperatures and on climatic conditions, both of which are discussed above for their reports in the *Bulletin* for 1916 and later. Both committees depended upon a number of observers to use similar equipment and methods for a significant period of time. Neither committee ever published their findings because they never achieved their goals well enough to be significant. Finally, during the 1980s, with establishment of NSF's Long Term Ecological Research Network (LTER), standardized meteorological measurements were established for ESA (Greenland 1986).

Heinrich K. H. Hoffmann had introduced the use of quadrats in plant ecology in his 1879 study on the middle Rhine valley (Tobey 1981:51–53). Quadrats were then used by Oscar Drude in *Deutschlands Pflanzengeographie* (1896). Roscoe Pound and Clements attempted to adapt Drude's methods to their needs in surveying *The Phytogeography of Nebraska* (edition 2, 1900). Clements addressed a broader need with his *Research Methods in Ecology* (1905), but it was an inexhaustible subject. For example, in the first issue of *Ecology*, Andrew E. Douglass (1920) reported on a correlation between tree rings and climate. He was an astronomer and an early paleoclimatologist (Webb 1983). In the same first issue, Edgar T. Wherry's article, "Plant Distribution around Salt Marshes in Relation to Soil Acidity" complained about authors who had determined soil acidity using misleading methods (1920a:42), and promised an article on methods to follow. That article on methods appeared in the July issue (Wherry 1920b). Implicit in his instructions was the possibility of everyone's data being comparable if all interested in the subject used Wherry's methods and equipment. Later in volume 1, Robert R. Hill, who was a grazing examiner for the U.S. Forest Service, explained "Charting Quadrats with a Pantograph" (1920). Two other U.S. foresters, Carlos Bates and Raphael Zon, published *Research Methods in the Study of Forest Environment* (1922). Their reviewer in *Ecology*, Burton Livingston, offered a mixed judgment (1924)—both praise and criticism.

Clements pioneered quantitative studies in ecology, which partly accounts for his influence in plant ecology. However, quantitative studies did not guarantee that his results were fully accepted (Gleason 1920). In December 1924, the Carnegie Institution of Washington published two studies of which he was the senior coauthor, *Experimental Vegetation: The Relation of Climaxes to Climates* (with Weaver, vii + 172 pages) and *The Phytometer Method in Ecology: The Plant and Community as Instruments* (with Glenn Goldsmith, vi + 106 pages). George E. Nichols at Yale reviewed them in *Ecology* (1926). He felt uneasy about "the apparent hiatus between the statement of facts and the drawing of conclusions," and he complained, "The average reader quickly becomes bewildered in reading through a mass of data such as that here presented, covering as it does, well over a hundred pages and including approximately seventy tables and twenty graphs." Nichols' implicit message was, maybe so, but let's wait for others to confirm or contest their claims.

Plant ecologists, however, did not usually wait for others to help them decide whether they were Clements' followers or opponents. Lucy Braun was willing to follow him to some extent, though her essay on Ohio (1926:354–372) in *Naturalist's Guide to the Americas* does not mention succession or climatic climax. Gleason was his most vocal opponent. Their positions were clearly evident during the 1920s, but their respective influences came later.

Plant geography attracted the interest of plant ecologists and other botanists (Harshberger 1911). Historical plant geography arose at the same time that ESA was founded (Good 1955:755–759), but failed to stimulate American research until 1930. Douglass' tree-ring research, discussed above, was on paleoclimate at particular locations, but not on past plant migrations. German meteorologist Alfred Wegener (1880–1930), began publishing his ideas on continental drift in 1912 and published *Die Entstehung der Kontinte und Ozean* in 1915; the third edition (1922) was translated into English (1924), and then into French, Spanish, Swedish, and

Russian (Bullen 1976; Schwarzbach 1986). Also in 1915, paleontologist William D. Matthew (1871–1930), at the American Museum of Natural History, published *Climate and Evolution*, which "reinforces the belief in the general permanence of continents," yet "the modern period in zoogeography must be dated from 1915" (Schmidt 1955:777–778). Only in the 1960s did Wegener's theory become the consensus of earth scientists.

Plant physiologist Burton Livingston (1875–1948), who had worked at the Tucson Laboratory, 1906 to 1909, before moving to Johns Hopkins University (Humphrey 1961:148-151), and desert ecologist Forrest Shreve had the Carnegie Institution in Washington publish *The Distribution of Vegetation in the United States, as Related to Climatic Conditions* (xvi + 590 pages, 1921), which "will remain the classic attempt to find the one environmental value which determines the distribution of vegetation types" (Ewan 1969b:120). They described in some detail Merriam's life-zone system (Livingston & Shreve 1921:519–529 + 3 maps), which they judged inadequate because he had not published all his data, though they felt that he had made a good start. They provided their own color map of the vegetation of the United States. A Michigan zoologist, Lee Dice, also criticized Merriam's system (1923; Schmidt 1955:783). Livingston and Shreve's monograph was obviously for scientists. Two U.S. Department of Agriculture (USDA) botanists, Homer Shantz (1876–1958), a physiologist in the Bureau of Plant Industry, and Raphael Zon (1874–1956), a forest economist in the Forest Service, collaborated on an *Atlas of American Agriculture: Natural Vegetation* (29 pages, 1924), which was for a broader audience. Although there were two maps that were a half-page each, the focus of the atlas was upon a single map of U.S. vegetation that spread across 2 pages. The remainder of the atlas consisted of two narratives with photographs, one with Zon describing forests (pp. 7–14) and the other with Shantz describing both grasslands (pp. 15–21) and desert shrubs (pp. 21–26).

Harvard botanist Merritt Fernald (1873–1950), who had trashed Harshberger's *Phytogeographic Survey of North America* (discussed above, 1911 for book and review), was invited to review for *Ecology* Douglas H. Campbell's *An Outline of Plant Geography* (ix + 392 pages, 1926). Neither of these American botanists was a member of ESA. Fernald (1926) concluded his 7-page review of Campbell's *Outline* by lamenting: "Many more such cases [as those already described] unfortunately mar the pages of a beautifully printed and illustrated and clearly written book," because, like Harshberger, Campbell had not used some crucial sources. That October issue of *Ecology* ended on a happier note and group photograph (with names) of 25 participants in a Field Symposium of Plant Ecologists who had come to the International Congress of Plant Sciences. The field trip in Colorado was led and reported on by ecologist Herbert C. Hanson (1890–1962) from the Colorado Agricultural College, Fort Collins (Burgess 1996:50–51). Frederic Clements was also a participant.

The January 1928 issue of *Ecology* has Edwin B. Fred's review of Selman A. Waksman, *Principles of Soil Microbiology* (xxviii + 897 pages, 19 plates). Both men published articles in *Ecology* (Aikman & Gates 1952), though neither Waksman nor Fred seemed to have joined ESA. Why is Fred's review noteworthy? It was not because Waksman would win a Nobel Prize in Medicine and Physiology in 1952

(Waksman 1954), or because Fred would later serve as dean of the graduate school and then president of the University of Wisconsin (Johnson 1974). But it was mostly because Waksman's book was a foundation for soil ecology and because Fred was such a good soil biologist that he could praise Waksman for carefully reviewing in this magnificent treatise some 2,500 original papers in English, French, German, and in Waksman's native Russian, but still criticized him for contradictions, lack of explicit conclusions, and some repetitions. Waksman and *Ecology* could hardly have found a better reviewer.

Norman Taylor (1883–1967), an English immigrant who had participated in the Mount Marcy study and as a curator at the Brooklyn Botanic Garden had arranged for the institution to publish *Ecology*, conducted two later studies that drew upon his Mount Marcy experience, both of which appeared in the *Brooklyn Botanic Garden Memoirs*. The first was a two-part study of "The Vegetation of Montauk: A Study of Grassland and Forest" (1923). Part I, 78 pages, is on the vegetation; part II, 23 pages, is on Montauk flora. Montauk is at the eastern end of Long Island, and its lighthouse is at the eastern extremity of New York State. In 1923, Montauk was very thinly settled, and much of its vegetation was little disturbed by man. Taylor's monograph is very well illustrated with a map, photographs, charts, diagrams, and a panoramic drawing stretching across 2 pages of Montauk's hills and vegetation, seen from the lighthouse. He found that forests could only grow behind dunes, sheltered from wind and where freshwater was available. Elsewhere, grasses dominated.

Taylor's second study in the *Brooklyn Botanic Garden Memoirs* was in collaboration with Barrington Moore, who was the senior author of "The Vegetation of Mount Desert Island, Maine, and Its Environment" (1927). It contains the same visual aids as in the previous study, but its map is a spectacular one in which different types of vegetation were indicated in different colors. The only indication of why they chose that island is that it is "the largest and most beautiful of the many small islands off the coast of Maine" (1927:1). However, they also mentioned that this island contained Lafayette National Park. At the time, the park covered less than half of Mount Desert Island. In 1929, the name was changed to Acadia National Park in compliance with a stipulation by John D. Rockefeller, Jr., for his gift of nearly 11,000 acres on the island to the park.

In 1929, Clements, Weaver, and Herbert Hanson published *Plant Competition* (xvi + 340 pages), which was an experimental monograph on plant physiology, possibly the most experimental large-scale ecological work published by then, which helped avoid Clements' propensity for dogmatic assertion instead of tentative exploration. It does begin with one of those historical surveys of the history of the concept of ecological competition which Clements did so well (Clements et al. 1929:3–36). Since *Ecology* published reviews of previous books by Clements and by Weaver, its lack of a review of *Plant Competition* perhaps was because an agreed reviewer never sent in the review.

During its initial 15 years, ESA was small by today's standards but still quite a respectable size for the time. Its bureaucracy functioned well. I found no evidence of any significant conflicts. From the start, it held annual meetings in December with one or more other organizations—either the American Association for the

Advancement of Science (AAAS) or other biological societies or both; and also one or more summer meetings, usually with another organization. The *Bulletin* began as a humble pamphlet with few pages and no illustrations. During 1915 to 1929, it gradually increased in the number of pages but was not yet very impressive. *Ecology* was a high-standard journal from issue one. ESA members were surely optimistic about their science and their society in 1929.

2 1930 to 1944
Challenges

The period 1930 to 1944 was one of stress for both Americans and Canadians: first, the Depression, then World War II (Miyanishi 2014). Most ecologists taught in academia, and during the Depression many young men (and some of the few young women who aspired to attend institutions of higher education) lacked the financial resources to pay tuition. That situation changed during World War II because the military services decided that it was important to educate some of their new personnel, and because wartime government expenditures had pulled the economy out of the Depression. (During the 1930s, my father taught physics at a small college [Louisburg in North Carolina] and was paid partly with meals at the college cafeteria. After the war began, Duke University hired him to teach engineering to students in uniform.) *Ecology* could hardly flourish during these years. The ESA membership had grown slowly but steadily until 1929, when it ceased to grow, and it actually contracted from a high of 645 to 546 by 1934 (Figure 2.1).

The esteem that ESA founder Victor Shelford enjoyed was indicated by two reviews in *Ecology* of his *Laboratory and Field Ecology: The Responses of Animals as Indicators of Correct Working Methods* (1929, xii + 608 pages). Charles Kofoid's review was the briefer one and was more descriptive of the book's contents than it was critical, but even he remarked upon Shelford's difficulty in separating autecology (which Shelford claimed was not the domain of ecology) from community ecology. William Cook's review was longer and more critical. He argued that "after defining ecology as the science of communities, and virtually excluding autecology from that science, Shelford proceeds to write an entire book on the environmental relations of single species" (1930:613). Both reviewers agreed that it was a valuable reference work with "an adequate bibliography of 1,012 titles" (Kofoid 1930:611).

During the 1920s, Alfred Lotka, Raymond Pearl, and Vito Volterra had founded mathematical animal demography, primarily as theoretical dynamics (Lotka 1925; Scudo 1971, 1984; Scudo & Ziegler 1978; Kingsland 1985:25–126). Their studies attracted the attention of entomologist Royal Chapman (1889–1939), a Minnesotan, who had received his bachelor and master degrees from his state's university before earning a Ph.D. at Cornell in 1917. He returned to the University of Minnesota to teach and developed a course on animal ecology—undoubtedly the first such course at that university and one of the first in America. He based his course "upon controlled experiment and quantitative field studies" (Graham 1941:522). When he published *Animal Ecology with Especial Reference to Insects*, he included as an appendix a reprint in English of a 40-page Italian article by Vito Volterra, "Variations and Fluctuations of the Number of Individuals in Animal Species Living Together" (1931:409–448). In her history of population ecology, Sharon Kingsland pointed out that in his own text, Chapman never used the sophisticated mathematics of Volterra's

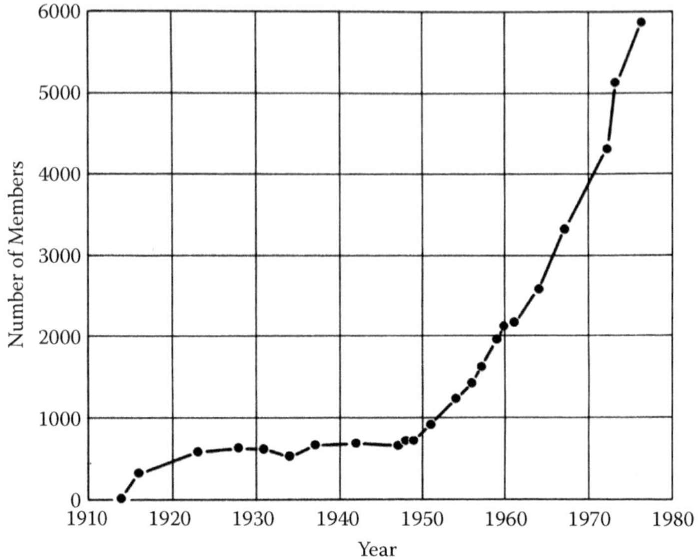

FIGURE 2.1 Growth in ESA membership, 1915 to 1976. (From Burgess 1977:3. With permission.)

article, and she suggested that Chapman reprinted it to add an aura of exactness to an insecure science (1985:128). Be that as it may, he had cited two of Volterra's publications in his *Ecology* paper, "The Quantitative Analysis of Environmental Factors" (1928), and to show his readers a worthwhile challenge which ecologists would do well to master.

During the Depression, ESA may have seemed daring in 1931, for creating a new journal, *Ecological Monographs*, for articles that were too long to squeeze into *Ecology*. However, this journal had been planned since 1928, and former ESA president Arthur Pearse who had been on the Duke Zoology Faculty since 1927, persuaded Duke University to publish it (Shreve 1929). There was a clear need, since the pages in *Ecology* had increased from 313 pages in volume 1 to 563 pages in volume 10, with a corresponding rise in the rejection rate (Burgess 1990:4). Furthermore, in 1931 no one knew how long the Depression would last. Ecologists carried on, meeting the challenge to persist during hard times. Its first article discussed the research of Edith S. Bowen, a student at the University of Chicago under Allee, "The Role of the Sense Organs in Aggregations of *Ameiurus melas*" (1931), which filled 35 pages.

The proto-scientific approach to regulating hunting and fishing was to set annual kill limits per hunter for land animals not considered pests, with unlimited controls of predators and prairie dogs to accommodate ranchers, and to raise fingerling fish from eggs in hatcheries for stocking rivers and streams to maintain fish populations (Leopold 1955:795–800). In the 1870s, some leaders in society were shocked at the slaughter, almost to extinction, of America's bison. William Hornaday, director of the New York Zoological Park, organized a Bison Society that in 1905 persuaded President Theodore Roosevelt to protect the bison by establishing the Wichita Mountains National Wildlife Refuge in Oklahoma, and he also persuaded Congress to provide a fence and guards (Laycock 1966:14–19; Dunlap 1988:7). In 1884, the American Ornithologists' Union asked Congress to establish a Division of Ornithology and Mammalogy, which Congress did in 1885, with C. Hart Merriam as director (Sterling 1977:58-66; Dunlap 1988:35). Merriam soon changed the name to Biological Survey. He had his staff collect life history information on species, and in 1905 it began rodent and predator control for farmers and ranchers. A Migratory Bird Treaty was approved in the United States in 1915 and in Canada in 1917 to protect species that migrated across their international boundary, and the U.S. Biological Survey was responsible for bird protection. In 1919, Biological Survey biologists formed the American Society of Mammalogists and elected a retired Merriam as president (Sterling 1977:179, 308–309; Dunlap 1988:36). By the 1920s, ecologists and government officials began to realize that more was needed to preserve hunted species than just seasons and bag limits.

Steps began to be taken then which led to the emergence of a game management profession in the 1930s (Leopold 1955:800–807). Three leaders of that new profession were also members of ESA: Aldo Leopold (1887–1948), Herbert Stoddard (1889–1970), and Paul Errington (1902–1962). They, like all wildlife managers I have ever known, were hunters. Progress in wildlife management ecology can be indicated during the period 1930 to 1945 by focusing upon their activities and contributions. Stoddard and Leopold began somewhat similar projects around the same time, rather independently. Stoddard and Leopold's meeting on October 8, 1928, in Cincinnati was arranged by the Biological Survey (Stoddard 1969:217–219; Meine 1988:264–265).

Leopold was the son of a businessman and avid hunter in Burlington, Iowa, along the Mississippi (Meine 1988, 1997). Aldo absorbed his father's fondness for hunting but not his business interest. His outdoor orientation influenced his decision to go to Yale University, Connecticut, to study forestry. He did read natural history books—he was impressed by Darwin's *Formation of Vegetable Mould, through the Actions of Worms* (1881)—but it is unlikely that Yale offered a course in ecology during the years 1905 through 1909, when he studied there. Neither his outdoors experiences nor his readings gave him insight into the role wolves played in the economy of nature. When he went to the Southwest as a U.S. forester, he decided that killing wolves would provide hunters with more deer to shoot, and when he found them, he killed the wolves. In 1909, he had second thoughts, which became the beginning of his ecological consciousness and the source of his famous essay, "Thinking Like a Mountain," which he did not write until 1944 (Meine 1988:93–94, 458–459, 543 note 10). In 1921, he published in the *Journal of Forestry* an essay, "The Wilderness and Its Place in Forest Recreation Policy," proposing to designate wilderness areas on government lands. His proposal became government policy in 1924 when Gila Wilderness Area was the first area so designated. In 1928 he left the Forest Service in order to make a game survey for the Sporting Arms and Ammunition Manufacturers' Institute in the upper Midwest. Based upon that research, he published a long series of essays and a *Report on a Game Survey of the North Central States* (1931). In 1932, he became America's first professor of Game Management at the University of Wisconsin, where he wrote his landmark *Game Management* (1933), which emphasized the maintenance of habitat. (When I took Professor Fred Barkalow's Wildlife Management course at North Carolina State University in fall 1958, Leopold's was our only textbook.) Leopold participated in founding The Arboretum of the University of Wisconsin–Madison in 1934; at its dedication he explained that it was an attempt to reconstruct "a sample of original Wisconsin," and he was its first research director (Jordan 1981). Leopold, who had never played an active role in ESA, was pleasantly surprised to be elected its vice president in 1946 and president in 1947. At the time, he was assembling the essays that became his best-known work, *A Sand County Almanac*, published posthumously in 1949 (Lorbiecki 1996; Knight & Riedel 2002; Newton 2006; Simberloff 2012a).

Stoddard was an ornithologist who was collecting and mounting birds for exhibits at the Milwaukee Museum in 1923 when he was invited to undertake research on quail near Thomasville, Georgia, where wealthy landowners wanted larger numbers to shoot (Stoddard 1969:174–180). They arranged a partnership with the U.S. Biological Survey to jointly fund this research, which Stoddard conducted in 1924 to 1930. His *Bobwhite Quail: Its Habits, Preservation, and Increase* (1931, xxix + 559 pages) was a landmark monograph that set a high standard for wildlife research (Way 2011). Leopold had begun research on July 1st for the newly founded Sporting Arms and Ammunition Manufacturers' Institute (SAAMI), which wanted him to study the decline of wildlife. SAAMI had decided by October to establish two research fellowships in game management at one or more universities. Stoddard and Leopold were to tour universities together in order to decide where the fellowships would be useful. In his autobiography, Stoddard commented: "I could see at once that we were kindred spirits" (1969:212). In three weeks, they visited universities in seven Midwestern states, meeting ecologists, ornithologists, mammalogists, entomologists, botanists, and horticulturists. They arranged for one fellowship at the University of Michigan, which went to Ralph Yeatter, and the other at the University of Wisconsin, which went to Paul Errington. Both Yeatter and Errington obtained doctorates and pursued careers in wildlife management (Errington 1973; Kohler 2011). Leopold could not match Stoddard's quail monograph, but he became the first professor of wildlife management in America at the University of Wisconsin, and wrote the first textbook, *Game Management* (1933, xxi + 481 pages). Leopold's influence has been broader and more lasting than that of either Stoddard or Errington, though both had important achievements and influence on wildlife management.

During the period 1930 to 1942, there were 21 ecological symposia held, and the papers presented were published together for 18 symposia, with five such collections in *Ecological Monographs* and one in *Ecology* (Park 1949:65–66). Park overlooked two which are included in the above tallies. One, on range ecology, was presented at the ESA meeting in Ames, Iowa (January 1, 1930); it had papers that had not been published together. John Weaver, Clements' protégé and coauthor, spoke there on "Underground Plant Development in Its Relation to Grazing" (1930) (Figure 2.2). Since Weaver was ESA president in 1930, it seems likely that

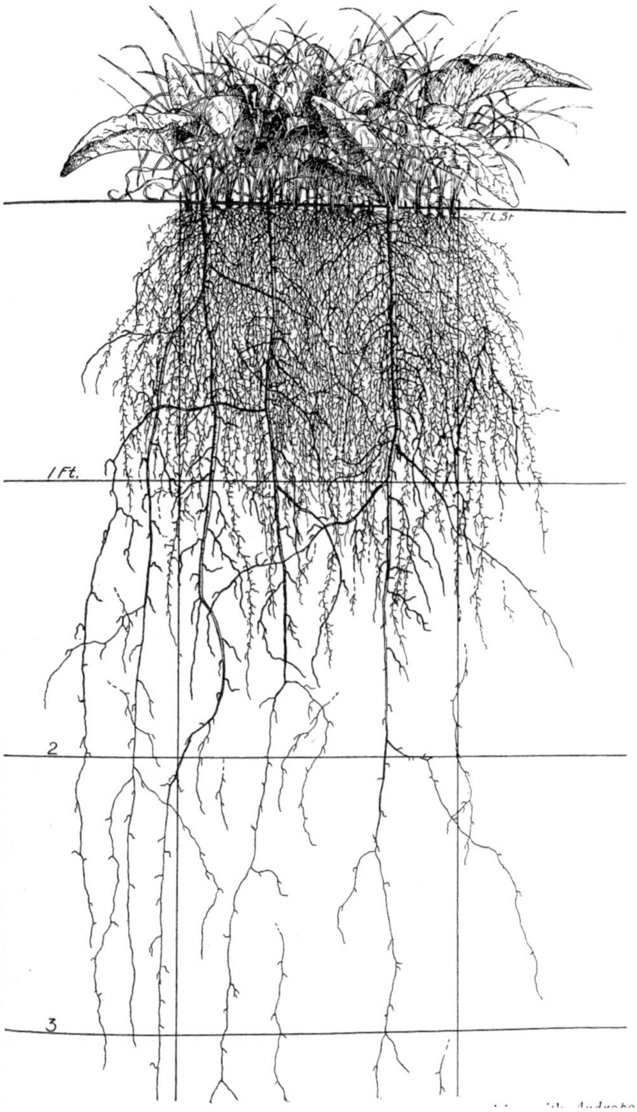

FIGURE 2.2 A weed in grassland, *Aretium lappa minus*, in competition with *Andropogon nutans*. (From Weaver 1930:555. With permission.)

he organized that symposium. The ESA symposium on range ecology would be relevant to the Dust Bowl, which, however, was most severe on plowed land. There was not another symposium relating to the Dust Bowl until 1940, and even then it had a more general title of "Relation of Ecology to Human Welfare." The other symposium that Park overlooked was on lake biology, with papers published together by AAAS (Moulton 1939).

Nevertheless, individual ecologists had begun responding to the Dust Bowl problem much earlier. There were already a good number of relevant papers in the 1930 volume of *Ecology* in addition to Weaver's, even though they were too early to have been inspired by the drought: T. L. Steiger, "Structure of Prairie Vegetation" (170–217); C. F. Shaw, "Potent Factors in Soil Formation" (239–245); A. E. Aldous, "Effect of Different Clipping Treatments on the Yield and Vigor of Prairie Grass Vegetation" (752–759); E. W. Nelson, "Methods of Studying Shrubby Plants in Relation to Grazing" (764–769); and others, depending upon how broadly one defines Dust Bowl problems. There were fewer such relevant studies in the 1932 volume.

The Dust Bowl conditions were not unique to the 1930s. There had been periodic droughts and dust storms on the southern plains throughout the 1800s and earlier 1900s (Hurt 1981:4–15; Lookingbill 2001:14–16; Schubert et al. 2004:1858). For nomadic Indians, dust storms were not a severe problem, because they could move their agriculture and hunting to better environments (Clements & Chaney 1937:40–43; Bonnifield 1979:8). American settlers, before and after the Civil War, were mostly ranchers, and although cattle can overgraze the land, with few settlements and no fences that was seldom a problem. Dust storms were caused by droughts, composition of the soil, and human settlement (Green 1973; Bonnifield 1979:9–12; Worster 1979:69–71; Hurt 1981:17–29; Lookingbill 2001; Schubert et al. 2004; Duncan & Burns 2012; Sylvester & Rupley 2012; Moon 2013:249–251). During the later 1800s, farmers brought barbed-wire fencing and farm machinery onto the Great Plains, and since the land was flat, they could plow many acres. During World War I, precipitation was greater than usual and so were wheat prices, which encouraged expansion of crop lands. The drought of 1931 lasted seven years—long enough for the Roosevelt Administration to respond with government assistance. A modified plow—basin lister—helped hold moisture in the soil, but there had to be moisture to hold, and contour plowing was introduced (Hurt 1981:68–74). A Federal Soil Erosion Service (Interior Department) and Soil Conservation Service (Agriculture Department) funded demonstrations before these services were combined. Hugh Bennett (1881–1960) joined the federal Bureau of Soils, USDA, in 1903 and crusaded for soil conservation. In 1935, as dust from Oklahoma and Kansas blew into the Capitol building in Washington, Congress provided funds to establish a more active Soil Conservation Service, under Bennett (Nash 1990:134–139; Egan 2006:125–127, 133–135; Armitage 2013).

Ohioan Paul Sears in 1935 was professor of botany and department chairman at the University of Oklahoma and also a botanist for the Oklahoma Biological Survey. In that year, he published *Deserts on the March*, which was the most important American environmental history since George Perkins Marsh's *Man and Nature; or, Physical Geography as Modified by Human Action* (1864). Marsh, from Vermont, focused mainly upon the impact by Americans upon the eastern forests. Sears focused on the use of arid lands by Americans. Although *Deserts on the March* was published by the University of Oklahoma Press, it was not a technical book. It explained the arid environment and how Americans were using it for grazing and farming, and the consequences of that usage. The book also invited Americans to reevaluate their policies and practices concerning land usage. Sears was the ecological public spokesman for this crisis. He suggested the appointment in each county of an ecologist to advise on land use.

Paul Bigelow Sears (1891–1990)

(From *ESAB* 46, 151, 1965. With permission.)

Ohioan Sears earned B.A. and B.S. degrees (1913) at Ohio Wesleyan College, an M.A. at the University of Nebraska (1915), and a Ph.D. at the University of Chicago (1922) under Cowles. He taught at Ohio State University (1915–1938) and at Oberlin College (1938–1950), when he became chairman of the Yale University Conservation Program (Patten 1994). His research interests in plant ecology were diverse, and he wrote conservation books for the public, beginning with Deserts on the March (1935), during the Dust Bowl. He served as ESA vice president in 1943 and president in 1948 (Burgess 1996:96–97, 2009; Stuckey & Sterling 1997; Harmond 1998; Shane 2009). He was one of 50 botanists awarded with a Certificate of Merit by a jubilee committee of the Botanical Society of America for "pioneering efforts in pollen analysis and continued interest in geochronological problems have made him the leader of all in this field, on our continent. The keenness of his mind, the warmth of his personality, the quality of his writing, and his capacity to relate all scientific problems to man have earned him the distinction of an exemplary figure in American science" (Meyer 1958:18).

A year later, Clements collaborated with paleontologist Ralph Chaney (1936) on a 53-page pamphlet, *Environment and Life in the Great Plains*, which traced the geologic and ecologic history of the region from the Pennsylvanian Period (Late Paleozoic) to recent times. The fact that readers were presented with 25 pages of geological and paleontological history before reaching a discussion of humans on the plains might help explain why this is one of Clements' lesser known works. When they discussed ecology, Clements' main point was about the same as Sears' (though the bibliography does not list *Deserts on the March*): that ecologists had accumulated a good understanding of the plains ecology since the 1890s (when Pound and Clements surveyed Nebraska's vegetation), but that ecological understanding had never guided human use of the land, and that it should in the future. Climatic cycles of rain and drought were fairly regular, correlated to an extent with sunspots, as Douglass had discovered, yet no significant contingent plans had developed to cope with this environmental cycle. Although Clements and Chaney's synthesis of knowledge was a positive achievement, it could not compete for public attention with Sears' book. Hugh Bennett's encyclopedic *Soil Conservation* (xvii + 993 pages, 1939) explained ways to avoid wind erosion.

The European windmill, used since the Middle Ages, was too cumbersome for use in American agriculture. In 1854, Daniel Halladay, part owner of a machine shop in Connecticut, began building windmills that were practicable for farmers (Baker 1980:38–40). Sales were slow, and so he moved his manufacturing to Batavia, west of Chicago, where sales became much better. Apparently independently, Rev. Leonard H. Wheeler, a missionary among the Ojibway Indians in 1867 invented another practical windmill, which was designed differently. These two kinds of windmills were available when farmers and ranchers moved onto the High Plains in the 1870s and later. A windmill could pump enough water—from 15 to 100 feet deep—for a homestead with some livestock, though S. E. Couch of San Angelo, Texas, claimed in 1890 that his windmill pumped water from a well over 360 feet deep (Baker 1980:41). In 1896, the U.S. Geological Survey sent hydrologist Nelson Darton (1865–1948) to the Great Plains to study groundwater (Aldrich 1974; Diffendal 2004; Ashworth 2006:84–86). He took a train to Ogallala, Nebraska, and examined the bluffs along the South Platte River. He found a Tertiary Age stratum containing water that seemed very promising, which he named the Ogallala Formation. He subsequently studied it elsewhere, and it became clear that it was a substantial resource that was too deep for many windmills to reach. By 1910, electrical and gasoline pumps were being used in a few places to pump water for irrigation. "In the mid 1920s, when Alma [Holmes] was in high school, she saw the Texas Land and Development Company drill a dozen deep irrigation wells to water alfalfa for cattle feed. It was the only landowner irrigating in a community of dryland farmers" (Cunfer 2005:173–174). Why? In 1935, her husband borrowed $1,950 to drill a well. "Alma wept when they made the decision because she thought they would never be able to repay the debt and would lose the farm."

The 1930 U.S. Census reported the existence of 170 Texas irrigation pumps, and the 1940 U.S. Census reported 2,180 Texas irrigation pumps (Green 1973:38–57, 143). After World War II, reliable pumps were developed, which could economically pump water from an aquifer for large-scale crop irrigation.

A different approach to paleo-ecology than correlating tree rings with past climates (Chapter 1) was the study of plant pollen in earth deposits—usually under bogs, ponds, and lakes—to determine past vegetation. Josias Braun-Blanquet (1932:338) cited G. Lagerheim's "Metoder för Pollenundersökning" (1902) as the origin of such pollen analysis. Nancy Slack (2010:406, note 28) suggested that the first study of pollen in peat was by a German, Carl A. Weber, in 1896, but she gave no citation. G. Erdtman possibly introduced the technique to Americans (1927, 1929, 1931), and Sears was among the first in the United States to adopt the technique for research (1930, 1931, 1932). His "Record of Post-Glacial Climate in Northern Ohio" (1930) cited two relevant studies published in 1929, one by P. Draper from his laboratory, who was possibly one of Sears' students, and the other was Ivey Lewis and E. C. Coke, "Pollen Analysis of the Dismal Swamp." He was likely in contact with Lewis and/or Coke, because otherwise he might not have discovered their papers in time to cite them in 1930. Lewis and Coke cited two papers by Erdtman as a source of guidance. Sears also attempted to determine the age of peat based upon the rate at which peat is laid down (Sears & Janson 1933), which was too uncertain to gain adherents. Edward Deevey, at Yale, wrote a history of land around Linsley Pond based on cores from the lake bottom containing pollen (Slack 2010:128). Raymond E. Janssen published *Leaves and Stems from Fossil Forests* (1939) that described and illustrated fossils from coal but without attempting to determine their age. The October 1939 issue of *Ecology* carried reviews of three studies on plant fossils—Janssen's and two others on fossil pollen—none of the studies or the reviews were by Erdtman, Sears, or Deevey. Radioactive isotopes to determine the ages of objects came into use after World War II. Erdtman published *An Introduction to Pollen Analysis* (1943), which also updated his previous bibliographies.

Ecology encompasses freshwater and saltwater ecology as well as terrestrial plant ecology and animal ecology. Wisconsin's limnologists, Edward A. Birge (1851–1950) and Chancey Juday, found it convenient to publish studies in both the *Transactions of the Wisconsin Academy of Science, Arts, and Letters* and the *Bulletins of the Wisconsin Geological and Natural History Surveys*; they surveyed many lakes, ponds, and streams in Wisconsin (Sellery 1956; Frey 1963b; Beckel 1987:2–29; Egerton 1987:86–94; 1999a; Harmond 1997a). To a lesser extent, and somewhat later, Evelyn Hutchinson and his students similarly published in the *Transactions of the Connecticut Academy of Arts and Sciences.* After the publication of *Ecology* began in 1920, terrestrial papers far outnumbered aquatic papers.

Chancey Juday (1871–1944)

(From Beckel 1987, cover. With permission.)

From Indiana, Chancey Juday obtained his B.A. and M.A. degrees from its university (1896, 1897). He studied and published on plankton in northern Indiana (1897, 1902). In 1900, he became the biologist for the Wisconsin Geological and Natural History Survey and began a collaboration with Professor Edward A. Birge that lasted for the rest of Juday's career, researching Wisconsin limnology (Frey 1963b; Beckel 1987:4–30; Egerton 1987:85–94, 1999b; Burgess 1996:59–60; Mitman 1997a). In 1931, the University of Wisconsin promoted him from lecturer to professor of limnology, and in 1933, Indiana University awarded him with an LL.D. degree. Juday was more responsible for the establishment of an informal limnological school at Wisconsin than Birge, who had ceased teaching in 1911. Juday supervised Ph.D. dissertations for 13 graduate students, including Arthur Hasler (Frey 1963b:43). Juday served as president of ESA in 1927. He was an important founder of the Limnological Society of America, and was its first president (1935–1937). He and Birge both received the Einar Naumann Medal from the International Association of Limnology in 1950. Two of the many studies that he coauthored with E. A. Birge were reprinted in *Limnology in Wisconsin* (Birge & Juday 1977).

In 1925, the National Research Council established a standing Committee on Aquaculture and Hydrobiology, which met annually with the AAAS (Lauff 1963:668-672; Needham 1941:10-11). At the 1934 meeting in Pittsburgh, limnologists formed a founding committee to organize the Limnological Society of America to be established at the AAAS meeting in 1935 (Lauff 1963). It more or less followed the pattern that had led to establishment of the ESA two decades earlier. The founding committee then discovered that an independent effort for the same purpose was underway on the Pacific Coast. The two committees joined for the 1935 meeting in St. Louis. The organizational meeting occurred there on January 1, 1936. There were 221 charter members, and by December 1940 there were 327 members. There was talk of a journal, but no action was taken until after World War II.

The papers from a symposium on lake biology were published by the American Association for the Advancement of Science (Moulton 1939). The first paper was by Canadian Professor Donald Rawson (1905–1961), who had a limnological school at the University of Saskatchewan that was comparable to the one that Birge and Juday ran at the University of Wisconsin (Hammer 2005). His paper was on "Some Physical and Chemical Factors in the Metabolism of Lakes," where he diagramed "A chart suggesting the interrelations of factors affecting the metabolism of a lake" (Rawson 1939:10) (Figure 2.3). It was an original kind of chart but resembled charts of food webs that were being diagramed at that time, such as August Thienemann's 1926 "Nahrungskreislauf im See" (reproduced in Egerton 2007:60). Rawson was equally skilled at discussing physico-chemical factors, trophic webs, and fishery ecology (Northcote & Larkin 1963:459–463). Such topics were discussed in the nine papers in the AAAS volume, but there was no summation or conclusion. The reviewer of the volume in *Ecology* (1940) was a bacteriologist, William Burrows (Chicago), who enthused over "The Distribution of Bacteria in Lakes" (1939) by Arthur Henrici (Minnesota), as being the first synthesis on the subject.

Juday was the first president of the Limnological Society from 1935 to 1936. From September 4 to 6, 1940, the University of Wisconsin and the Wisconsin Alumni Research Foundation sponsored a Symposium on Hydrobiology, at which 32 invited papers and 16 "volunteered" papers were presented. The volume, which the University of Wisconsin Press published, is entitled *A Symposium on Hydrobiology* (1941). The

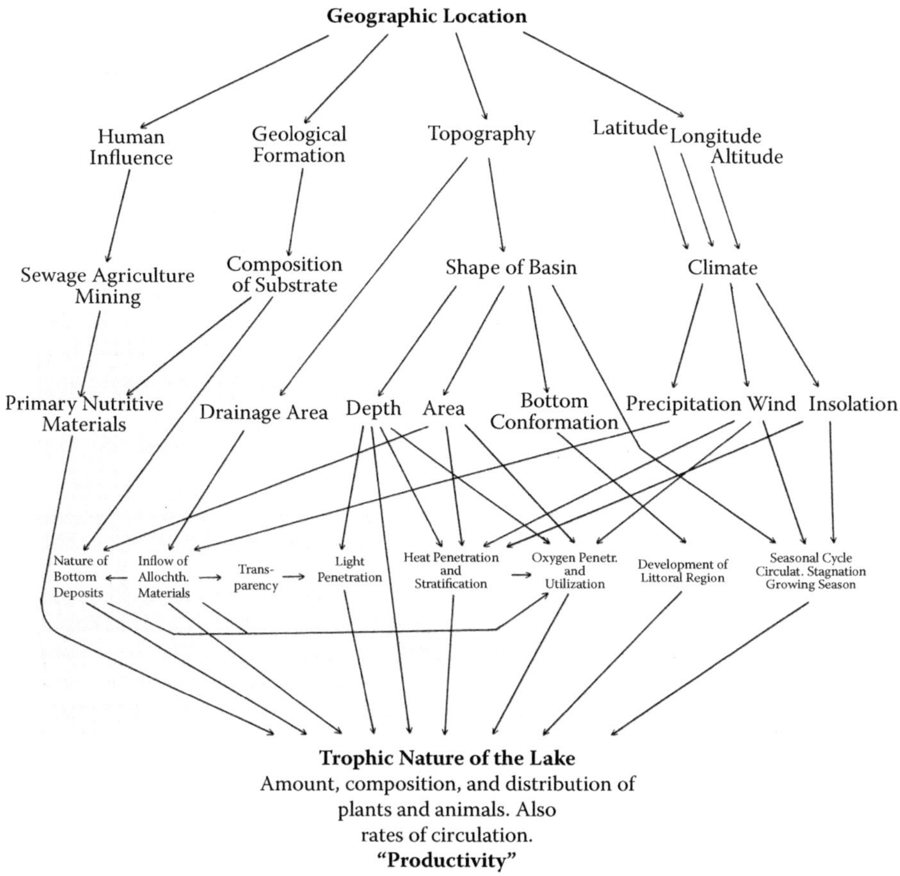

FIGURE 2.3 A chart suggesting the interrelations of factors affecting the metabolism of a lake. (From Rawson 1939:10. With permission.)

title page lists the names of 38 speakers, with no indication of an editor; the introduction has no name attached to it and may be by Juday, who did not provide a paper for the volume. All invited papers were printed in full, but the volunteered papers have only an abstract, with most being a page or less, and a few slightly more than a page. There is no mention of the Limnological Society of America (LSA), except that the frontispiece is a collection of six photographic portraits of the first six LSA presidents. The invited papers were very well distributed by subject matter, and the 405-page book (no index) could have been used as an up-to-date textbook for a limnology course.

Wisconsin limnology research, which also included colleagues of Birge and Juday and students of Juday, culminated around 1940 in a new understanding of the heat budget of lakes and phytoplankton productivity (Morris 2009:769). In relation to the annual movement of copepods, Birge had studied and named the "thermocline," which he noted was called "Sprungschicht" in German (Birge 1898:295). In 1910, he also introduced the terms "epilimnion" and "hypolimnion" (Mortimer 1956:177). He then distinguished between lakes that have circulation of water throughout the summer and others that are thermally stratified (Frey 1963b:24). The species of plankton found in American lakes were known before the studies done by Birge and Juday. However, they measured the productivity of phytoplankton in four lakes near Madison as indicators of potential productivity of the grazing zooplankton (Birge & Juday 1922). Two of these lakes became thermally stratified during the summer and two did not. Arthur Hasler was the only one of 13 students who earned Ph.D.s under Juday who remained at the University of Wisconsin–Madison, beginning as an instructor in zoology (1937) and becoming professor in 1948. Juday retired from teaching in 1937 and from the directorship of the Trout Lake Limnological Laboratory in 1942. The "Birge–Juday Limnological School" was an informal network, and there was no real school for Hasler to inherit. During World War II, there was no Wisconsin limnological school.

Hutchinson did not run a limnological school at Yale. However, he did not need one, because he had students and resources by virtue of his teaching ability and reputation, and he did not limit his interest or research to limnology. Yet, his first three graduate students during the 1930s were in limnology.

George Evelyn Hutchinson (1903–1991)

(From *ESAB* 43, 113, 1962. With permission.)

Hutchinson was the son of an English mineralogist at Cambridge University. Those around him promoted his early interests in natural history (Slack 2010:15–41). His undergraduate studies focused on invertebrates, and after graduating he gradually moved into animal ecology (Slack 2010:42–86). He received an M.A. degree from Cambridge in 1928 and was hired that year as an instructor in biology at Yale University. He spent the rest of his career at Yale, where he became one of the most important ecologists in the world (Burgess 1996:56–57). He was considered to be just as an effective teacher as Cowles (Figure 2.4, Kohn 1971), and he was also a productive author on a broad range of animal ecological

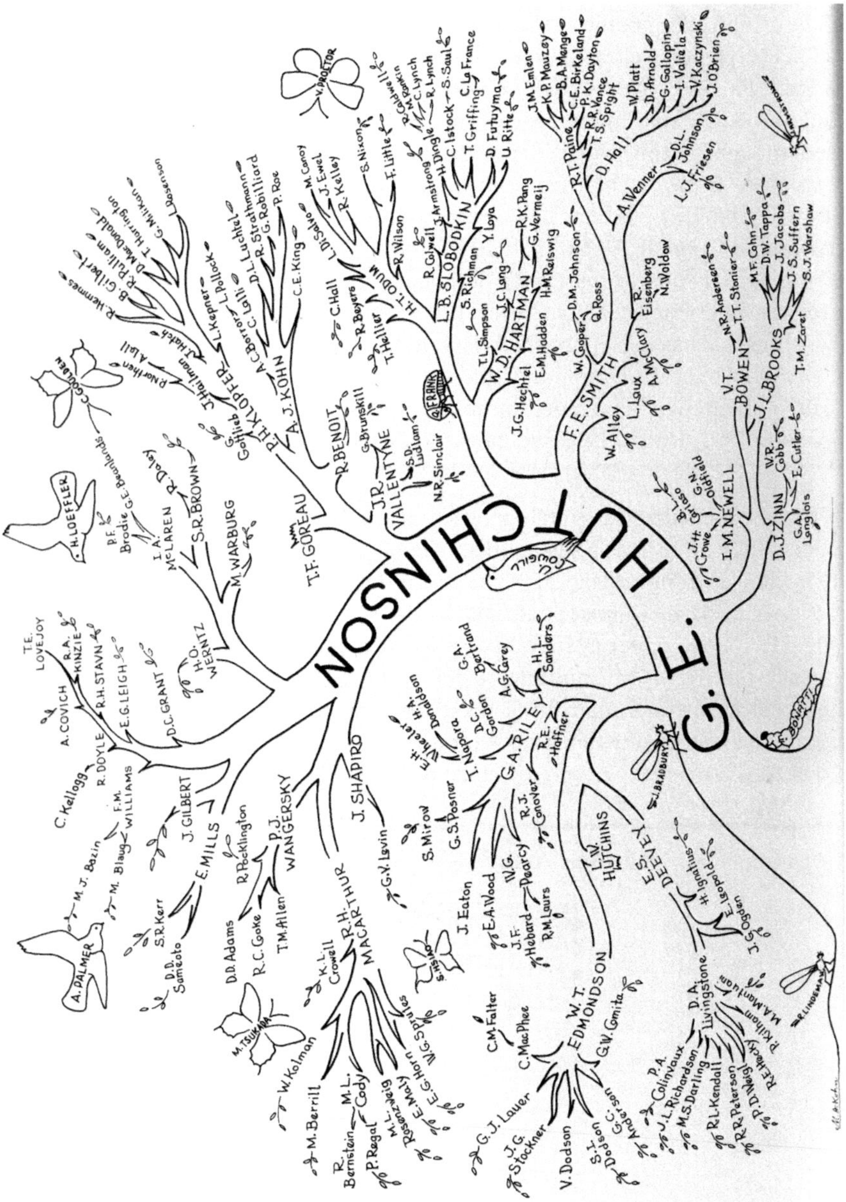

FIGURE 2.4 Evelyn Hutchinson's academic descendants. (By Marion Kohn. From Edmondson 1971:163. With permission.)

subjects. Hutchinson's documented involvement with ESA only began in 1939, when he published a study on fish from an expedition to Kashmir and Tibet in which he had participated in *Ecological Monographs*. In 1940, he published in *Ecology* his famous review of Clements and Shelford's *Bio-Ecology* (1939), which he criticized for being descriptive, and not mathematical in treating ecological subjects. His encyclopedic four-volume *Treatise on Limnology* (1957–1993) rivals Ruth Patrick's *Rivers of the United States* as being the longest ecological treatise by one person, his is general and abstract, hers is more geographically focused (though her third volume was also general and abstract). His *Introduction to Population Ecology* (1978) was an important contribution to that subject, on which two of his own students had already published textbooks (Slobodkin 1961; MacArthur & Connell 1966). In 1979, Hutchinson published an autobiographical *The Kindly Fruits of the Earth: Recollections of an Embryo Ecologist*. He received the Eminent Ecologist Award in 1962. There is also an anthology of his best writings that has been published (Hutchinson 2010). Nancy Slack's *G. Evelyn Hutchinson and the Invention of Modern Ecology* (2010) is an excellent biography. His bibliography is included in both of these last works (Edmondson 1971; Edmondson 1991; Hagen 1992:see index; Slobodkin & Slack 1999).

In 1935, Gordon Riley was Hutchinson's first graduate student, who had "defected" from embryology after hearing Hutchinson lecture for a week (Slack 2010:121–127). Since Riley had a strong education in chemistry, Hutchinson suggested a dissertation on the copper cycle in Connecticut, and Riley chose Linsley Pond as his primary study site. He found copper availability varied seasonably, with the highest concentration in the fall. He used "multiple correlation techniques to test the adequacy of hypotheses about factors controlling the copper cycle. This approach became the most significant feature of Riley's work for nearly the next decade" (Mills 1989:266). Riley "defected" to oceanography when he needed a job.

Gordon Arthur Riley (1911–1985)

(Courtesy of Eric Mills [as in part 51], 1989. With permission.)

Riley was a Missourian who earned a B.S. degree in biology from Drury College (1933), an M.S. from Washington University (1934), he then went to Yale, where he switched from embryology to limnology after hearing Hutchinson lecture

(Hutchinson 1982; Slack 2010:122–123). He became Hutchinson's first graduate student, and since he had strong training in chemistry, Hutchinson had him write his doctoral dissertation on the copper cycle through the environment and aquatic organisms in Linsley Pond (Mills 1989:266). "For his freshwater research, Riley borrowed some techniques used previously in oceanography at the Plymouth Laboratory in England. One was a quantitative plant pigment analysis that H. W. Harvey had developed. The phytoplankton were filtered, the pigment extracted with acetone and measured against a color standard. This an estimate of total phytoplankton more easily and perhaps more accurately than the usual laborious cell counts" (Slack 2010:125). After completing a draft of his dissertation, he joined an oceanographic cruise—his first—for two weeks to study plant pigments and nutrient chemistry. Hutchinson taught him "that populations needed to be studied in terms of dynamic processes—rates of production and consumption and the way these are affected by ecological factors" (Riley 1984, quoted from Slack 2010:127). Riley received a Ph.D. at Yale in 1937 and remained there from 1937 to 1938; he then became a marine biologist at the Bingham Oceanographic Laboratory at Yale (1938–1965); and for the remainder of his career, he was professor and director of the Dalhousie University Institute of Oceanography. "Gordon's view of phytoplankton ecology in the early 1940's was quite revolutionary….based on the premise that in biologically, chemically, and physically complex medium such as seawater, the only approach that held a reasonable expectation of return was one which was strongly quantitative" (Fournier 1982:332). Riley's most significant ecological paper was "Factors Controlling Phytoplankton Populations on Georges Bank" (1946). From 1961 to 1962, Riley served as president of the American Society of Limnology and Oceanography.

Hutchinson's second graduate student was Edward Deevey, who also started graduate studies in 1935 (Slack 2010:128–132). He had earned a B.A. degree in botany at Yale in 1934, and intended to continue graduate studies in botany. He was interested in researching the history of the plants around Linsley Pond with the cores from its bottom. No one in the Botany Department was interested in this subject, but Hutchinson was, and he convinced Deevey to switch to the Zoology Department and do his research under him. Paleolimnology had been pioneered by a Swede, G. Lundquist, in the 1920s, but no comprehensive study of pollen analysis from the cores of a pond had preceded Deevey's study.

Milwaukeean Tommy Edmondson's mother moved her family to New Haven, Connecticut, when his older brother entered Yale (Slack 2010:115–121). There, he went to an excellent high school, where a biology teacher let him use a microscope after school to study rotifers. While still in high school, in 1931, he met Hutchinson, who showed him a colonial species he had not seen before. In 1932, when Hutchinson returned from Yale's North India Expedition, he gave his rotifers to Edmondson to study. As an undergraduate, Edmondson assisted Riley and Deevey in their fieldwork. After Edmondson graduated from Yale in 1938, Hutchinson sent him to Madison for a year to work under Juday. There he met one of Juday's students, Yvette Hardman

(Ph.D.: 1940), whom he would marry in 1941. He returned to Yale in 1939 and wrote a doctoral dissertation on rotifer ecology (1942).

In 1948, the Limnological Society of America expanded and became the American Society of Limnology and Oceanography (ASLO). By 1951, its membership had increased to 1,103, and at the 1952 meeting, the society began taking steps to establish a journal (Lauff 1963:673). The National Science Foundation (NSF) provided funds for three years to launch *Limnology and Oceanography*, which published its first volume in 1956. Yvette H. Edmondson served as the assistant editor (1966–1968), then as editor (1968–1986). Since Hutchinson's health was declining, she then stepped forward to edit the last volume of his *Treatise on Limnology*, which appeared posthumously in 1993. She received ASLO's Distinguished Service Award in 1999 (Twombly 1999).

Terrestrial ecologists tended to be either botanists or zoologists, and so the great majority of studies published were either on plant ecology or animal ecology. However, beginning with the Mount Marcy study (Adams et al. 1920), some American ecologists thought that ideally, plant and animal ecology should be studied together rather than separately. Booker Klugh provided encouragement with his "Common System of Classification in Plant and Animal Ecology" (1923). The *Naturalist's Guide* (Shelford & Shreve 1926) was another step in that direction. Yet, one animal ecologist complained that plant ecologists tended to see plant ecology as "ecology," ignoring animal interactions (Taylor 1927). Since Taylor cited a work by William Cooper as an example, Cooper (1927) assured him that no plant ecologist thinks like that. Maybe, but nevertheless Shelford thought it worthwhile to discuss "Some Concepts of Bioecology" (1931), exploring the implications of plant–animal interactions.

This concern culminated into two works published in 1939: Clements and Shelford's *Bio-Ecology* and a collection of 10 articles, "Plant and Animal Communities," published in *The American Midland Naturalist* (Just 1939). These latter papers (reprinted by the University of Notre Dame Press as a book) were the proceedings of a Conference on Plant and Animal Communities organized by Stanley A. Cain and held at the Biological Laboratory, Cold Spring Harbor, Long Island, New York, August 29 to September 2, 1938. One paper in 1939 by J. Richard Carpenter, "The Biome," focused directly upon the conference's concern of integrating plant and animal ecology. Carpenter's article and Clements and Shelford's book both begin with historical reviews of previous studies, and both overlooked the three-part study on Mount Marcy (Adams et al. 1920, discussed in Chapter 1).

The conference papers included one by C. E. MacGinitie on littoral marine communities, one by Frank E. Eggleton (1893–1970) on freshwater communities, four on plant ecology, and three on animal ecology. The Estonian Theodor Heinrich Lippmaa (1892–1943) spoke on "The Unistratal Concept of Plant Communities," and Dutchman Niko Tinbergen (1907–1988) spoke on "The Analysis of Social Organization among Vertebrates, with Special Reference to Birds." Following the presentations, there were conference discussions on all the papers. So comments on one another's presentations contained some brief thoughts on synthesis; participant Thomas Park regretted that the "facts have not been adequately assembled into principles and concepts"

(1939a:250); it was up to the reader to reach a general synthesis. Allee, from the University of Chicago, reviewed the reprinted book (1939) for *Ecology* and criticized Eggleton's article stating that it was not as well done as the others, but he recommended that all ecologists read these articles. Thomas Park from the University of Chicago, despite having an article in the volume, reviewed it for the *Quarterly Review of Biology*, and commented that "ecology is seriously in need of a thorough house-cleaning followed by a well-planned program of refurbishing" (1939b:332). Charles Elton dismissed the book as "Scholasticism in Ecology" (1940).

Neither Clements nor Shelford attended the conference (they might have declined, being too busy writing their own book on the subject), though their ideas and publications were frequently cited. The advantage of their *Bio-Ecology* was that they did much more of the synthesis work themselves than did those at the conference. Animal ecologist Walter P. Taylor (1927) had defended use of the term "bio-ecology" on grounds that plant ecologists used the term "ecology" as being virtually synonymous with "plant ecology." Two plant ecologists (Cooper 1927; Nichols 1928) protested that bio-ecology was redundant. Clements and Shelford's *Bio-Ecology* resembled a continent-wide version of the Marcy study (Adams et al. 1920), even if they did not consult that study when writing their book. Only *Bio-Ecology* included a reference to a study in which herbivores were excluded from an experimental plot to show their impact on vegetation. Since "bio-ecology" really is a redundant term, its usage did not persist.

Perhaps the most famous review in *Ecology* is Hutchinson's review of Clements and Shelford's *Bio-Ecology* (1939). Their book was a commendable uniting of plant and animal ecology between two covers. Hutchinson (1940) acknowledged the book's virtues, then complained that it was obsolete, because it was entirely descriptive and had no biogeochemistry or mathematical reasoning on metabolic aspects of communities. (It sold well anyway, for it is difficult to defeat something with nothing.)

Another taking stock of ecological concepts was Warder Allee and Thomas Park's "Concerning Ecological Principles" (1939). They thought that ecologists and others imagined that ecology had fewer theoretical principles than it did. Perhaps because they did not claim that their set of principles was necessarily definitive, they did not enumerate them, but Robert McIntosh (1976:358–359) did so for them, listing nine: law of the minimum, adaptation, community, succession, population growth, cooperation and competition, niche separation, geographic distribution, and emigration and dispersal. However, McIntosh's listing is incomplete, as it does not explicitly include their discussions of gregariousness, web of life, biotic cycles, or balance versus unbalance. Regardless of whatever influence their analysis had on others at the time, it had a lasting influence upon them and three of their zoologist colleagues: Alfred Emerson, Orlando Park (Thomas' older brother), and Karl Schmidt. These five Chicago colleagues, three at the University of Chicago, Orlando Park at Northwestern, and Schmidt at the Field Museum of Natural History, were members of an informal Chicago Ecology Club, where they debated their ideas (Allee et al. 1949:viii–ix; Mitman 1992:134–136). The eventual result of those interactions would be a landmark encyclopedia, *Principles of Animal Ecology* (Allee et al. 1949).

Warder Clyde Allee (1885–1955)

(From Schmidt 1957 or *Ecology* 37(2), 1956, frontispiece. With permission.)

Allee was born into a Quaker farm family near Bloomingdale, Indiana. He earned a B.S. degree from Quaker-supported Earlham College (1908), and M.S. and Ph.D. degrees, under Shelford, at the University of Chicago (1910, 1912). He taught in four other Midwestern institutions before returning to the University of Chicago in 1921 (Shelford had left in 1914), where he remained until 1950, when he moved to the University of Florida (1950–1955). During the summer in 1914 to 1921, he taught at the Marine Biology Laboratory, Woods Hole, Massachusetts. Allee's Quaker concerns for peaceful human communities influenced his ecological interests, which focused on community and cooperation among animals, beginning with freshwater isopods he found in the Indiana dunes—a favorite site for ecological studies at the University of Chicago. He published many ecological articles and also three books: *Animal Aggregations: A Study in General Sociology* (1931), *Animal Life and Social Growth* (1932), and *The Social Life of Animals* (1938). He joined four other animal ecologists in the Chicago region—Alfred Emerson, the Park brothers Orlando and Thomas, and Karl Schmidt—to write a landmark encyclopedic, *Principles of Animal Ecology* (Allee et al. 1949). Allee was ESA president in 1929 (Emerson & Park 1955; Schmidt 1957; Engel 1983:172–182; Banks 1985; Kimler 1988; Mitman 1988, 1992, 1999; Collias 1991; Burgess 1996:11; Sterling 1997a). A symposium in September 1954 held in Gainesville, Florida, on the Social Organization of Animals was held to honor Allee; and the April 1956 *Ecology* issue was named the Warder C. Allee Memorial Issue, with a collection of animal behavior papers (Scott 1956a,b). He was elected president of AAAS for 1960 (Wright 1960).

Animal ecologist Charles C. Smith, University of Oklahoma, addressed a timely topic in his study on "The Effect of Overgrazing and Erosion upon the Biota of the Mixed-Grass Prairie of Oklahoma" (1940). His use of the term "biota" in his title was accurate, as he studied the impact upon both the flora and fauna. In studying the impact upon plant species, he could draw upon the published works of Clements and Weaver, and he confirmed their conclusion, that overgrazing in the spring when grasses were growing was much more damaging to the total amount of vegetation produced than if cattle were rotated to allow grasses to grow to maturity before heavy grazing. His findings concerning animals were variable, because grazed vegetation created a more favorable environment for some species while creating a less favorable environment for other species. Counterintuitively, hemipteran and homopteran insects increased in abundance in overgrazed lands than in less-grazed lands, but other orders of insects declined in numbers with overgrazing. Birds eating insects increased when the insects did, but moderately grazed land contained more meadowlark nests than undisturbed land, and overgrazed land had fewer nests than moderately grazed land. Small mammals that tunnelled under the prairie were not necessarily affected badly by overgrazing, but larger species found less cover in overgrazed land and were more vulnerable to predators.

One of the main concepts of ecology is the ecosystem, which British scientist Roy Clapham named and defined in 1930, at the request of plant ecologist Arthur Tansley (Morris 2009:768–769), who made it famous in his polemical essay, "The Use and Abuse of Vegetational Concepts and Terms" (1935), which appeared in *Ecology*, despite the fact that Tansley was editor of the British *Journal of Ecology* (Hagen 1992:86–87; Golley 1993:33–34). "Ecosystem" was not the main point of his essay and occupied less than 4 pages in his discussion (Tansley 1935:299–303). He was busy with other projects and never developed the concept further himself. Three historians of the ecosystem concept (Major 1969:12; Hagen 1992:7–11; Golley 1993:36–37) have argued persuasively that the idea goes back at least to Stephen Forbes' address on "The Lake as a Microcosm" (1887).

"Ecosystem" first attracted attention when Raymond Lindeman (1915–1942) used the term in his landmark theoretical essay, "The Trophic-Dynamic Aspect of Ecology" (1942b:400; Cook 1977; Hagen 1992:87–95; Golley 1993:48–56). Minnesotan Lindeman received a Ph.D. at the University of Minnesota, where he was influenced by Professor William Cooper (Cook 1977). His dissertation had focused upon nutritional relations among the inhabitants of a shallow lake in succession to a terrestrial habitat. He and his wife had taken extensive samples at Cedar Creek Bog (Minnesota) of aquatic plants, phytoplankton and zooplankton, and animals, which he related to the community with a trophic-dynamic viewpoint. He published two articles derived from his dissertation (Lindeman 1941a,b) and went to Yale for postdoctoral studies under Hutchinson. Lindeman's trophic-dynamic concept also owed an acknowledged debt to the German limnologist, August Thienemann (1882–1960). A diagram in Thienemann's "Der Nahrungskreislauf im Wasser" (1926:57) influenced the development of Lindeman's famous trophic-dynamic food cycle diagram (Lindeman 1942b:401; Egerton 2007:60–61). After Lindeman's early death, Hutchinson prepared Lindeman's paper for publication and submitted it to the *Ecology* editors, Thomas Park (zoology) and Charles E. Olmsted (botany), both at the University of Chicago. They sent it to two limnologists to referee: Juday at the University of Wisconsin and Paul Welch at the University of Michigan. Both rejected it. Juday commented that lakes are "stubborn about fitting into mathematical formulae," to which McIntosh (1985:197) commented that this seems similar to Haeckel's criticism in 1890 of Hensen's quantification studies on plankton (Lussenhop 1974; Egerton 2012a:199–200). Welch objected to a lack of data. However, Lindeman had already published several articles, including the lead one in the January 1942 issue of *Ecology* (Lindeman 1942a). Park was not convinced by his referees, and Hutchinson strongly defended Lindeman's article, and by 1941 Hutchinson was already as respected a limnologist as Juday and Welch were, if not more so (Slack 2010:149–151). This was another example of Hutchinson pushing for a "new" ecology and limnology against the old guard. It was published in *Ecology* (Lindeman 1942b). During World War II, the ecosystem concept did not progress further.

However, American ecologists were not generally resistant to theoretical and quantitative papers: the first paper in the January 1940 issue of *Ecology* was Edward Haskell's "Mathematical Systematization of 'Environment,' 'Organism' and 'Habitat.'" In the same issue, Thomas Park, *Ecology*'s zoology editor, enthusiastically reviewed *Quantitative Zoology* (1940) by George G. Simpson and his wife Anne Roe. ESA had a committee on quantitative ecology.

Zoologist Orlando Park of Northwestern University, at the 1944 ESA Annual Meeting in September, gave a speech as retiring president, "Observations concerning the Future of Ecology." With World War II winding down—D-Day had been June 6th—it was an occasion in which one might optimistically envision the future of ecology during peacetime. Instead, Park compared ecology to other biological disciplines, such as genetics and physiology, and found that "ecologists have 'missed the boat' so many times that the process is becoming habitual" (Park 1945:1). Specifically, he compared progress in ecology and genetics, since their modern histories had begun at about the same time. Mendel (1822–1884) and Haeckel (1834–1919) were contemporaries. Around the turn of the century, Mendel's work in genetics was rediscovered and Hugo de Vries' mutation theory was published. In ecology, both Warming and Cowles developed a theory of succession. Then, geneticists bred *Drosophila* and mice and Morgan developed his gene theory. Geneticists were consolidating their science, but for ecology, that was impossible. Variety and complexity made early ecological consolidation impossible. Park thought ecologists should have responded to this situation by collaborating with biologists in other disciplines, such as geneticists, taxonomists, and evolutionists, but failed to explain what might have come from doing so. He then offered 10 recommendations for the future, including teaching, cross-discipline collaborations, and developing a program in tropical ecology at Barro Colorado Island in the Panama Canal Zone.

W(alles) Thomas Edmondson (1916–2000)

(From [facing t.p.] *Limnology & Oceanography* 33, November 1988) or *ESAB* 84(11), 2003. With permission.)

Although he was born in Milwaukee, Edmondson's widowed mother moved her three sons to New Haven, Connecticut, in hopes of them going to Yale (Hutchinson 1988:1231). In high school, he became interested in studying rotifers, and his biology teacher took him over to Yale, where Hutchinson provided a place in his own laboratory to study them. Hutchinson and others provided rotifers for him to study, and he published his first scientific paper at age 18: "Investigations of Some Hispaniolan Lakes. I. Rotatoria" (1934). A steady succession of other papers was only interrupted for two years during World War II, when he worked for the U.S. Navy. He earned his B.S. degree (1938) at Yale, and then studied for a school year under Juday at the University of Wisconsin, where he met Yvette Hardman, his future wife, who would become editor of *Limnology and Oceanography* for 19 years. After earning a Ph.D. at Yale (1942) and conducting research for the U.S. Navy, he became an instructor at Harvard University. In 1949, he moved permanently to the Zoology Department, University of Washington (Burgess 1996:38; Hairston 2000a). In 1955, he began a study of eutrophication in Lake Washington, which became "the greatest single case study of eutrophication and recovery yet conducted" (Lehman 1988:1237; Klingle 2007:211–222). Edmondson's paper on the "Reproductive Rate of Rotifers in Natural Populations" (1960) explained that "the ratios between eggs and females in plankton collections could be converted into quantitative estimates of birth rates" (Lehman 1988:1237). He became ESA's Eminent Ecologist in 1984, and he also won the Hutchinson Award of the Society of Limnology and Oceanography and the August Thienemann-Einar Naumann Medal of the Societas Internationalis Limnologiae (Hairston 2000b).

Wartime sometimes affected the quality of the paper used, but it did not stop the publication of scientific books. In 1944, Professor Stanley Cain, Botany from the University of Tennessee, published *Foundations of Plant Geography*, which received a detailed, well-informed review in *Ecology* from Carl Sharsmith, at the University of Minnesota. Sharsmith explains (1945) that Cain necessarily followed in the footsteps of Ronald Good in England and H. L. Mason in America, but Cain also went beyond them in trying to synthesize the two traditions of floristic and vegetational phytogeography. This was an interesting challenge that Sharsmith explained was reconciling one school that could not see the forest for the trees with the other that could not see the trees for the forest. The synthesis could not be fully achieved, but Sharsmith judged Cain as having been considerably advanced for the discipline.

World War II had various impacts upon ecologists (Miyanishi 2014:105).

John Curtis, at the University of Wisconsin, was offered a directorship of a research program in Haiti to produce a rubber substitute from *Cryptostegia* (Burgess 1993a:11–13). That project was unsuccessful, but while there, he became interested in the tropical flora and corresponded with Gleason at the New York Botanical Garden. Gleason sent Curtis reprints of some of his papers, and Curtis studied the vegetation in a Haitian forest and published a paper on it in 1947, after he returned to Wisconsin. Eugene Odum, with a recent Ph.D., arrived at the University of Georgia in September 1940, he had received a deferment from the draft in order to teach zoology to nursing and pharmacy students (Craige 2001:33). His younger brother, Howard, after studying for two years at the University of North Carolina (UNC), served three years in the Air Force as a "tropical meteorologist, where undoubtedly he gained his basic interest in large systems and the energetics behind them" (Brown, Hall, & Jorgensen 2004:4). He served in Puerto Rico and the Panama Canal Zone, then returned to UNC and graduated in 1947. A group of Hutchinson's former students worked at Woods Hole (Massachusetts) on war-related marine projects (Slack 2010:131). Catherine Keever (1985) was teaching high school during the war, while working on graduate degrees at Duke. She was able to obtain summer scholarships at the Highlands Biological Station in western North Carolina since there were few male students with whom to compete. There was no annual meeting of the ESA in 1945.

A few ESA members have published autobiographies—including Beebe (in many books and popular articles), Harshberger in 1928, Sumner in 1945, Pearse in 1952, Errington in 1973 (posthumous), Hutchinson in 1979, and Keever in 1985—which shed some light on their careers. (Gleason and Riley left unpublished ones, which have been consulted by historians.) Francis Sumner (1874–1945) had a varied career and published *The Life History of an American Naturalist* in his last year. He died on September 5, 1945, and likely lived to see his book published. It received a friendly, substantial review in *Ecology* by Graham DuShane (1946), who had known him. Sumner wrote it during wartime when there was little concern in society for the environment, and he made clear his low opinion of American culture. DuShane came to his defense, writing that Sumner had sold himself short by not making clear his warm feelings for his family and others. *American Naturalist* lacks illustrations, bibliography, and index. One can, however, find photographs and Charles

Child's biographical memoir of him (1947) on the Internet. Child's biography, for the National Academy of Sciences, includes Sumner's complete bibliography.

Sumner was born in Pomfret, Connecticut, grew up outside Berkeley, California, and earned a B.S. degree from the University of Minnesota (1894) and a Ph.D. from Columbia University (1901). He taught at City College of New York officially from 1899 to 1906, but he did not enjoy doing so and took a two-year leave of absence. He became the director of the U.S. Bureau of Fisheries Laboratory at Woods Hole until 1911, when he moved to the Scripps Institution of Oceanography, La Jolla, California, for the remainder of his career. He had written his Ph.D. dissertation on fish physiology and continued that research for awhile at Woods Hole, and then began collaborating with three others on *A Biological Survey of the Waters of Woods Hole and Vicinity* (Sumner et al. 1913). The other three only conducted research in the summers, when dredging on the *Fish Hawk* at 458 stations was a bit like being on vacation. It was up to Sumner to do the tabulations and other work during the winter and to write up the results. In California, he conducted similar collaborative research with two others in San Francisco Bay, but that study was limited to physical conditions (Sumner et al. 1914). However, he had not given up his research on animals. In 1910, during a visit to the Stazione Zoologica in Naples, Italy, he became interested in the protective coloration in fish, and he continued that research for years. He also successfully studied the geographical races of deer mice (*Peromyscus*) in order to shed light on evolutionary theory. His findings were widely appreciated and cited by evolutionary biologists (Provine 1979).

The Depression of 1929 to 1940 was emotional as well as economically devastating, and World War II was a constraint on noncombatants at home. American and Canadian ecologists at least had an easier time than did ecologists in Britain, Europe, and the Soviet Union. The ESA continued publishing *Ecology* and *Ecological Monographs*, and ecologists in the United States and Canada were in positions to flourish after the war, where the war damage was absent; more easily than ecologists in Europe and Russia, where the war damage was substantial.

3 1945 to 1959
Expansion

The allied victory in World War II left the United States in a commanding position, economically and politically, and Canada also prospered. Other participant countries—whether winning or losing—faced repairing lives, buildings, infrastructure, and economy. American government and commerce were eager to assist other countries in recovery, while America simply adjusted to peacetime prosperity. The G.I. Bill enabled many U.S. veterans to attend institutions of higher education. However, faculty at those institutions still had to compete for students, even within departments. By 1949, ESA membership began a rapid increase that has continued to the present day (Burgess 1981a,c and see Figure 2.1 in Chapter 2). As the ESA expanded, it began presenting two awards for outstanding contributions (Burgess 1977:19–21). Dr. Frank W. Preston established the Mercer Award in memory of British ecologist George Mercer, who was killed in World War I. It is awarded for an outstanding paper in an ESA journal and was first presented in 1949, to Henry P. Hansen. The Eminent Ecologist Award was first presented in 1953, to Gleason, for a lifetime contribution to ecology. These awards are presented at annual meetings and are announced in the *Bulletin*.

During the war, the federal government had funded extensive research in universities and within government agencies for a wide range of topics, and many of those involved wanted to see this funding continued into peacetime. The subject was widely debated by government advisors and congressional members during the war and afterward, and new funding began in 1951 (Penick et al. 1965:72–89, 179–190; Kevles 1977; England 1982). Until the National Science Foundation (NSF) became a significant source of research funds, wartime agencies continued to support scientific research.

Orlando Park was ESA president in 1944, and his presidential address as his term ended was on the future of ecology, and it became the first article in the January 1945 issue of *Ecology*. Since ecology was poised on the verge of a successful expansion, his worries now seem overly pessimistic and his suggestions seem rather obvious. However, at the time, he lacked our hindsight, and his concerns likely found a receptive audience. Perhaps one of the reasons why ecology did expand successfully was because ecologists heeded his advice. He wanted to see ecologists collaborate with geneticists and taxonomists to advance the understanding of evolution. Ecologists should do more to expose undergraduates to ecology, in general biology courses, in laboratory exercises, and in textbooks. Ecology textbooks should be written for graduate students, ecologists should collaborate in placing such students in graduate programs that best meet their needs, and ecologists should collaborate with government agencies to solve conservation problems. His final recommendation was that American ecologists should make a strong commitment to advancing an understanding of tropical ecology. He had himself done research at the Institute for Research in

FIGURE 3.1 Profile diagrams of tropical montane formations. (From Beard 1944:144. With permission.)

Tropical America on Barro Colorado Island in the Panama Canal Zone (1942) and recommended it as a place to start such research.

Park was an animal ecologist, and he may have overlooked a study that British plant ecologist J. S. Beard had published in the April issue of *Ecology*: "Climax Vegetation in Tropical America" (1944). Beard had conducted research on Trinidad and Tobago Islands. He was interested in the classification of vegetation, and he supplemented verbal definitions with photographs and vegetation diagrams (Figure 3.1). He continued these studies and offered a revised and more comprehensive version of his classification in the January 1955 issue of *Ecology*, this time dispensing with photographs.

In the same January 1945 issue of *Ecology* as Park's address, Carl Sharsmith had a 2-page favorable review of Stanley Cain's *Foundations of Plant Geography* (1944). Cain attempted a synthesis of two traditions, of floristic and vegetative plant geography. Sharsmith commented (1945:101) that "the proponents of one school have failed to see the forest for the trees, and conversely that those of the other have failed to see the trees for the forest." Sharsmith acknowledged that a synthesis was a difficult challenge, and he thought that Cain had sometimes fallen short, but that he had nevertheless achieved a "masterful and stimulating" work.

In the second issue of *Ecology* for 1945, Alfred Redfield had a more restrained 1-page review of Frank R. Lillie's *The Woods Hole Marine Biological Laboratory* (1944). Actually, Redfield praised Lillie's book; his restrained enthusiasm was for the institution itself, which he explained had a misleading name. It is indeed at the seashore, but its research participants came from all aspects of biology, with marine ecology being a negligible field of research there. Redfield lamented (1945:208) that "Marine ecology has developed much more slowly than the ecology of lakes, and the land in this country, and still occupies a minute share in the total biological effort." This, despite the establishment of marine research stations beginning in the 1870s (Egerton 2014c:348–374).

One slight retrenchment during this period of expansion occurred in 1945. A referendum decided that ESA should not be directly involved in saving natural areas for research (Dexter 1978). Two preservation committees were ended. Shelford was unhappy with this decision and organized an Ecologists' Union to continue advocating for preservation of natural areas (Birchard 2005:19–20, 235–238). Ironically, in 1950 the ESA and the Ecologists' Union held a joint session at an American Institute of Biological Sciences (AIBS) meeting on "Ecological Researches in Natural Areas," and afterward the Union held its annual meeting and changed its name to Nature Conservancy in order to broaden its appeal for nonecologists. The new name was borrowed from Britain, where the Nature Conservancy was established by the government to administer natural areas and where ecological research was encouraged. The American Nature Conservancy is a nongovernmental organization that, despite a potential confusion of two different organizations sharing a name, has become the largest environmental organization in the world (Birchard 2005).

In the late 1950s, the Federation of Ontario Naturalists (FON) began to have the same preservationist anxieties that had led Shelford's group to found the Nature Conservancy. In 1961, FON established a committee that studied both the British and U.S. Nature Conservancies that in 1962 recommended that Canadian naturalists establish a similar land-trust organization. On November 28, 1962, the Nature Conservancy of Canada "became duly constituted as a trustee-governed, no-share capital corporation (a not-for-profit corporation), incorporated under the Canada Corporations Act, with charitable status under the Income Tax Act" (Freedman 2013:39). It has been an equally successful organization as its British and U.S. predecessors.

In 1946, (Rand) Aldo Leopold (1887–1948) became ESA vice president, and in 1947, president. He was not an active member and wrote to a fellow member on January 11, 1947: "I am astonished by my election. I feel deeply the responsibility implied in my being elected despite this failure on my part" (Meine 1988:493, 1997, 1999). He was known for advocating an understanding of the environment of game animals in order to maintain game stocks for hunting. However, Leopold also had a genuine interest in natural history, influenced somewhat by his reading of Thoreau (Lannoo 2010). One of Thoreau's projects had been keeping phenological records (before the word was coined), and in 1935 Leopold began his own phenological records. After a decade, he decided to publish them, with the aid of a plant ecology graduate student, Sarah Elizabeth Jones. He sent the manuscript to the editor of *Ecological Monographs*, where it was published (Leopold & Jones 1947). It seems likely that the phenological manuscript, being read by other ecologists for the editor, may have given someone the idea of nominating him for vice president, and a year later for president. His election reflected well on both him and ESA. He was the best-known professional ecologist, going back to the popularity of his *Game Management* (Leopold 1933) and other writings on the subject. He died on April 21, 1948, before he could deliver a past president's address at the ESA annual meeting, and before his now-famous *Sand County Almanac* (Leopold 1949) was published. However, he had expressed an intention to base his address upon that book, and his colleague and former student, Joseph Hickey selected "The Land Ethic" as the most appropriate for the address, which his successor as president (and friend) Paul Sears read on Leopold's behalf at the annual meeting (Meine 2014).

In 1947, Hutchinson published in *Ecology* "A Note on the Theory of Competition between Two Social Species," which covers about 2 pages. This seems to be his first foray into population ecology, though he had raised doubts about Charles Elton's demographic understanding in his review (Hutchinson 1942) of Elton's *Voles, Mice, and Lemmings* (1942). Hutchinson persisted in this new field, in which two former students had already published textbooks (Slobodkin 1961; MacArthur & Connell 1966), and his only textbook was *An Introduction to Population Ecology* (1978). He had a strong background in mathematics, which was already evident in this first note. In 1948, he published "Circular Causal Systems in Ecology," a very different subject and a major contribution to the foundation of systems ecology. His thesis was that groups of organisms "may be acted upon by their environment, and they may react upon it" (1948:221). This is a commonplace concept now because of that article.

Two years after Redfield vented his frustration in his review of Lillie's *The Woods Hole Marine Biological Laboratory* at the comparatively slow progress in marine ecology, and he reviewed Robert Coker's *This Great and Wide Sea* (1947) in *Ecology.* Coker was a marine ecologist, and Redfield approved of the book, somewhat, for that reason. But, it was not quite what Redfield wanted; its defect was that Coker wrote it for two audiences, undergraduates in a course on marine ecology and the public. In one respect, Coker met this challenge with "abundant and superb" photographs, but Redfield thought that Coker had skimped on maps, diagrams, and tables of numbers needed by students out of deference to the taste of the public.

In 1947, Carl (Leavitt) Hubbs (1894–1979) and Karl F(rank) Lagler (1912–1985), both on the University of Michigan faculty, published the second edition of their *Fishes of the Great Lakes Region*, a work that never went out of print, with a recent revision done by Gerald R. Smith (Hubbs, Lagler, & Smith 2004). Reviewing the second edition were Loren Woods and Robert Miller, both at the U.S. National Museum. They were very pleased with the improvements over the first edition (published in 1941). It was thoughtfully designed for convenience to carry afield, yet it had 26 color plates, a map, and numerous drawings. It was not just a manual for identification, but also included a summary of the development and drainage history of the Great Lakes, "ecological information concerning life histories," and references (Wood & Miller 1948:222). The origin of the fauna was correlated with the retreat of glaciers, with species coming from the northwest, south, and east. Hubbs brought 35 years of study to the project.

As ESA expanded, it began giving two awards for outstanding contributions to ecology (Burgess 1977:19–21). Dr. Frank W. Preston established the Mercer Award in memory of British ecologist George Mercer, who was killed during World War I. It is awarded for an outstanding paper in an ESA journal. The first Mercer Award was made in 1948 (reported in the August 1949 *Bulletin*) to Henry P. Hansen for his paper "Postglacial Forest Succession, Climate, and Chronology in the Pacific Northwest" (1947). The Eminent Ecologist Award, for a lifetime of contributions to ecology, was first awarded in 1953, to Henry Gleason (who received it again in 1959). Computer-based citation statistics has made it possible for the non-ESA periodical,

Current Contents—Life Sciences, to identify ecological citation classics since 1947. The titles are limited to the English language but are inclusive worldwide and are based on publications cited 55 or more times in the scientific literature. Robert (Mac) McIntosh (1989) explained the background of this selection method and compiled a list of the Ecological Citation Classics (ECC) (1947–1977), with authors' names, an abbreviated title of each ECC, and the journal in which it appeared. The list identifies important articles, chosen by citation statistics rather than by a committee of judges.

A kind of celebration of the war's end was the achievement of five Chicago animal ecologists who published an encyclopedic *Principles of Animal Ecology* (1949, xii + 837 pages)—Warder C. Allee, Alfred E. Emerson, Orlando Park, Thomas Park, and Karl P. Schmidt—since this large publication during wartime would have been impractical. The idea for the book arose from discussions at the Chicago Ecology Club. They felt the need for "a study of the underlying principles together with a sampling of the evidence on which they are based" (Allee et al. 1949:vii). Each coauthor assumed responsibility for a portion of the text:

> Preface, Introduction, Chapter 1: K. P. S. (based on drafts by W. C. A. and T. P.). Chapters 2, 4 to 16, inclusive, and 23: W. C. A. Chapters 3 and 18 to 22, inclusive: T. P. Chapter 17: W. C. A. and K. P. S. Chapters 24 and 31 to 35, inclusive: A. E. E. Chapters 25 to 29, inclusive: O. P. Chapter 30: K. P. S. and O. P. General editing: K. P. S. The four junior authors here acknowledge the leadership of Dr. Warder Clyde Allee and their indebtedness to him throughout the preparation of the present work.

Its review in *Ecology*, of almost 3 pages, was by LaMont C. Cole, at Cornell University, New York. He readily agreed (1950:153) with the publisher, Saunders' promotional claim that it is "without question the definitive work in the field." However, Cole doubted the wisdom of Saunders' additional claim that elementary biology students could understand a large part of it. On his review's third page, Cole summarized his "conviction that *Principles of Animal Ecology* supersedes all existing volumes on general or animal ecology, that it is an indispensable work for those who seek a broad understanding of biological phenomena generally, and that it will be a source of inspiration for research, speeding ecology toward a more mature and analytical future." And, "it is a reference book which every ecologist must possess." Cole did not mention that he had earned B.S. and Ph.D. degrees at the University of Chicago, where he had studied under three of these five authors.

That January 1950 issue of *Ecology* also had reviews or summaries of other books of general interest, though none were on the same heroic scale as *Principles of Animal Ecology*. Lagler's *Studies in Freshwater Fishery Biology* (1949) had gone through two previous mimeographed editions for his students before he gave it to a publisher. Its reviewer, Albert Hazzard, called it a pioneering work because of its broad scope and its value for courses in fisheries management. Two book notices by T. P. (Thomas Park) were about a half-page each: Marston Bates' *The Natural History of Mosquitoes* (1949) and Edward A. Steinhaus' *Principles of Insect Pathology* (1949). Park highly praised both as important contributions to entomology. Two books which received only a paragraph each and no indication of who summarized them were Albert H. Wright and A. A. Wright's *Handbook of the Frogs and Toads of the United States and Canada* (1949) and Asa C. Chandler's *Introduction to Parasitology* (1949). The reason for their brief treatment was that it was the third edition of the Wrights' book and the eighth edition of Chandler's.

The October 1950 issue has an article, with a map, "Vegetation Zones of Mexico," by Starker Leopold (one of Aldo Leopold's sons). He was a wildlife biologist, and he explained that "The objective has been to divide the country into natural bio-geographical units, thereby simplifying considerations of the problems of wildlife distribution and wildlife conservation" (Leopold 1950:507). He drew upon his own observations, available literature, and the advice of three botanists: Sharp, Gilly, and Gentry. He acknowledged that his map did not meet the standards of botanists, but that it was basically useful for practical purposes. His sources included P. C. Standley's *Trees and Shrubs of Mexico* (5 parts, 1920–1926, 1721 pages); E. W. Nelson and E. A. Goldman's "Mexico" (1926) with its own vegetation map of Mexico (Figure 3.2) in the *Naturalist's Guide to the Americas*; and six articles by Forrest Shreve (1934–1944). Shreve died in July 1950, so he would not have seen Leopold's map, unless he had read Leopold's article for the *Ecology* editors.

A German ethologist, Karl von Frisch (1886–1982), would become one of three winners of a Nobel Prize in 1973 for studies on animal behavior. In 1950, he published *Bees—Their Vision, Chemical Senses, and Language* (Cornell University Press). Can an American reviewer recognize a future Nobel Laurate? Yes! The book is actually lectures that he had delivered at Cornell, which would have been a clue that he should be taken seriously. The reviewer, ethologist Theodore C. Schneirla, of the American Museum of Natural History, fully appreciated Frisch's work and wrote an enthusiastic 2.5-page review (1950). Schneirla may never have joined ESA, but its editors sought the best reviewer for a book. In 1955 when Frisch published *The Dancing Bees*, William Dreyer (1955) also wrote an enthusiastic review but did not indicate how it differed from Frisch's *Bees* (1950).

A Dutch ethologist, Niko Tinbergen (1807–1888), would be another winner of that 1973 Nobel Prize. His *Social Behaviour in Animals, with Special Reference to Vertebrates* (1953a) was reviewed for *Ecology* by Nicholas E. Collias, a behaviorist (1953) with enthusiastic appreciation similar to Frisch's reviewers. When Collias also reviewed Tinbergen's *The Herring Gull's World* (1953b) for *Ecology* (Collias 1955), there was no problem separating it from Tinbergen's other book since their scopes differed. This book summarized observations and experiments that Tinbergen and his students had conducted over a number of years, and Collias thought the book was a progress report on breeding behavior that was continuing. He praised Tinbergen's work for inspiring research by others.

Schneirla's colleague at the American Museum of Natural History, plant ecologist Frank Egler, wrote the most provocative essay review ever published in *Ecology*: "A Commentary on American Plant Ecology, Based on the Textbooks of 1947–1949" (1951). In acknowledgment of incendiary aspects of his review, he stated that his manuscript had been read by 13 prominent plant ecologists, whom he listed and whose comments ranged from high enthusiasm to high censure. He reviewed four books that he said "are a fitting culmination to the first half century of American

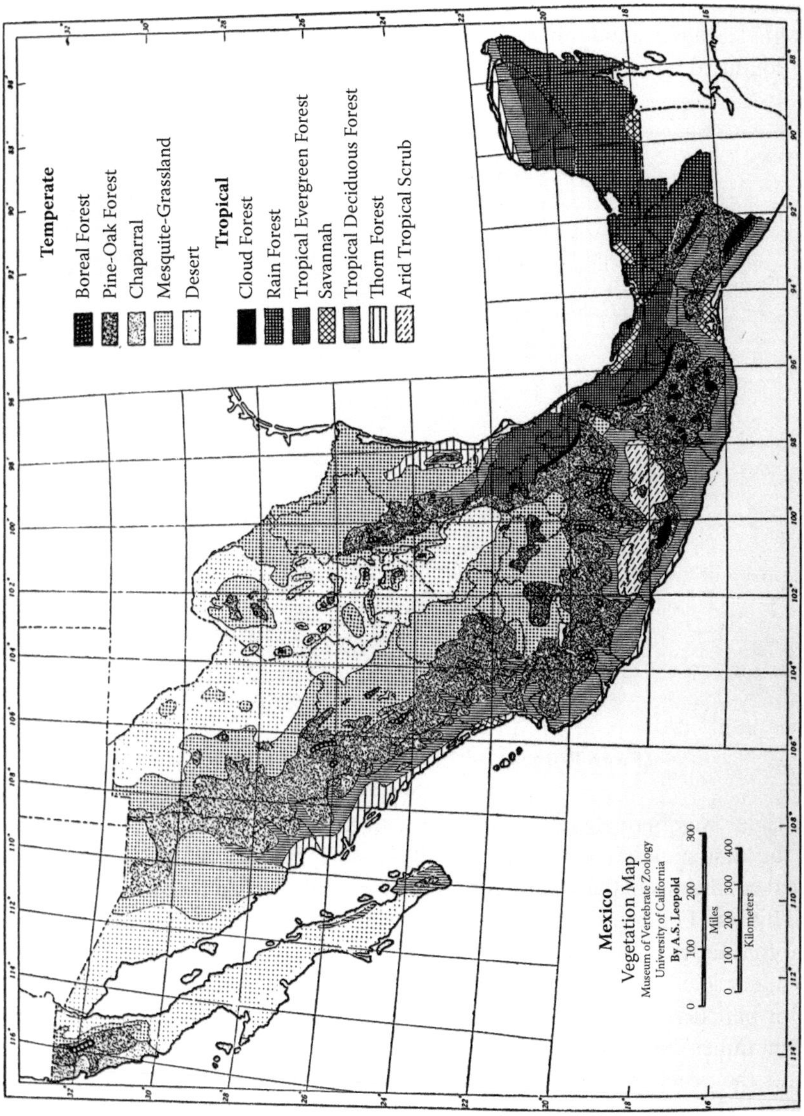

FIGURE 3.2 Vegetation zones of Mexico. (From Leopold 1950:508. With permission.)

Plant Ecology, and were they written to order for that purpose, they probably could not be improved" (Egler 1951:675). He further explained, "The comments on these four books are not to be considered primarily a review of their factual contents and a spotlighting of the good and bad in omission and commission … comments are in the nature of a critical analysis of certain methodologic and epistemologic foundations of American Ecology" (Egler 1951:676). He deplored the blind following of traditional dogma as against the spirit of science. He disavowed the notion of propagating his own dogma to replace that of others. His goal was Socratic, not Aristotelian.

Frank E(dwin) Egler (1911–1996)

(From Burgess 1997:193. With permission.)

When Egler was born there was no one with a tape recorder, but if there had been, the first sounds recorded would likely have been a complaint about the crowded conditions in the womb. And that imagined complaint seems characteristic of his life. For example, here is C. E. Cushing's editorial comment prefacing a letter that Egler published in the ESA *Bulletin* 67 (September 1986:233): "the following article by Dr. Frank Egler has been published exactly as he desired in terms of punctuation, spacing, grammatical usage, etc. …I am publishing it in this form rather than rejecting it outright, to provide members with his perceived views of the problems with the field of ecology." Former ESA *Bulletin* editor William A. Niering then wrote a letter complaining about Cushing's editorial comment (*ESAB* 68:139–140), to which Cushing replied that he had published his remarks with Egler's permission.

If Egler were merely a crank, he would not merit space here. He was a New Yorker who received a B.S. degree from the University of Chicago under Cowles

(1932), an M.S. from the University of Minnesota under Cooper (1934), and a Ph.D. from Yale under Nichols (1935) (McCormick 1994b). He received the 1978 Distinguished Service Citation and the awards committee (A. Lindsey, chair; C. Muller; and J. Tanner) explained that it was granted (ESA *Bulletin* 67 December 1978:169), "Because of his important and long-sustained work as book reviewer, seminar-leader and advisor, and his research in the science of vegetation and the art of management," and that he had "written five books and over 200 papers in fields from ecology and conservation to electrical engineering and physics." In an obituary, Robert Burgess wrote that Egler's "A Commentary on American Plant Ecology, Based on the Textbooks for 1947–1949" in *Ecology* "set a standard for scientific critique for over 40 years" (Miller et al. 1978b; Burgess 1997:193; Niering 1998).

The first of these books was a posthumous collection of Frederic E. Clements' writings, *Dynamics of Vegetation* (1949). Egler thought that Clements' influence was because he had a talent for understanding and communicating the practical facts of land management, and that talent had therefore persuaded readers to accept his idealist philosophy, which was "as neatly organized and arranged as the components of Dante's *Inferno*" (Egler 1951:677).

A second book was by Rexford Daubenmire (1909–1995)—who would serve as ESA president in 1967 (Burgess 1996:34–35)—*Plants and Environment: A Textbook of Autecology* (1947). Egler (1951:680) could complement as readily as criticize:

> The seven chapters on separate factors (soil, water, temperature, light, atmospheric, biotic and fire) are without doubt admirably written, well organized, thoroughly annotated, and highly suitable as a source of reference for the great mass of detailed information that can be segregated under these headings.

By now, however, the reader anticipates that any Egler complements are a prelude to accompanying criticisms. His foremost criticism was that single-factor chapters lead to minimizing the complexity of multiple causes operating upon plants. Another Egler criticism was that Daubenmire only discussed an average of any factor's magnitude, with no consideration of the impact of droughts, hurricanes, and other irregularities in nature.

The third text that Egler evaluated was *The Study of Plant Communities: An Introduction to Plant Ecology* (1948) by Henry J. Oosting (1903–1968), who served as ESA president in 1956 (Anon. 1955; Billings 1970; Burgess 1996:83). Anyone expecting to read about Egler's enthusiasm for synecology after his condemnation of autecology was overly optimistic. Egler acknowledged (1951:681) that Oosting had "organized his excellent command of the pertinent literature, to draw for all of us a detailed portrait of that speculative philosophy that can be known as Synecology." Egler judged that Oosting's synecology actually borrowed the single causation outlook of autecology and also Clements' dynamic ecology. Oosting's *truth* "follows the party lines closely, and is supported by such a wealth of accessory data and excellent illustrations that even the deepest disbeliever will read the book with pleasure and profit" (Egler 1951:681). Egler devoted just over 2 pages to a discussion of Clements' book, almost 2 pages to Daubenmire's, and almost 3 pages to Oosting's. His complaints about Oosting's book consisted of a series of smaller complaints than he had for Clements' and Daubenmire's books.

Egler coined the term "symbiontology" to characterize Walter Byron McDougall's *Plant Ecology* (edition 4, 1949). McDougall (1883–1980) was never an ESA officer, but he was designated Eminent Ecologist for 1977 (Niering 1978; Burgess 1996:75). Egler welcomed McDougall's introduction of symbiosis into plant ecology but thought McDougall took the concept too far, while also overlooking pollination by animals, wind, and water. Egler devoted only about a page to this text, possibly because it was not a new work. After considering all four books, Egler went on to expound his general critique of American ecology, including praise for the work of Gleason, Cain, and the Wisconsin group: Fassett, Curtis, and Cottam.

An interesting historical exercise is to compare Egler's review of these four texts (1950) with Stuart Allison, William Ehmann, Aaron Ellison, and John Mull's review (1996) of six comparable textbooks almost a half-century later. They had no philosophical agenda, and so their perspectives were more similar to those of the authors of those texts.

Emma Lucy Braun (1889–1971)

(From *ESAB* 34(9), 1953. With permission.)

Braun was a lifelong resident of Cincinnati, Ohio, and she earned degrees from the University of Cincinnati: B.A. (1910), M.A. (geology, 1912), and Ph.D. (botany, 1914). As an undergraduate, she was interested in both botany and geology, and it was only when she read Harshberger's "Ecologic Study of the Flora of Mountainous North Carolina" (1903) that she realized that plant ecology was where she wanted to devote her career (Braun 1994:63). She joined the university's faculty in 1910, was professor of botany from 1927 to 1946, and was professor of plant ecology from 1946 to 1948 (Stuckey 1973, 1994, 1997a, 2001; Bonta 1991:241–243; Braun 1994; Burgess 1996:18). Her plant ecology studies focused on Ohio and Kentucky vegetation in the Cincinnati region, leading to her greatest work, *Deciduous Forests of Eastern North America* (Braun 1950). That synthesis, however, did not end her research. Braun's later studies included three substantial publications: "The Phytogeography of Unglaciated Eastern United States" (1955), *The Woody Plants of Ohio* (1961), and *An Ecological Survey of the Vegetation of Fort Hill State Memorial* (1969). She was vice president of ESA in 1938 and president in 1950—the first woman to hold either office. She was one of 50 botanists awarded a Certificate of Merit by a jubilee committee of the Botanical Society of America "for her contribution to our knowledge of the origin and structure of the Eastern American deciduous forest. Her critical evaluation of the works of others, her capacity to observe correctly in the field and to interpret forcefully have given biogeographers a new point of departure" (Meyer 1958:16). One of her papers is in the BSA's jubilee volume (Braun 1958). A symposium at the August 1994 meeting of the BSA and AIBS was on "The Eastern Deciduous Forest Since E. Lucy Braun 1950," for which Ronald Stuckey compiled an 83-page booklet of biographical accounts, maps, and photographs. In it, he judged her as (Stuckey 1994:5) "One of the most original thinkers for the developing field of plant ecology during the first half of the twentieth century in North America."

Lucy Braun's research career culminated in her *Deciduous Forests of Eastern North America* (1950, xiv + 596 pages). Harvard's Bullard Professor of Forestry, Hugh Miller Raup (1901–1995), published a 3-page review of it in *Ecology* (1951). He welcomed it as a guide to more intelligent management of the eastern forests. American forestry had begun in the late 1800s, guided by European practice. The vast Northeast and Lake States' white pine-spruce forests had been logged, and hardwood forests grew there. Braun's book included a general survey of soils. Most of her research was in the Cumberland Mountains and Plateau, but she also traveled widely in the East, though she depended extensively upon the forest ecology literature. She reconstructed the original forests settlers encountered, to the extent possible from historical sources and paleontology. "If she had accomplished this alone, her work would have been of monumental value" (Raup 1951:304–305). She subdivided eastern forests into nine regions, based upon topography. She concluded that the other regional forests derived from mixed mesophytic forests, which varied according to topography and climate. Raup thought that American plant ecologists, who built upon the legacy of Clements and Cowles, were so preoccupied with the past and future of a forest that they neglected its present attributes. While Raup thought Braun's descriptions of forests were superb, he regretted her effort to place all forests either in a climax or in a stage evolving to a climax (1951:306). She made substantial contributions to phytogeography. She supplemented her text with well-chosen photographs, tabular data, maps, and a 538-title bibliography.

Animal ecologist Professor Lee R(aymond) Dice (1877–1977), the University of Michigan, published *Natural Communities* (1952, x + 547 pages), which seemed important enough to warrant two reviews in *Ecology* by Ralph Dexter (1953) and Orlando Park (1953). Although both reviewers had minor complaints, both welcomed it as an original contribution that could well serve the profession. Dexter called it a textbook and considered it suitable for undergraduate ecology courses. Dice was critical of succession studies, but Dexter did not think Dice had resolved controversies concerning succession. Park did not call it a textbook and considered it met Dice's purpose as stated in his Preface of describing "the more important methods and concepts…dealing with natural communities." Interest in his book was helped by Dice being elected ESA president for 1953 (Collins 1994b; Burgess 1996:36–37; Chambers 1997). It seems poignant that Dice's book could easily have been modified under the guidance of a textbook publisher (rather than the university press which he chose) to become a textbook that might have captured the high demand that was met by Eugene Odum's *Fundamentals of Ecology* (1953), which cited Dice's book as a source.

The Odum brothers, Eugene and Howard Thomas (later, "H. T."), were the sons of a sophisticated professor of sociology, Howard Washington Odum, and they had grown up watching campus politics and hearing their father's comments about how it played out at the University of North Carolina (Craige 2001:3–12). Eugene and H. T., therefore, would enter the academic profession more sophisticated than most of their peers. Both sons graduated from that university, but they went in different directions for graduate school. Eugene went to the University of Illinois to study under Shelford and Shelford's earlier student, Charles Kendeigh. Eugene was well grounded in animal ecology, and after several other positions, settled at the University of Georgia in 1940 (Craige 2001:22–32). H. T., being 11 years younger than Eugene, after two years at the University of North Carolina, had joined the Air Force during World War II and had learned meteorology (Craige 2001:17–19). After the war, they continued professionally in different directions, but that was not a major detail, because they constantly communicated and shared an ecological perspective.

The prestige of the Atomic Energy Commission (AEC) was enormous after two atomic bombs ended the war with Japan. AEC had ample funds to explore peaceful uses of atomic power, and many Americans expected peaceful uses to solve all sorts of problems. When the Cold War emerged in 1946, this only added to the AEC's prestige and also funding for research. Before his contact with AEC, E. Odum was a small fish in a small pond. When the dozen members of the University of Georgia biology faculty met in 1948 to plan a new core curriculum, his suggestion that it include ecology was rejected, with the excuse that ecology "had no basic principles" (Craige 2001:37–39; E. Odum 2001:16)—that is, ecology had no prestige. He was already thinking of writing an ecology textbook before that snub, and it spurred him on. He had earned tenure with scientific publications and could afford to undertake a new kind of challenge.

Eugene P(leasants) Odum (1913–2002)

(From Barrett 129, 1964. With permission.)

The Odum brothers were sons of a prominent professor of sociology at the University of North Carolina, Howard Washington Odum (Craige 2001:3–12). Both brothers attended the University of North Carolina as undergraduates, and Gene received his B.A. degree in 1934, and his Ph.D. under Charles Kendeigh at the University of Illinois in 1939, with a dissertation on bird ecology. After teaching at three other universities, he settled at the University of Georgia in 1940. Gene Odum became one of the most influential American ecologists during the second half of the 1900s. His ESA publications were only two articles in *Ecological Monographs*, one coauthored with his brother. He served as ESA president in 1965. His influence came partly from his publications, including his very influential textbook, *Fundamentals of Ecology* (1953, edition 3, 1971). Odum was also an effective teacher, and he obtained early funding from the AEC to fund his Savannah River Ecology Laboratory which studied radiation effects upon ecosystems (Craige 2001:48–54). Gradually, he built a very important Georgia Institute of Ecology at the University of Georgia (Hagen 1992:see index; Burgess 1996:82; Barrett and Barrett 2001; Craige 2001:79–143; Golley 2001; Odum 2001; Barrett 2003).

H. Odum's wartime experience as a meteorologist in Puerto Rico was influential on his ecological thinking. His fascination with tropical rain forests arose then and culminated in a gigantic book, *A Tropical Rain Forest* (Odum & Pigeon 1970, discussed in Chapter 4). Before that encyclopedia appeared, however, he had to finish his undergraduate degree at the University of North Carolina (1947) and obtain his Ph.D. under Hutchinson at Yale (1951). H. T.'s education from his father and older brother was as important as what he learned at the University of North Carolina. When he arrived at Yale, therefore, he was already skilled far beyond an average new graduate student, even at Yale.

He immediately appreciated Hutchinson's broad vision and sophistication. He accepted a dissertation topic that Hutchinson suggested, even though it might have sounded pretty humdrum coming from anyone else: *The Biogeochemistry of Strontium*, which he completed in 1950 and published in six articles in 1951 (Limburg 2004). H. T.'s dissertation built upon the theoretical foundations from Lotka (1925), Tansley (1935), and Hutchinson (1943, 1948). Both the United States and the Soviet Union were testing atomic bombs in the atmosphere, and scientists discovered that radioactive strontium (^{90}Sr) was a dangerous by-product from the explosions; it entered the food chain and became concentrated as it traveled up the food chain, which could make milk radioactive. However, radioactive strontium does not occur naturally and did not concern H. Odum, who traced the natural cycle. Even so, ^{90}Sr behaves chemically like other isotopes of strontium and so follows the same cycle in nature (Figure 3.3).

In 1951, the AEC announced that it had acquired 278,000 acres along the Savannah River, near Aiken, South Carolina, to produce material for hydrogen bombs (Bebbington 1990). The AEC needed an ecological assessment of the area and requested proposals from the universities of Georgia and South Carolina. Biologists at both universities responded positively, as they could be patriotic while obtaining research funds. The biologists at the University of South Carolina used the

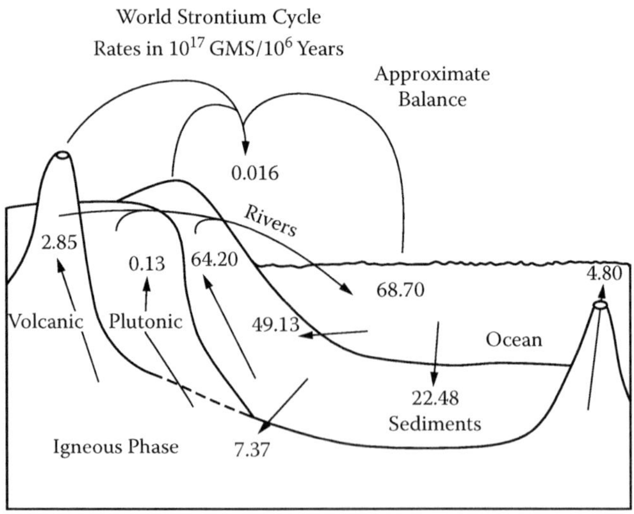

FIGURE 3.3 Global flux of strontium. (From Odum 1951:408. With permission.)

funds for faculty summer salaries to conduct the surveys they agreed to make, and E. Odum used the University of Georgia funds to fund graduate student research. For the University of South Carolina, it was a one-time event, but for E. Odum it was a foot in the door to AEC funds for ecological research at the Savannah River Site. The AEC established its own team of ecologists doing research similar to that of the University of Georgia at its Oak Ridge Laboratory, and the two laboratories were close enough geographically to communicate and learn from each other. That collegial situation would ultimately lead to the development of systems ecology, but there were missing ingredients before that happened.

H. Odum had learned from Hutchinson to focus upon both the environment (his strontium dissertation an example) and the dynamics of ecological communities (as in Lindeman's 1942 article). Eugene picked up the message also, for he was already committed to a broad ecological understanding. His own understanding, enriched by H. Odum, went into the first postwar textbook. In 1949, while H. T. was busy with his dissertation, Eugene had invited him to coauthor *Fundamentals of Ecology* (1953). H. Odum declined this "most generous" invitation on grounds that he could not provide an equal amount of substance (Craige 2001:40). In the Preface, however, Eugene thanked H. T. for his critique of the whole manuscript and his major contributions to Chapters 4 through 7, and he even gave H. T. a portion of the royalties. After publication, however, H. T. decided that he had earned mention on the title page of the second edition (1959), and Eugene added to the title page: "In collaboration with Howard T. Odum" and again gave him part of the royalties. Most book publishers in 1953 had not yet realized that American ecology had already begun its rapid expansion, illustrated by Eugene having offered it to several other publishers before W.B. Saunders accepted it. *Fundamentals* had the dynamism that Hutchinson, in a 1940 review, had found lacking in Clements and Shelford's *Bio-Ecology* (1939). Incidentally, the influence of both Clements and Shelford on Eugene's thinking was quite evident in the book.

Eugene began with "Basic ecological principles and concepts" which his Georgia biology colleagues had claimed ecology lacked. The first principle was "ecosystem and biochemical cycles" and the second was "energy in ecological systems." *Ecology*'s book review editor, LaMont Cole, recognized *Fundamentals*' importance and arranged to have it reviewed by both a plant and an animal ecologist. Plant ecologist Edwin Moul (1954) fully appreciated Odum's treatment of the dynamics of ecosystems, its illustrations, and its bibliography. He praised the sections on freshwater and marine ecology and only regretted that the section on terrestrial plant communities was too brief. Zoologist Peter Frank's review in *Ecology* suggested that "The possible influence of the book on the field is substantial" (1954:297). *Fundamentals* had virtually no competitor in America or abroad and was translated into a dozen languages and got Eugene's picture in *Life, Newsweek*, and *Time* as America's leading ecologist (replacing Leopold, who had died in 1948, and Sears) (Craige 2001:47).

After receiving a Ph.D., H. Odum moved about from one Southeastern university to another until he finally settled at the University of Florida in Gainesville, for a second time, in 1970. However, the first time he taught there (1950–1954), he had begun research at Silver Springs, a popular vacation destination. The attraction for him was an ecosystem that had a constant flow in an environment that was also virtually constant (Odum 1957; Kemp & Boynton 2004:43–45). He was already far enough into that research to give a talk at the annual ESA meeting at Madison, Wisconsin (September 7, 1953) on "Community Metabolism of Silver Springs, Florida" (*ESA Bulletin* 34:67). H. Odum produced a seminal article on trophic energy flow and productivity, published in *Ecological Monographs* (1957). Yet, this was only one of a blizzard of publications he produced in the 1950s, some being coauthored with colleagues.

A notable publication coauthored by the Odum brothers and supported by the AEC was the "Trophic Structure and Productivity of a Windward Coral Reef Community in Eniwetok Atoll" (1955), in *Ecological Monographs.* It drew a quick response in *Ecology* from Cadet Hand (1956), who had been studying corals in the Caroline Islands at about the same time they were at Eniwetok. He questioned their claim that corals are herbivores. The Odums responded that they appreciated Hand's interest in their study, but stated that he had oversimplified their views (Odum & Odum 1956):

> Our viewpoint was summarized on page 305: "the animal part of the coral is partly divided in trophic classification between two trophic levels, herbivores, and carnivores" and again on page 310: "The single coral is first a producer, to a lesser extent (in many cases) a herbivore, and somewhat of a carnivore." We suggest on page 299 that corals must obtain some food and especially critical nitrogen from plankton.

The Odum brothers received the Mercer Prize in 1956 for their coral paper.

A major AEC facility was at Oak Ridge, Tennessee, where the Manhattan Project had built an atomic bomb. By 1955, the Oak Ridge National Laboratory (ORNL) was "one of the largest research institutions in the world" (Bocking 1997a:63). The AEC also developed an ecology program there, which stayed in contact with Odum's group at the Savannah River Site. The first Oak Ridge ecologist was Stanley Auerbach, who had studied under Orlando Park and was hired in 1954 upon Park's recommendation (Bocking 1997a:66–69). White Oak Lake had been used to store low-level radioactive waste, and in 1955 it was drained. In 1956, Auerbach began studying the movement of radionuclides in soil, plants, and animals in the lake bed of White Oak Lake. Auerbach read Odum's *Fundamentals of Ecology* and oriented his research toward ecosystem ecology.

In the 1950s, two ecological schools flourished at the University of Wisconsin–Madison. The limnological school under Hasler was a continuation of the tradition that had developed in the earlier 1900s under Birge and Juday, though it differed from theirs. After publication of his major work, *The Vegetation of Wisconsin* (1959), the plant ecology school that Curtis had developed suffered a serious blow, when he died in 1961 at the age of 47. Both professors were active in ESA, with Hasler serving as president in 1960, and Curtis likely serving as president had he lived longer. Let us first examine Curtis' school, where his leadership lasted for more than a decade.

John Thomas Curtis (1913–1961)

(From Stearns 1961:167. With permission.)

From Waukesha, near Milwaukee, Wisconsin, John Thomas Curtis was an undergraduate at Carroll College, majoring in biology, with a strong interest in botany (Burgess 1993a:1–8). In the fall of 1934, he entered graduate school at the University of Wisconsin–Madison's Botany Department to study under plant physiologist Benjamin Duggar. Curtis studied plant physiology and apparently minored in plant pathology. He wrote both an M.S. thesis and Ph.D. dissertation on orchid seed germination and included a study of symbiosis of fungal mycorrhiza. He received a Ph.D. in 1937 and joined the department faculty as an instructor. He published nine papers before he was 25 (Curtis Bibliography: Burgess 1993a:36–41), and he married another student studying orchid-mycorrhizal symbiosis, Jane Ann Kurtenacker, in 1938. Duggar had taught a plant ecology course that became Curtis' course, and he also collaborated with other faculty on courses that had ecological aspects. During World War II, he ran an agricultural experiment station in Haiti to see if *Cryptostegia* could become a latex substitute for rubber. He studied the species and grew it on a plantation, but the invention of synthetic rubber was the eventual solution to rubber scarcity (Burgess 1993a:8–13). However, he found time to study tropical vegetation while there, and published a "seminal paper" on a Palo Verde (*Cercidium praecox*) forest in 1947 (Burgess 1993a:13). Returning to Wisconsin in 1946, Curtis began a long-term study of Wisconsin vegetation, which would engage him and his many students for the rest of his 15 years of life (Fralish et al. 1993; Martin 1994). That research culminated in his *Vegetation of Wisconsin: An Ordination of Plant Communities* (1959), which "has had a lasting influence on the development of modern ecology. It is an ecological classic" (Cottam 1971). Had he lived longer, he likely would have been an ESA officer, most likely president.

Curtis was a Wisconsin native, who received a Ph.D. in botany from the University of Wisconsin (Burgess 1993a, 1996:33; Nicolson 2001). He had some interest in ecology from the start, but he had mainly focused upon plant physiology and orchid reproduction in his early research. Only after World War II did his orientation shift to Wisconsin vegetation. Curtis stated (1959:4) that

> A definite attempt was begun at the Plant Ecology Laboratory of the University of Wisconsin in 1946 to survey the entire vegetation of the state and to learn the geographical limits, species compositions, and as much as possible of the environmental relations of the communities composing that vegetation.

Beginning in 1947, many graduate students who earned degrees under Curtis undertook research contributing to the knowledge of Wisconsin vegetation (Burgess 1993a:42–43). Curtis and a few others completed a revolution begun by Gleason (Barbour 1996); Gleason had argued that plant communities were illusions, because every species has a different range. It was one thing to argue this in theory, quite another to demonstrate it in the field all over Wisconsin. It required new quantitative methodologies (Curtis & McIntosh 1950; McIntosh 1958:116–118, 1974:135–137, 1993:97). Tragic though Curtis' early death was, the plant ecology school he founded persisted. His students stepped in and continued his work (Mladenoff & Burgess 1993), with his former student and colleague since 1949, Grant Cottam (1918–2009), replacing him at the University of Wisconsin–Madison (Davidson 1994; Allen et al. 2009) (Figure 3.4).

How did Hasler's limnology school at Wisconsin differ from that of Birge and Juday's school? In 1879 to 1910, Birge began to study invertebrates in lakes, but by

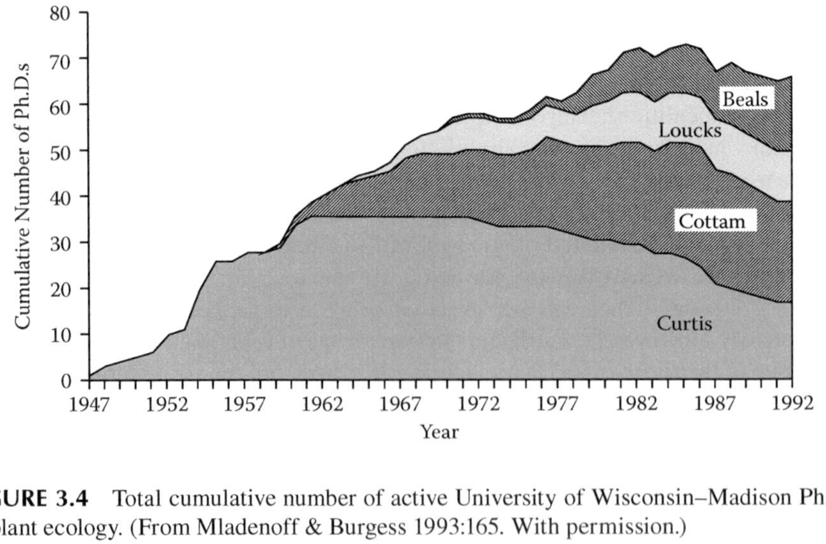

FIGURE 3.4 Total cumulative number of active University of Wisconsin–Madison Ph.D.s in plant ecology. (From Mladenoff & Burgess 1993:165. With permission.)

1911 his interest had shifted to environmental factors (Frey 1963b; Beckel 1987:1–30; Egerton 1987:86–94). Most of the 13 Ph.D. students whom Juday advised (1928–1940) wrote dissertations on some form of aquatic life (Egerton 1987:116), showing that organisms had not been entirely neglected, but organisms were not emphasized in Juday's own research. When Birge and Juday wanted collaborators, they obtained them from University of Wisconsin–Madison faculty or graduate students. Birge and Juday wrote descriptive studies, as did some of Juday's students, which was typical of the times. Juday wondered if plankton could use dissolved organic matter and suggested to Hasler that he study digestion in plankton Crustacea. Hasler did so, but he was supervised by physical chemist H. C. Bradley, for whom it was routine to duplicate experiments and to use controls (Beckel 1987:32–33; Egerton 1987:95). These practices became routine in the Hasler limnology school, when practicable. Controlled experiments were not limited to a laboratory. Hasler's field experiments were developed independently and were not part of a tradition going back to experiments by Clements in the 1920s. Clements' influence in this respect seemed not to survive the Depression and World War II.

An early doctoral dissertation that Hasler supervised, written by Sarah Elizabeth Jones (Leopold's coauthor), had four experimental silo-ponds 3.6-m diameter by 1-m deep, erected in a large fish hatchery pond to test the impact of rooted aquatic vegetation upon algae. All four ponds had algae, with two also having rooted vegetation, which produced a duplicated controlled experiment, showing that rooted vegetation limited algal growth (Jones 1947; Hasler 1963:55). Hasler then condensed Jones' 166-page dissertation into a 5.33-page article for *Ecology* (Hasler & Jones 1949). A footnote to the article indicated that by 1949 she had married limnologist David Frey (who had earned his Ph.D. at the University of Wisconsin in 1940).

A less rigorous experiment—no duplication—was conducted at Cather Lake in northwest Wisconsin. It had brown-stained water that apparently inhibited photosynthesis. Hasler had two students, Oscar Brynildson and William Helm, study Cather Lake's physical characteristics and photosynthesis from 1948 to 1949, and in 1950 they added hydrated lime to it which cleared the water, allowing greater photosynthesis that they then measured (Hasler et al. 1951). Brynildson continued these studies for his Ph.D. dissertation (1958) to determine what impact increased photosynthesis had upon the productivity of trout and largemouth bass; Helm decided to study yellow bass in a small lake close to the Madison campus for his Ph.D. dissertation (1958).

A sort of duplication of the Cather Lake experiment was conducted at northern Peter and Paul Lakes, by two other graduate students. Peter and Paul Lakes together were shaped like an hourglass, with a narrow connection between them. Hasler decided that an ideal controlled experiment would be to close the neck between Peter and Paul and use one for the experiment and the other for a control. The ponds belonged to the University of Notre Dame, which gave permission to fill the neck for this experiment (Hasler 1963:55–57; Egerton 1987:99–100). All these experiments constituted a small part of the scope of limnology under Hasler (1963:57–93).

Arthur Davis Hasler (1908–2001)

(From Beckel 1987, cover. With permission.)

From Utah, Arthur Davis Hasler graduated from Brigham Young University (1932) and then attended graduate school at the University of Wisconsin, where he studied limnology under Chauncey Juday for his Ph.D. (1937). Juday suggested his doctoral research, but his dissertation supervisor was a chemist, from whom he learned to conduct controlled experiments. Hasler worked for the U.S. Fish and Wildlife Service for two years (1935–1937) during which he began using controlled experiments in streams. He joined the Zoology Department at the University of Wisconsin in 1937, the same year that Juday retired from teaching. In the summer of 1946, Hasler returned to the streams on Mount Timpanogos in Utah, where he had fished as a youth. He wondered how salmon found their way back to their home stream to spawn. A breeze brought the fragrance of mosses and columbine to him, and the image of the falls immediately returned to his mind, with childhood associations. He wondered if salmon might also use smell as a memory aid, and proving that became the major research of his career. This led to two research books: *Underwater Guideposts* (1966) and *Olfactory Imprinting and Homing in Salmon* (coauthored with Scholz, 1983). Hasler also greatly expanded the limnological school at the university that Birge and Juday had informally begun. He obtained funds for a Limnology Laboratory from the NSF, which was built in 1963. He served as the president of ESA in 1961, the American Society of Zoologists in 1971, and the International Association for Ecology from 1967 to 1974. He became a member of the National Academy of Sciences in 1969 (Hasler 1963; Beckel 1987:31–56; Egerton 1987:94–122, 2008; Burgess 1996:52; Carpenter & Kitchell 2001; Likens 2002).

Hasler, like E. Odum, knew how to obtain research grants from federal agencies, for graduate student research fellowships, equipment, and, finally, a new limnology laboratory funded by the NSF and built on Lake Mendota's shore on campus in 1963 (photo: Beckel 1987:77; Egerton 1987:107–108). Because of his funding successes, the university administration asked him to obtain funding for a new building at the university's Trout Lake field station. He agreed, if the university would fund half, to which the administration agreed. The NSF was willing to fund the other half in 1966. Hasler supervised Ph.D. dissertations of 64 students (1946–1978) (Egerton 1987:117–122). The most famous investigations by Hasler and his associates were on the homing abilities of salmon, which after years in the ocean return to their native streams to reproduce (Egerton 1987:102–106). Their investigations on homing began in 1947 but most famously occurred later than this chapter's dates. Many other noteworthy limnological advances that occurred during the period of 1945 to 1959 in North America are described elsewhere (Frey 1963a; Egerton 2016).

(Samuel) Charles Kendeigh (1904–1986), who served as ESA president in 1951 and received ESA's Eminent Ecologist Award in 1978 (Muller et al. 1978a; Odum 1987; Burgess 1996:61), wrote a "History and Evaluation of Various Concepts of Plant and Animal Communities in North America" (1954). It included eight maps—seven encompassing 1823 to 1890, plus his own. Use of the term "community" in his title seems odd, since none of the maps, including his own, referred to the areas identified as communities, though "Biotic Communities" is the heading of the final section of his article (1954:166), and previously he had published "Bird Populations and Biotic Communities in Northern Lower Michigan" (1948). He quoted (1954:164) Lee Dice's definition of a biotic province but did not embrace Dice's *The Biotic Provinces of North America* (1943). He thought his own term, "biociation" was better than "biotic province," because it recognized the importance of successional stages and because it focused upon vegetation forms rather than plant species. If he had begun his historical survey with Humboldt (1807) instead of Schouw (1823), he would have discovered the ultimate forerunner to his thinking. Kendeigh published a slightly different and much larger version of his own map as the front endpaper in his *Animal Ecology* (x + 468 pages, 1961) book, which discussed "biomes" and "biotic communities," as well as "biociations."

Angus M. Woodbury (1886–1964) published *Principles of General Ecology* (1954, viii + 503 pages), which—unlike Odum's *Fundamentals*—received only one review in *Ecology*, by Canadian zoologist James Bendell, though his review was actually longer than Moul and Frank's reviews combined (which might be why only one review was published). Although most of Bendell's review merely describes the book's content, he found aspects of it to praise. Nevertheless, the key comment is as follows (Bendell 1954:586): "Woodbury's approach is the antithesis of Odum's and, while ably demonstrating that life does not exist in a vacuum, fails to move beyond natural history and general biology to clear cut principles of ecology." In other words, Woodbury's *Principles* would not be a strong competitor of Odum's *Fundamentals*. Following Bendell's review was a brief review of John Storer's *The Web of Life* (1953, xii + 144 pages); William Dreyer warmly praised it (1954) as an introduction to ecology for the general public. Professor Emeritus John Weaver's *The North American Prairie* (1954, xi + 348 pages) received a similarly brief, positive review from B. W. Allred (1954), who advised that it could interest scientists as well as laymen.

George L. Clarke (1905–1991), from Providence, Rhode Island, was educated at Harvard University and spent his career there and at Woods Hole Oceanographic Institution. He published *Elements of Ecology* (1954, xiv + 534 pages), for which Australian animal ecologist Louis Birch wrote a 66-page review in *Ecology*. Why it did not merit two reviews like Odum's *Fundamentals* or a long review like Woodbury's *Principles* is unclear. However, the length of these reviews was possibly left up to the reviewer. Clarke began the book with chapters on what he considered elements of ecology: physics, chemistry, plant physiology, animal physiology, pedology, oceanography, and animal behavior. Next, he discussed the biological effects of these elements, and finally discussed communities. Birch offered a good number of small corrections, especially concerning population ecology, Birch's specialty. Nevertheless, "these are minor faults and will not prevent the book from being accepted as a substantial contribution to the science and teaching of ecology" (Birch 1955:369). Although Odum's *Fundamentals* remained more popular with teachers of ecology than Clarke's *Elements*; *Fundamentals* did not drive *Elements* from the market. No doubt Clarke corrected the errors in later printings that Birch had identified, and *Elements* was translated into French and Japanese.

Birch's review appeared in the April 1955 issue of *Ecology*. The July issue featured a review of *The Distribution and Abundance of Animals* (1954), which Birch coauthored with fellow Australian Herbert Andrewartha. LaMont Cole was both the book review editor of *Ecology* and the reviewer of Andrewartha and Birch's book. The idea to have Birch review Clarke's book apparently came to Cole after he received a review copy of the latter book. Both books likely arrived at Cornell at about the same time, because Andrewartha and Birch's book did not have to come from Australia—it was published by the University of Chicago Press.

Their volume "is probably the most significant ecological book to appear since the publication of Allee et al. [1949]" (Cole 1955:538). Since Cole, like them, was an animal population ecologist, this might not have seemed an objective evaluation, but as Cole observed, their book has 792 pages versus Allee et al.'s 849 pages, and no other rival comes to mind. The authors argue there are two "sorts" of ecology, and they brush aside synecology versus autecology as inappropriate for their context. Instead, they focus upon niche versus distribution and abundance. Their own extensive research was on insect populations, which left them skeptical of much traditional thinking about populations: "no need for density-dependent factors to halt unlimited population growth or decline, and an alternative classification of ecological factors into 'biotic' and 'physico-chemical' is also rejected as unnatural" (Cole 1955:538). Cole agreed with some of their arguments, but others he thought were redundant. However, regardless of how much one accepts or challenges, they had written a provocative book that "must be on every [animal?] ecologist's bookshelf" (Cole 1955:538).

Comparable progress was being made in the quantitative study of plants (Goodall 1952; Greig-Smith 1957; McIntosh 1974:135–136).

The existence of a U.S. Fishery Laboratory at Woods Hole had influenced the building of the Marine Biology Laboratory and the Woods Hole Oceanographic

Institution there. Similarly, the existence of an Ohio State Fish Hatchery at South Bass Island in Lake Erie influenced Ohio State University to build its Franz Theodore Stone Laboratory nearby, for use by its biology faculty and students. Thomas Huxley (Hux) Langlois (1898–1968) had been the head of the fish hatchery since 1930, and when the Stone Laboratory first hired a director in 1936, Langlois became its assistant director, and in 1938 its director until 1956. He retained his directorship at the hatchery until 1946 (Langlois 1949; Egerton 1985:13–14). The Stone Laboratory was oriented toward limnology (Beeton & Chandler 1963:545), but its findings were also useful for fishery management, which was the focus of the hatchery. A substantial number of research publications from the laboratory were published in *Ecology* (Herdendorf et al. 1974; Prantner et al. 1974). Langlois synthesized many such studies in *The Western End of Lake Erie and Its Ecology* (xx + 4,479 pages, 1954), which received a rather neutral, slightly negative, review in *Ecology* (Verduin 1956). Nevertheless, it was a kind of "ancestor" to a later, larger survey, *Lake Erie and Lake St. Clair: Handbook*, Stanley Bolsenga and Charles Herdendorf, editors (1993).

According to plant ecologist Dwight Billings (1957:375), "The fields of physiology and ecology are so closely interrelated that it is often difficult to detect any boundary between them," and hence, the title of his essay review: "Physiological Ecology" (1957). This was not a review of a few books. His "Literature Cited" carries citations to 147 works, published from 1940 to 1956, with few from the 1940s. The topics he surveyed included germination, microclimate and phenology, photoperiodism and plant distribution, water relations, edaphic factors, competition, life histories, ecology of related taxa, and ecosystem productivity. Billings commented (1957:376–377) on the importance of new instruments and techniques for some kinds of studies. One topic that he neglected was mathematical computation of data. Coincidentally, British botanist P. Greig-Smith published his *Quantitative Plant Ecology* also in 1957, which was used by botanists throughout the English-speaking world.

A Cold Spring Harbor Symposium on Quantitative Biology in 1957 brought together population ecologists and human demographers, where they discovered they had little in common (McIntosh 1985:190–192; Slack 2010:276–279). The saving grace for the conference was Hutchinson's "Concluding Remarks" (1957), which did not try to knit together the disparate presentations but instead offered a new niche theory without even mentioning Grinnell or Elton. Hutchinson (1957) defined the niche in terms of an abstract multinational hyperspace in which the organism's needs and properties were defined as dimensions (Slobodkin & Slack 1999:28). Tansley's ecosystem paradigm integrated the study of biotic and abiotic factors in the environment, and Hutchinson's theory did the same for a more abstract entity, his niche concept.

Robert MacArthur had two degrees in mathematics when he entered Yale to study under Hutchinson. In a memoir on him, Wilson and Hutchinson (1989:320) generously claimed that MacArthur's unique blend of mathematics and ecology could be credited to the influence of David Lack and MacArthur's older brother, John MacArthur, Jr., a physicist. He did spend the last school year before receiving a Ph.D. (1957–1958) with Lack, but by then he had also spent five years, more or less (interrupted by U.S. Army service), at Yale with Hutchinson, whose mathematically sophisticated ecological reasoning was far greater than anything Lack ever published.

(MacArthur wrote his first published paper [1955] in Arizona, while serving with the Army.) Others, including Lindeman (1942b) and Hutchinson (2010:87–131), had published ecological papers that used mathematical tools. MacArthur's starting point was Hutchinson's "The Concept of Pattern in Ecology" (1953; Kingsland 1985:184). Where MacArthur exerted a strong influence was by showing how mathematical reasoning could greatly enhance reasoning on long-standing topics, such as intraspecific and interspecific competition and other aspects of community dynamics. His reasoning was often daring, and "it energized a generation of young population biologists and transformed a large part of ecology" (Wilson & Hutchinson 1989:321).

The postwar years (1945–1959) were favorable to progress in ecology because of a substantial increase in research funds that became available from federal agencies and because of many veterans going to college on the G.I. Bill, with a portion of them studying to become ecologists. Eugene Odum at the University of Georgia and Arthur Hasler at the University of Wisconsin were especially adroit at obtaining research funds and students in order to develop ecological research institutions at their universities, and both also served as ESA presidents. Wartime had been unfavorable for publishing textbooks, but afterward, general ecology, plant ecology, and animal ecology textbooks reappeared. The ESA began offering its annual Mercer Award in 1949 and Eminent Ecologist Award in 1953. The ESA had only about 800 members in 1949, but its membership began increasing steeply in that year, and by 1959 it had almost 2,000 members. In the late 1940s, Hasler had introduced field experiments into limnology, which soon spread to other ecological sciences. In the late 1950s, MacArthur introduced quantitative thinking into community ecology, and quantitative studies began to replace purely descriptive studies.

4 1960 to 1974
The International Biological Program (IBP), The Institute of Ecology, and Others

In 1930 and in 1945 there were sweeping changes in American society, which were also turning points in the history of ESA that defined the beginnings of Chapters 2 and 3. From 1960 to 1974, the period discussed in this chapter, the West and the U.S.S.R. were locked in a Cold War that influenced the sciences that seemed strategically important, but its impact on ecology was real but rather indirect: the emergence of radioecology. (My wife and I attended an International Congress for the History of Science in Moscow in 1968 and easily made friends with a Russian colleague and his wife. The only incident that smacked of the Cold War was that after I forgot my camera on a tour bus, when I retrieved it a few hours later, my film had been confiscated.) The International Biological Program (IBP), a momentous worldwide development for ecologists, belatedly began in the United States and Canada in 1967, three years later than in Europe. The IBP officially ended in June 1974, which coincides with the end of this chapter, though some IBP programs continued.

The ESA membership continued to expand throughout the 15 years discussed in this chapter, from 2,000 members in 1960, to 3,000 in 1966, to around 4,200 in 1974 (Burgess 1977:3). This inevitably led to an annual increase in ecological literature. For example, the number of pages in *Ecology* averaged 476 from 1945 to 1949, with an average of 58 articles; from 1970 to 1974, the number of pages in *Ecology* averaged 1,279, with an average of 150 articles (Burgess 1981a:75). Furthermore, in 1954 the printed part of a page increased from 13 by 20.5 cm to 15.5 by 22.5 cm, an increase of about 30% of print per page. In 1965, the number of issues per year increased from four to six. The size of the *Bulletin* also grew, from 44 pages in 1924, to 156 pages in 1974, as there was progressively more and more news that a growing membership wanted to read.

Thomas Park was president of ESA during 1959, and his presidential address was published in the January 1960 issue of the *ESA Bulletin*. When his brother Orlando had been president in 1944, his presidential address had been the first article in the January 1945 issue of *Ecology* (see Chapter 3). At that time, World War II was winding down and the future of ESA seemed uncertain, and he advised the fruitful paths that the ESA should follow. From 1959 to 1960, ESA was flourishing, and Thomas Park was optimistic about ESA's future, but nevertheless he offered further advice on fruitful paths to follow. A glance at molecular biology and genetics revealed that ecology had not reached their level of theoretical advance, and therefore there was no room for complacency. Two areas of research which he identified as flourishing were population ecology (his specialty) and community energetics. He thought ecology specialists were drifting apart, when instead they should be collaborating and learning from one another. He felt that natural history studies were losing popularity but were still needed to provide insights and knowledge so that ecologists could ask cogent questions and increase their capacity to interpret their findings. He also thought that, although some ecologists were already modifying environments to see what impacts occurred among the inhabitants, these studies needed to expand. The same was true for theoretical models in ecology—already in use, but wider use would be more fruitful.

Thomas Park (1908–1992)

(From Buell et al. 1971:15. With permission.)

Illinoian Thomas Park was the younger brother of ecologist Orlando Park (1901–1969), who taught at Northwestern University (Englemann 1970) and was a coauthor with Allee et al.'s landmark *Principles of Animal Ecology* (1949). Both brothers earned B.S. and Ph.D. degrees from the University of Chicago under W. C. Allee (Buell et al. 1971; Mertz 1992:248). After receiving a Ph.D. (1932), Thomas received a National Research Council Fellowship (1933–1935) that enabled him to conduct postdoctoral research at Johns Hopkins University under Raymond Pearl, where he remained, teaching at Johns Hopkins (1935–1937), after which he joined the University of Chicago faculty (1937–present). Park's own work became an influential model of laboratory experimental ecology for many other ecologists, his students and beyond, who followed in his footsteps. In addition to the influence of Allee and Pearl on his thinking, Park greatly admired the experimental work of his colleague, then at Chicago, Sewall Wright. Allee studied cooperation, Thomas Park studied competition; both studied insect ecology. Leslie Real and James Brown, for their sourcebook, *Foundations of Ecology* (1991), reprinted Park's "Experimental Studies of Interspecies Competition: Competition between Populations of Flour Beetles" (1948). Park's former student and collaborator David B. Mertz (1992:249) judged "Struggle for Existence: The *Tribolium* Model" (Neyman, Park, & Scott 1956) as Park's most important scientific contribution. Park taught the first course in population ecology, and he introduced that term into the literature (1946). Aside from teaching, research, and advising graduate students, Park served as an editor for several professional journals, including *Ecology* (1940–1950). The best remembered episode during his editorship of *Ecology* was defying his two limnology referees' negative evaluation of Raymond Lindeman's "The Trophic-Dynamic Aspect of Ecology" (1942b), which was strongly backed by Lindeman's posthumous champion, Evelyn Hutchinson. Park was also ESA president in 1959 and Eminent Ecologist in 1971 (Burgess 1996:85), and president of the American Association for the Advancement of Science (AAAS) (Wright 1960).

In 1960, three animal ecologists at the University of Michigan—Nelson Hairston, Frederick Smith, and Lawrence Slobodkin—published "Community Structure, Population Control, and Competition," which was the only ecological citation classic for 1960 (McIntosh 1989:34). It was in response to six outstanding papers published in 1957 from a Cold Spring Harbor Symposium on Quantitative Biology dealing with animal populations. Their 4.5-page paper intended "to demonstrate a pattern of population control in many communities which derives easily from a series of general, widely accepted observations" (Hairston et al. 1960:421). They argued that populations of species at different trophic levels experience different controls. Their arguments were based upon published literature, and their three conclusions were as follows (Hairston et al. 1960:424):

1. Populations of producers, carnivores, and decomposers are limited by their respective resources in the classical density-dependent fashion.
2. Interspecific competition must necessarily exist among the members of each of these three trophic levels.
3. Herbivores are seldom food-limited, appear most often to be predator-limited, and therefore are not likely to compete for common resources.

(Eric Mills stated that number 3 was the most influential among us at the time.)

Professor Marston Bates (1906–1974) at the University of Michigan was an entomologist specializing in mosquitoes (Kimler 1990; Tino 1997). He taught an undergraduate course, "Zoology in Human Affairs," and he turned his course lectures into *The Forest and the Sea: A Look at the Economy of Nature and the Ecology of Man* (1960). With no illustrations, it did not compete with introductory ecology textbooks. Since Bates had collected mosquitoes around the world, he had a grasp of the natural history of many lands. His book seemed to answer Thomas Park's call for more natural history studies, though it was already in press when Park made his plea. Oddly, Frank Egler was its reviewer for *Ecology* (1960). Odd, because Egler was a botanist, but he was enthusiastic about Bates' book. He displayed none of the knives with which he had slashed away at textbooks by fellow plant ecologists. With warm praise for Bates' book, Egler was apparently glad to show that he was not a professional curmudgeon. *The Forest and the Sea* was Bates' seventh book, and with its favorable reception, Time-Life Books gladly persuaded him to write the text for its *Land and Wildlife of South America* (1964).

Rachel Carson's three books on life in the sea (1941, 1951, 1955) were not reviewed in *Ecology*, probably because the publishers did not send the journal review copies. *Ecology* gladly reviewed books on ecology for the public, and since she was an ESA member, it is unlikely she would have been snubbed, had copies been sent to *Ecology*. At any rate, *Silent Spring* (1962) received a full-page review by Francis Fosberg (1963). He wrote a very favorable and thoughtful review and urged all ecologists to read her book, and certainly many of them did. It is commonly said that this book founded the environmental movement (Murphy 2005; Lytle 2007; Souder 2012), and if the emphasis is on "movement" this is a fair statement. Environmentalism as a perspective predated the book but had been an inconspicuous aspect of conservation. With the rise of the environmental movement, the public seemingly discovered "ecology" for the first time—as a synonym for environment (Egerton 1993). One heard people urging others to help save our ecology. Donald Cox published an attractive book for teens, *Pioneers of Ecology* (1971), in which the only "card-carrying ecologists" he discussed were Rachel Carson and Paul Ehrlich (however, I also called Thoreau an ecologist in Egerton 2012a:151–156).

According to ESA member and entomologist Kenneth Raffa (1991:127): "The field of plant-insect interactions took ecology by storm in the late 1960s through mid 1980s, greatly increasing our understanding of population regulation, evolutionary processes, and interspecific mediation of life history parameters." ESA member and entomologist Nancy Stamp agreed (1992:28) and traced the stimulus for this research to three articles, by Dethier (1954), Fraenkel (1959), and Ehrlich and Raven (1964). Dethier had argued that plants gradually develop biochemical defenses against insect feeders, and that when an insect species shifts its diet from one plant species to another it is because the previous victim species had evolved new chemical defenses. Another entomologist, Gottfried Fraenkel, had investigated secondary biochemicals that plants synthesize to inhibit potential insect feeders, and he traced the insight that these substances are insect inhibitors back to E. Stahl (1888). Animal ecologist Paul Ehrlich and plant ecologist Peter Raven had studied the coevolutionary process among butterfly larvae and their food plants. Although Raffa and Stamp discussed this research in the *Bulletin*, had textbook authors noticed? Stamp (1996) searched seven ecology textbooks published from 1990 to 1994, and found little interest in this progress.

Professor John Cantlon (b. 1921) was provost at Michigan State University in 1969 when he served as ESA president. His retiring presidential address, "Ecological Bridges" (1970), confronted the role of ecologists in an "Age of Ecology." Ecologists could publicize and assist the public in understanding ecosystems and the biosphere. Critics had suggested that ecological concerns were a diversion from more serious problems of peace, racism, poverty, and disintegrating central cities. Ecologists could acknowledge the seriousness of these social problems but explain that one does not have to choose between social and environmental problems, as they are linked. Ecologists, therefore, should reach out to leaders in both the humanities and the technologies to explore ways to bridge the gap that C. P. Snow had earlier identified as *The Two Cultures* (1959).

Eugene Odum published the second edition of his *Fundamentals of Ecology* (with H. T. Odum's name also on the title page) in 1959, and, considering that it was the second edition of a textbook, it received a long review (.75 of a page) in *Ecology* (1960). That was tacit acknowledgment that this was the leading introductory ecology textbook in the world. Its world domination was also tacitly illustrated when it was reviewed by a Danish ecologist, Overgaard Nielsen. Perhaps a foreign reviewer could also be more objective than an American reviewer of this book. His geographical distance did not inhibit Nielsen from highly praising it and even suggesting that it be considered a "catechism." A new feature of the second edition was a chapter on radioecology, which Eugene Odum was well positioned to discuss, since he had used Atomic Energy Commission (AEC) funds to establish an Institute of Radiation Ecology at the University of Georgia in 1958 (Odum 2001:20), and he was in close contact with radioecologists at the AEC's Oak Ridge facility. Since the Cold War was a potential menace to the whole world, Nielsen's review devoted more discussion to this addition than to other aspects of the book.

E. Odum's ecological empire flourished and expanded during the years 1960 to 1974. The modest foundations had been laid during the 1940s and 1950s, and during the 1960s, the several strands of an Institute of Ecology were woven together around the Institute of Radiation Ecology and around Odum's ability to keep those strands moving forward (Barrett & Barrett 2001). The Radiation Ecology Laboratory became the Savannah River Ecology Laboratory. Central and South American tropical research began, Coweeta Watershed research began (for Georgia ecologists; the U.S. Forest Service had founded a station there for research in 1927), the Savannah River Site became a National Environmental Research Park, and Sapelo Island became the University of Georgia Marine Institute. A new Ecology Building was completed on the University of Georgia campus in 1974. All of the Institute's efforts focused upon achieving a holistic ecological science.

Stanley Auerbach's adoption of the ecosystem concept (as a result of E. Odum's influence) was advantageous ideologically at Oak Ridge, Tennessee, as it focused upon both organisms and the environment while tracing pathways of radioactive elements. "By 1960 his ecology program had become one of the largest in the United States, with twenty-two permanent and temporary employees and visitors" (Bocking 1997a:75). They pursued a variety of ecological and environmental projects on the 15,000-hectare AEC reservation (Bocking 1995). In 1961, this Radiation Ecology Section of the Health Physics Division acquired its own building. In 1956, Auerbach had recruited his first ecologist, D. ("Dac") A. Crossley, Jr., with a Ph.D. from the University of Kansas, who stayed until 1967, when he moved to the University of Georgia as Professor of Entomology and became a member of Odum's Institute of Ecology. David Reichle, with a Ph.D. from Northwestern University in ecosystem

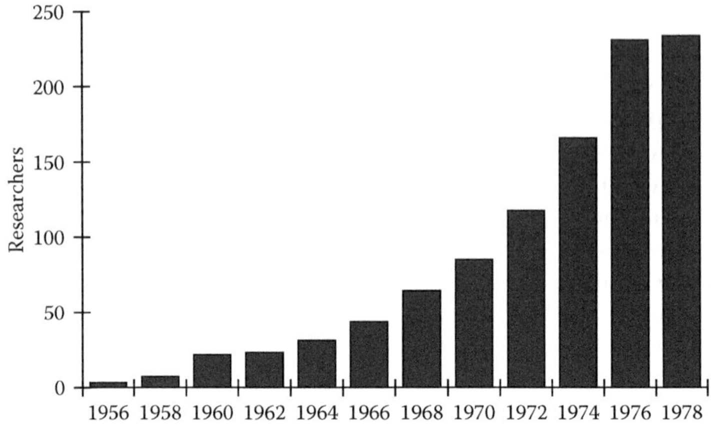

FIGURE 4.1 Permanent and temporary researchers in Oak Ridge Ecology/Environmental Science Program, 1956 to 1978. (From Bocking 1997a:75. With permission.)

analysis, had joined Auerbach's team in 1964, and he continued work that Crossley had begun at Oak Ridge (Figure 4.1).

Henry Cowles had died in 1939, but his influence lingered at the University of Chicago, where Jerry Olson obtained his Ph.D. in 1951, with a dissertation on the plant communities of the Indiana Dunes beside Lake Michigan. In 1958, Olson joined Auerbach's team and studied nutrient cycling in forests, using cesium 137 as a trace element (Bocking 1997a:77–78). That process freed him and his associates from only studying contaminated areas of the Oak Ridge Reservation. The first electrically powered computers were built to compute problems concerning the building of atomic bombs during World War II. Auerbach and Olson were well positioned to utilize that technology when ecosystem studies produced lots of data they wanted to manipulate. Alvin Weinberg, Director of Oak Ridge National Laboratory (ORNL), coined the term "big science" in 1961, and in 1963 historian of science Derek Price published his little book *Little Science, Big Science*. One indication of big science was that scientists had begun to gather in institutions that had computers (Price 1963:86); this was years before personal computers, when even mainframe computers were not plentiful.

Radioecology became a popular fad, partly because research funds were available for its research and partly because it brought a new technique and technology to bear upon ecological questions. Testimony to its popularity is evident in the astonishingly large volume edited by Vincent Schultz and Alfred Klement, *Radioecology: Proceedings of the First National Symposium on Radioecology* (xx + 746 pages, 1963). The symposium had been held in 1961 in Colorado, sponsored by the Division of Biology and Medicine, AEC. *Radioecology* contains 87 reports by 111 participants. Each report carries its own bibliography, but the references are also collected into three bibliographies, pages 649 to 744. The reports included these by ecologists already mentioned: Crossley's "Movement and Accumulation of Radiostrontium and Radiocesium in Insects," Eugene Odum and Edward Kuenzler's "Experimental Isolation of Food Chains in an Old-Field Ecosystem with the Use of

Phosphorum-32," and Olson's "Analog Computer Models for Movement of Nuclides through Ecosystems."

Other ecologists found new questions to ask. Herbert Bormann earned his Ph.D. in plant ecology under Oosting at Duke (1952) and received ESA's Mercer Award in 1953 for the published version of his dissertation. Gene Likens is the most prominent of Hasler's limnology students at Wisconsin (Ph.D. 1962). Bormann and Likens were on the Dartmouth faculty—Bormann from 1956 to 1966 and Likens from 1963 to 1969—after which Bormann went to Yale and Likens to Cornell. Their long collaboration (details below) reminds one of that between Birge and Juday, except that neither had Birge's stolid personality.

While at Duke, Bormann had visited the Coweeta Hydrologic Laboratory in southwestern North Carolina, a U.S. Forest Service facility where ecological studies have been conducted since Forest Service ecologist Charles R. Hursh established it in 1926 (Swank et al. 2001:144). However, Hursh had published in a Forest Service publication and in a geographical publication and had had little influence in ecology. Bormann, after receiving his doctorate, taught at Emory University and took his students to Coweeta (Bocking 1997a:117). E. Odum's students began research at Coweeta in the mid-1960s (Swank et al. 2001:150), after Bormann and Likens' Hubbard Brook (New Hampshire) research began.

F(rederick) Herbert Bormann (1922–2012)

(From *ESAB* 51(4:2), 1970. With permission.)

Bormann's German father immigrated to New York City to escape the Prussian draft. His father became a waiter at the Astoria Hotel, and he was born in New York, but his family later moved to a suburb, Westwood, New Jersey. His mother loved nature and gardening and transmitted these interests to her son. After graduating from high school, he went to the University of Idaho because it was much less expensive than eastern universities. There he took a general botany course under Rexford Daubenmire. After the Japanese attack on Pearl Harbor, he joined the Navy and worked as a pipe fitter on submarines until he took an exam for officer training and was sent to Princeton. In 1946, he enrolled at Rutgers University on the G.I. Bill, and studied plant ecology under Murray Buell, where he learned Clementsian ecology. Bormann valued Clements' "emphasis on processes of vegetation change, biophysical factors that produced community development, and a dynamic approach that promoted development of hypotheses and observational and experimental methods to test them" (Bormann 1996:4). For his Ph.D., Buell recommended that Bormann study under Henry J. Oosting. Bormann was uncomfortable about ecologists estimating community variables, and he discovered that the Duke forestry had methods of measuring them not being used by ecologists. The published version of his dissertation won the Mercer Award for 1954. A forestry professor also introduced him to the hydrology installation at the U.S. Forestry Service (USFS) experimental station at Coweeta, North Carolina. Bormann saw Coweeta as his Rosetta Stone (1996:7). He was also impressed with Eugene Odum's presentation of nutrient cycling in *Fundamentals of Ecology* (1953). In 1956, Bormann joined the Botany Department at Dartmouth College, and in 1957 he met Victor Jensen, USFS, who showed him newly established hydrologic studies being conducted at the Hubbard Brook Experimental Forest, where Bormann in 1959 began bringing his plant ecology students from Dartmouth on field trips. In 1961, the Dartmouth Zoology Department temporarily hired Gene Likens, who was finishing his doctoral studies in limnology. Bormann and Likens found common interests, and in 1963 Likens joined the Zoology Department. They applied for a National Science Foundation (NSF) grant to study the Hubbard Brook ecosystem, which was funded then, and continuously for 32 years. By 1983, dozens of ecologists and graduate students had studied there, producing 373 research papers, 36 Ph.D.s, and 28 masters and honors theses, and two books. Bormann served as ESA president in 1971, and he and Likens were ESA's Eminent Ecologists in 1995, and they also shared the Tyler Award for 1993, and the Blue Planet Prize (2003) for "outstanding scientific research achievements and help to solve global environmental problems" from the Asahi Foundation (Hagen 1992:181–188; Burgess 1996:18; Hedin 2012; Likens 2013a, 2013b; Likens & Hedin 2014).

In 1956, Bormann left Emory for Dartmouth College, where he learned that the U.S. Forest Service had accumulated several years of hydrologic data on the Hubbard Brook Watershed. He realized that the data could be used in ecological studies, and in November 1960 he laid out a research program in a letter to Robert Pierce at the U.S. Forest Service Research Center, Laconia, New Hampshire (quoted in Golley 1993:143–144). When Likens arrived at Dartmouth, he and Bormann gradually developed a plan for ecological studies at Hubbard Brook Watershed (Figure 4.2). Likens had already learned to think of limnology as an experimental science. The NSF funded their study. They began their watershed study in 1963, assisted by geologist Noye M. Johnson who had also arrived at Dartmouth in 1963, and U.S. Forest Service hydrologist Robert S. Pierce (Bocking 1997a:116). The site is in the White Mountain National Forest in New Hampshire. Their first challenge was to gain an understanding of the natural watershed nutrient cycles (Bormann & Likens 1967), and then to modify portions of the watershed—experimental deforestation, herbicides (Likens et al. 1970)—to learn the impacts of such manipulations upon the ecology of both a manipulated small-stream watershed within the Hubbard Brook Watershed and then upon the whole watershed (Likens & Bormann 1975b).

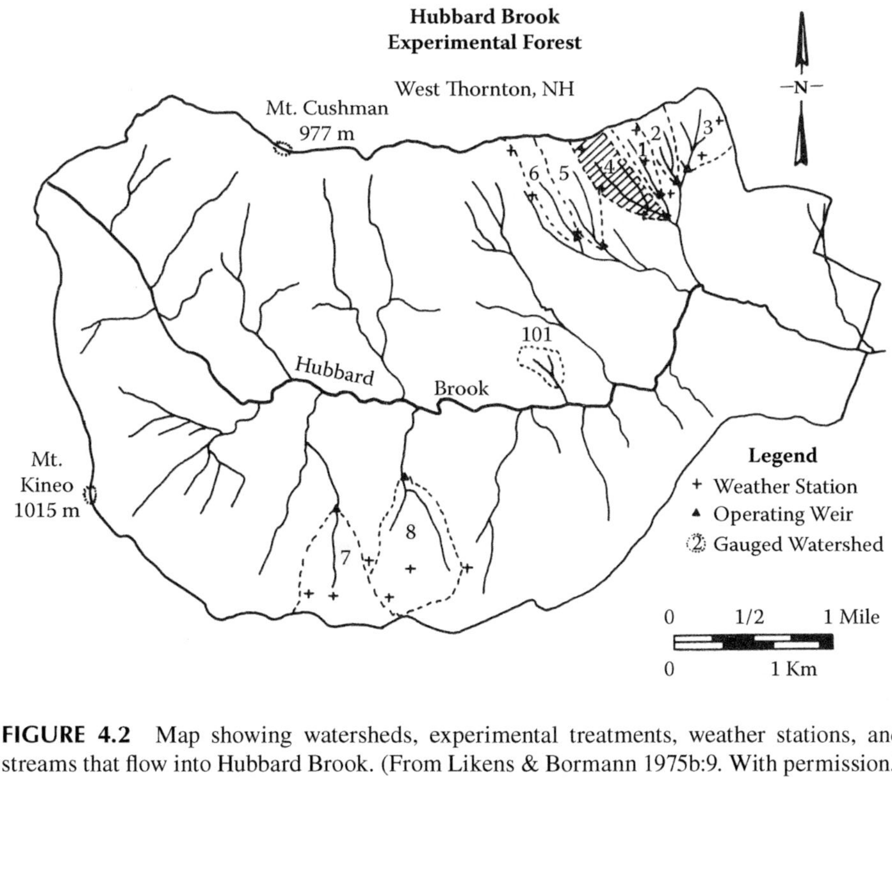

FIGURE 4.2 Map showing watersheds, experimental treatments, weather stations, and streams that flow into Hubbard Brook. (From Likens & Bormann 1975b:9. With permission.)

Gene Elden Likens (1935–)

(From Gilbert 62(197), 1981. With permission.)

Indianian Likens earned a B.S. degree in zoology from Manchester College (1957) and master's and doctoral degrees from the University of Wisconsin (1959, 1962). He is the most illustrious graduate student ever supervised by Professor Arthur Hasler, who taught limnology as an experimental science. Likens taught at Dartmouth College in the fall semester 1961 and joined the Dartmouth Zoology Department in February 1963. He soon collaborated with Herbert Bormann to apply for an NSF grant to study the Hubbard Brook ecosystem, where much of his subsequent research was conducted. There they conducted their landmark study, discovering acid rain in North America (Likens, Bormann, & Johnson 1972; Jenkins et al. 2007:41, 99, 134). Likens moved to Cornell University in July 1969 and continued his research on pH measurements of precipitation around the Finger Lakes (New York), providing evidence that acid precipitation was regional (Likens & Bormann 1974). Likens served as ESA president from 1981 to 1982 (Gilbert 1981; Hagen 1992:181–188; Burgess 1996:66). In September 1983, he became the director of the Cary Institute of Ecosystem Studies (with a slightly different institution title at first). Likens rivals Ehrlich, Hutchinson, and E. O. Wilson for having the most impressive curriculum vitae among ecologists (see his web page). He is the sole author of two books, the sole editor of nine books, coauthor or co-editor of 14 books, and author or coauthor of 574 articles. He has received 10 honorary doctorates, was elected an honorary member of eight learned societies, served as a member of the board of trustees of nine societies, has received 55 awards, has given 182 invited lectures and five commencement addresses, and lists 178 professional activities. He and Bormann jointly became ESA's Eminent Ecologists in 1995, and they shared the Tyler Award for 1993 (Likens 2013a, 2013b). In 2002, President Bush presented him with the National Medal of Science for 2001.

Systems ecology predated IBPs in America, as seen in the volume that Kenneth Watt edited, *Systems Analysis in Ecology* (1966). Ecologists picked up systems methodology from physical sciences and applied it to their study of ecosystems (McIntosh 1985:209–213; Jamison 1993). Systems analysis was a means of describing the functioning of a complex system in which changes of any component affected the other components. Watt's example was a hypothetical forest composed of trees, man, insect defoliators, warblers, spiders, hymenopterous parasites, insect diseases, and weather, in which the change in the abundance of any of these components has a ripple effect through the whole system (excepting impact on weather, which he did not mention). When ecologists began worrying about this complexity, along came computers, which were able to handle such complexities.

George (Mason) Van Dyne (1932–1981) claimed that the first systems ecology paper was Jerry Olson's "Energy Storage and the Balance of Producers and Decomposers in Ecological Systems" (1962, cited from Kwa 1989:41, 1993a:213–214). However, the term "systems ecology" was first published by E. Odum (1964). Two ecologists led the way in developing systems ecology: E. Odum and Auerbach. Both were oriented toward radiation ecology (another term E. Odum coined, in 1956) and received research funds from the Atomic Energy Committee (AEC). E. Odum became an advisor for the ecology program that Auerbach developed at ORNL. In the 1950s, both research groups were peripheral in ecology, but the AEC soon provided both groups with substantial funding, and their status quickly changed (Kwa 1993a:214–215). The AEC had shown little interest in radioactive pollution until after the hydrogen bomb test at Bikini in 1954 that caused worldwide concerns over radiation pollution of the Marshall Islands and a Japanese fishing boat (Kwa 1993a:218). AEC's funding of E. Odum and Auerbach's teams was meant to make up for previous imprecise knowledge. In 1956, before systems ecology, four ecologists at the Hanford National Laboratory (Washington) reported that despite low levels of radioactive P-32 in the Columbia River, organisms living in the river contained up to 350,000 times more P-32 than the water, and birds preying on those organisms had even higher amounts in their tissue (Kwa 1989:50–51). This finding helped orient both E. Odum and Auerbach's research teams toward tracing radioactive isotopes through food chains, and food chain studies led to ecosystem studies.

Auerbach was interested in bringing mathematically inclined ecologists to Oak Ridge, even when their previous research did not seem relevant to the subjects studied at Oak Ridge. Two examples were Bernard Patten (b. 1931) and Van Dyne. Patten was a marine biologist, and Van Dyne a grassland ecologist (Bocking 1997a:79–80). They increased the mathematical sophistication of Auerbach's team and increased their own knowledge of ecosystems. Both stayed only a few years each, but when they left they were prepared for leadership in the IBP.

The idea for the IBP came from the success of the International Geophysical Year (IGY) from July 1, 1957, to December 31, 1958 (Aronova, Baker, & Oreskes 2010:183–196; Coleman 2010:1–3, 16–17). The idea for IBP, which was developed in Britain and Western Europe since 1959 and got underway there in 1963 (Golley 1993:110–114). Congress had generously funded the U.S. IGY and presumably would do the same for a well-planned U.S. IBP. The IBP was announced in the ESA *Bulletin* (December 1963:146–147; March 1964:39), but it required lobbying Congress and much discussion among U.S. ecologists on how to organize the research programs (Golley 1993:1–2, 114–119; Coleman 2010:24–26).

America's IBP began in July 1967 and lasted until June 1974. Although the IBP stimulated much new research (McIntosh 1976:363–365, 1985:213–215; Burgess 1981a:78), it was less controlled from the top than IGY had been (Aronova et al. 2010:198). The March 1968 issue of the ESA *Bulletin* (p. 4) reported that six major IBP programs were approved: aerobiology, large ecosystems, Eskimo populations, terrestrial life in the Hawaii Islands, phenology, and migration of persons from rural to urban areas. However, these research programs were modified as implementation began. In the United States, five biome programs emerged: the Grasslands, the Coniferous Forest, the Eastern Deciduous Forest, the Desert, and the Arctic-Alpine. University of Illinois plant ecologist Larry Bliss became the general supervisor and representative sites for each biome were selected (Aronova et al. 2010:200):

> The tundra program (part of the Arctic-Alpine Biome) conducted its research at Point Barrow, Alaska; the Desert Biome projects were carried on at several sites, one of which, at Curley Valley, Utah, functioned as a coordinating center; the Coniferous Forest Biome program had two sites, the Cedar River basin near Seattle and J. J. Andrews Experimental Forest in Oregon; the Deciduous Forest Biome program had five sites; and the Grassland Biome, based in Fort Collins, Colorado, was directed by Van Dyne and was supposed to function as a model and coordinating center for other biome studies.

E. Odum, the Oak Ridge ecologists, and limnologist Art Hasler were prominent in developing the guidelines for the IBP studies on ecosystems (Golley 1993:116–119). Odum thought it would be interesting to study ecosystems that contained both freshwater and terrestrial habitats. Hasler edited such a volume (1975, see below). These biome studies became the dominant feature of the IBP in North America. Both the U.S. and Canadian IBP programs began in 1968. Auerbach at Oak Ridge became head of the Eastern Deciduous Forest Biome (EDFB), which only began in 1970, but by 1974 had some 29 ecologists at Oak Ridge alone, and there were EDFB researchers at four other sites from lakes Wingra and George in the north to the Carolinas in the south (Bocking 1997a:93–94). Congress was persuaded to provide the NSF with unprecedented funding for the IBP. Other branches of biology feared it would come from their accustomed funding, but that did not happen.

Planning for the IBP coincided with advances in computer technology, leading to systems ecology (Aronova et al. 2010:212):

> A systems approach in ecology drew from cybernetics, information theory, thermodynamics, physical equilibrium theory, statistical ecology, and computer science in developing models of large ecosystems with the information about its components and linkages.

In other words, computers are built to deal with complexity, ecosystems are extremely complex, so put enough data on ecosystems into a computer and something wonderful will come out—Is it a new theory? Is it similar to what happened when the *Challenger* Expedition mapped the floor topography of the Atlantic Ocean (1872–1876), leading to speculation in the early 1900s about the movement of the continents (Skinner 1986:405)? Or is it maybe (as we say now) only "garbage in, garbage out?"

Optimism was not a bad thing when a new page was being turned in the history of ecology, with strong impacts on the history of ESA. Complex studies with many collaborators yielded problems as well as solutions, which took time to sort out. The first and most dominant biome study was of the Grassland Biome (1967–1976), under the dynamic leadership of Van Dyne, age 35 (Blair 1977:169; Golley 1993:120–130; Kwa 1993b; Auerbach 1995; Patten 1995; Coleman 2010:29–32). His stay at Oak Ridge had given him a vision of big science, and after he returned to Colorado State University in Fort Collins in August 1967, he took an IBP official, Frederick E. Smith, on a seven-day tour of America's grasslands, using a small airplane. (Smith was far from being a mere bureaucrat; he had earned a Ph.D. under Hutchinson at Yale, and in 1969 he moved from a professorship at the University of Michigan to one at Harvard.) That convinced Smith that Van Dyne was the right person to head the Grassland Biome, which became the model for other U.S. biome organizations. The combination of being first and having an aggressive administrator ensured that the Grassland Biome would have the largest budget and the largest research program. An example of Van Dyne's managerial thinking was his response in September 1969 to an invitation from Canadian IBP co-investigators at the University of Saskatoon for him and others to come to a conference. Van Dyne chartered a DC3 airplane to fly two dozen investigators from Denver to Saskatoon (Coleman 2010:35–36; on Saskatoon, see below). By December 1969, there were some 130 researchers involved with the Grassland Biome, including a systems engineer. To an invitation from *Bulletin* editor William Niering to write an informal progress report, Van Dyne and eight colleagues responded with 3 pages on the "Grassland Ecosystem Evaluation" (Coleman et al. 1973). It included a photograph of three bison wearing fecal collection bags; its caption stated: "Bison are less selective in their diet than domestic cattle, and yet are more efficient in digesting this non-specific diet—one of the strategies evolved for survival on the prairie."

Van Dyne was interested in constructing ecological mathematical models, and development of such models was one of the main achievements of the U.S. IBP. He wrote a brief history of his team's development of a model of the grassland ecosystem (Van Dyne 1978:v–xxii) and co-edited the 974-page *Grasslands, System Analysis and Man* (Breymeyer & Van Dyne 1980:9), which is volume 19 in the IBP reports. There was also a series of U.S. IBP synthesis reports, of which *The True*

Prairie Ecosystem (Risser et al. 1981) was volume 16. Frank Blair, a U.S. IBP participant, believed that no other country benefited as much as the United States. He thought that the U.S. IBP was responsible for "a quantum advance in ecological science as a basic science and for significant steps toward the development of the linkages necessary for the application of ecological knowledge to the solution of major environmental problems" (Blair 1977:162). The U.S. IBP's main achievement was "advancement of the concept of the ecosystem as a unit for ecological investigation," because an "ecosystem provides a unit for management of the world's resources" (Blair 1977:163; Coleman 2010:15–88).

In a memorandum of May 7, 1973, Frederick Smith commented: "To me the weakest aspect of the [grassland] program is a willingness to accept and incorporate a variety of ecological tenets without vigorously and systematically challenging their validity" (Kwa 1993b:133, n. 30). The NSF did not provide the $7.6 million requested, but only $4.6 million, as the Grassland Biome could no longer justify receiving a disproportionate amount of IBP funding (Kwa 1993b:148–149). When Van Dyne tried to achieve a grand synthesis of the grassland project for the official ending of the IBP in 1974, even though other investigators felt it was premature, it created a crisis, and he was relieved of his leadership role but allowed to continue working on a synthesis, which he did until he died of heart failure on August 2, 1981, at age 48 (Kwa 1993b:149–151).

The Deciduous Forest Biome began with an excellent international workshop organized by the IBP and the Oak Ridge National Laboratory, in July 1968, in Gatlinburg, Tennessee (Reichle 1970:v). This meeting brought together 11 researchers from Canada, England, France, Belgium, Germany, Russia, Israel, and Japan, who had already begun their IBP studies, with 13 American ecologists who were just embarking upon theirs (Reichle 1970:xi–xii). Although the Americans would only study temperate forests in the United States, this meeting provided a worldwide summary in 18 chapters of what was already known about temperate forest ecology, with chapters on ecosystem analysis, primary producers, consumers, decomposers, nutrient cycling, and the hydrologic cycle. In his chapter on "Secondary Production of Birds and Mammals," Dale McCullough (1970) of the University of Michigan, included a discussion of "Energy Flux and Biomass Distribution between Trophic Levels," which relied on research by Briton P. J. Newbould and Polish K. Petrusewicz. American ecologists were already rather familiar with relevant international literature, and they gained greater familiarity at the meeting.

A trans-Atlantic collaboration reported on "The Role of Soil Invertebrates in Turnover of Organic Matter and Nutrients" (Figure 4.3). C. A. Edwards was from the Entomology Department, Rothamsted Experimental Station, Harpenden, England; David Reichle was from the Ecological Sciences Division, Oak Ridge National Laboratory, Oak Ridge, Tennessee; and D. A. Crossley Jr. was from the Department of Entomology, University of Georgia. Decomposition of organic remains in a forest is a two-stage process, begun by invertebrates and completed by microorganisms. These investigators' report (1970) only studied the first stage, and that was limited to plant decomposition. Their goal was to determine energy flow and nutrient cycles through the soil populations of invertebrates.

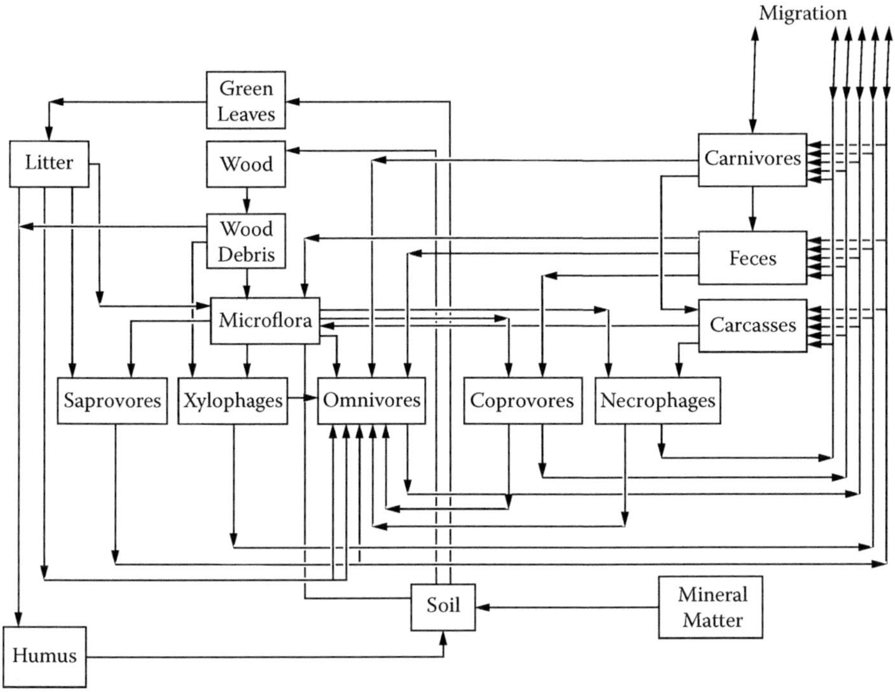

FIGURE 4.3 "A trophic model of the decomposer community of forest floor and soil. This diagrammatic description of the detritus food web illustrates the innate complexity of the trophic relationships." (From Edwards, Reichle, & Crossley 1970:156. With permission.)

However, the ecosystem models were much less developed than had been anticipated, and an anticipated central data bank never materialized (Aronova et al. 2010:209). Since computers were still in the early stages of development, computer models were difficult for ecologists to develop, and data from different biomes were not comparable (Mitchell, Moyer, & Downower 1976:865). A handbook on *Systems Analysis and Simulation in Ecology* appeared too late to guide the initial modeling, though it assisted in reporting the results (4 volumes, Patten 1971–1975).

One decision was to simultaneously study terrestrial and aquatic systems within drainage basins. This was not a minor detail (Golley 1993:109):

> Ecosystem theory was constructed from thermodynamics, from physical equilibrium theory, from information theory, from evolutionary theory, from field natural history, and so on. In 1965 it did not yet form a coherent, organized body of knowledge.

Golley's last comment may seem surprising, coming from an ecologist at the University of Georgia where E. Odum had written a famous textbook that emphasized the ecosystem concept. However, Odum's discussion in *Fundamentals of Ecology* emphasized theory and programmatic statements but was brief in describing results of such studies. A coherent theory, yes, but a coherent body of knowledge, not so much.

By enormously increasing the number of ecological researchers worldwide, the IBP also enormously increased the number of ecological publications. However, it was only a modest increase in ESA publications, because most of the increase was in separate publications. One of the earliest was a symposium volume, volume one of Springer-Verlag's *Ecological Studies*, most of which were IBP volumes. The series itself had four editors, only one of whom was an American, Jerry Olson, at Oak Ridge. This symposium volume, *Analysis of Temperate Forest Ecosystems*, was edited by David Reichle, at Oak Ridge (1970, xii + 304 pages). It contains the 18 revised papers delivered by 24 participants (13 from the United States and 2 from Canada) at an international workshop–seminar that the U.S. IBP Temperate Production Section and ORNL held in Gatlinburg, Tennessee, in July 1968 (Reichle 1970:vii).

Its reviewer for *Ecology*, Robert Wright, thought it useful, but he was less than excited by it (1971):

> While *Analysis of Temperate Forest Ecosystems* could scarcely be called a landmark volume, it will have real value in the doing of many kinds of ecology. It leaves to the reader the job of integrating the several approaches into the best combination for his study. It is by no means complete, either in breadth or depth. But it does give the serious experimenter a good start toward knowing some theoretical and practical aspects of ecosystem research in temperate forests.

This was the conclusion of Wright's full-page review that methodically covered the volume, describing its contents in detail. The authors were all very capable specialists who described their discoveries about ecosystem analysis or particular primary producers or consumers or decomposers or nutrient cycles or hydrologic cycles. At least *Analysis* had the virtue of integrating terrestrial plant ecology and animal ecology into one volume. Who could make a grand synthesis of all this? The introduction by two France-based ecologists, François Bourlière and Malcolm Hadley, did this to the best extent possible in 6 pages, but for more than that, it was up to authors of textbooks or of synthesis monographs written by specialists for other specialists (Figure 4.4).

Herbert Bormann, who would become ESA president in 1971, expressed concern for global ecological problems in an essay, "Subtraction by Multiplication" (1970). An initiative that president Bormann took was to appoint an ad hoc Committee on Urban Ecology to study how ecologists could respond to challenges of urban life and planning. That committee's report was written by C. S. Holling and Gordon Orians (1971). They explained the challenges and offered four suggestions for ecologists and two for ESA. A note to the report stated that the ESA Council was taking actions to implement their recommendations. A symposium on urban ecology was held on June 22nd at an AAAS meeting on urban ecology, and abstracts of the six papers that were presented appeared in the March 1972 issue of the *Bulletin*. Concurrently, the NSF asked some 200 scientific organizations to explain how their scientific research had helped solve human problems. Bormann (1971:2) in turn asked about 20 prominent ecologists to respond, and he later learned that the NSF considered ESA's response among the best.

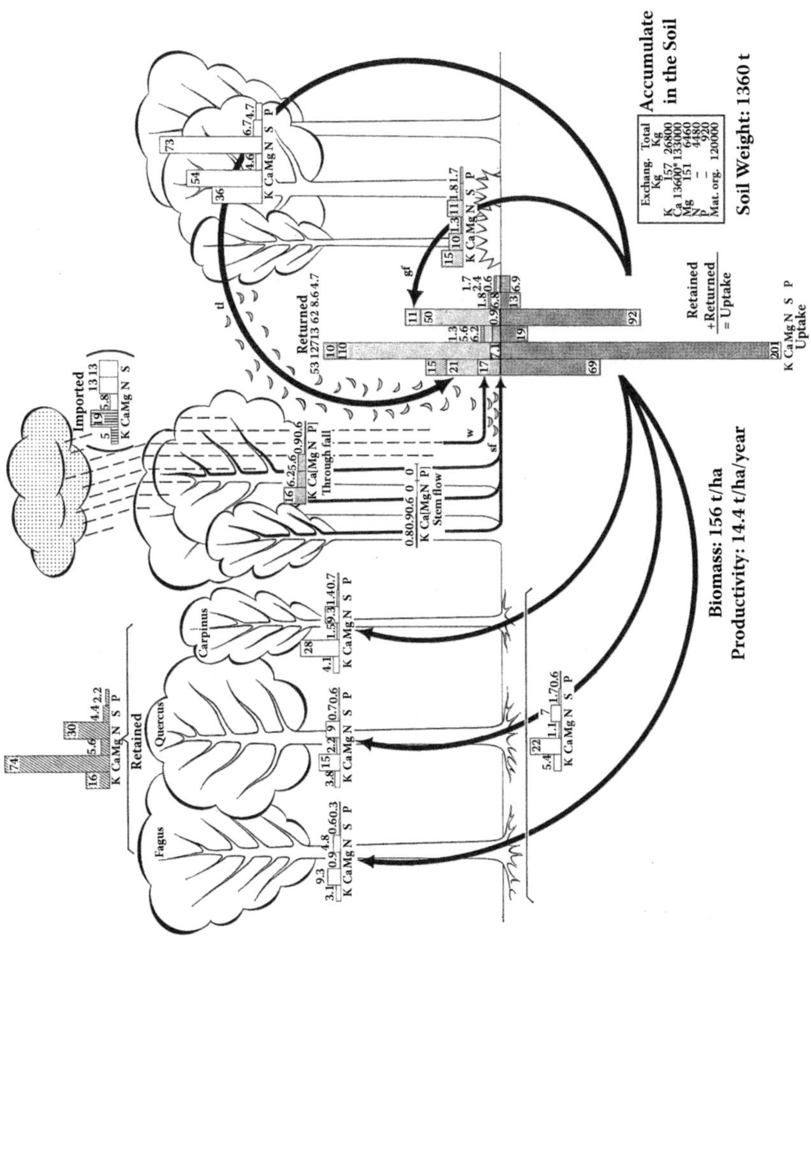

FIGURE 4.4 Annual mineral cycling of macronutrients in a mixed oak ecosystem in Virelles, Belgium. (From Duvigneaud & Denaeyer-De Smet 1970:216. With permission.)

The tenth volume in the Springer-Verlag series of the IBP volumes was the *Coupling of Land and Water Systems*, edited by Arthur Hasler, who had served as ESA president in 1961. Although this volume appeared in 1975, I discuss it in this chapter because all of its papers were written by 1974 and all the literature cited was published by 1974. It features 18 papers by 25 authors, 14 of whom were American, 8 were European, 1 was from the USSR, 1 was Argentine, and 1 was Malaysian. It is a hybrid volume in the sense that some of its 18 reports were IBP sponsored and funded, and others were International Hydrological Decade (IHD) sponsored and funded, all of which were presented at the XVIII International Congress of Limnology in Leningrad, USSR (Hasler 1975:vii, xi).

It begins with three reports at IBP sites in the eastern United States. The book's introduction and first report were by Bormann and Likens; they updated their ongoing experimental ecosystem studies, since 1963, at Hubbard Brook Experimental Forest, U.S. National Forests, New Hampshire. They began that research while on the Dartmouth University faculty, assisted by Dartmouth geologist Noye M. Johnson and U.S. Forest Service hydrologist Robert S. Pierce (Bocking 1997a:116). Their 1975 bibliography cites 14 previous published studies by them and coworkers, allowing one to study the details of their work to whatever extent desired.

The third report in this volume was "Models Linking Land–Water Interactions around Lake Wingra, Wisconsin" (1975), written by Orie Loucks (b. 1931), a Canadian who held two degrees from the University of Toronto, and who went to the University of Wisconsin to earn a Ph.D. in forest ecology under John Curtis (1960). At the time that Loucks wrote this 1975 report, he was at the Institute of Ecology's Office of Ecosystem Study, in Madison, Wisconsin. Five others on his team included students or former students from both Curtis and Hasler's Wisconsin schools. Lake Wingra is south of Lake Mendota and is much smaller (22 km^2 or 6 $miles^2$ for Wingra, 39.4 km^2 for Mendota; for maps of both, see Frey 1963a:12, 76). Two-thirds of Wingra's shoreline is alongside the University of Wisconsin Arboretum, and the other third is urban, in Madison.

The goals of the Wingra study were to document the movement of specific materials between the land and water systems and the impact of urban versus natural land uses, simulate with computer models the land and water ecosystems, and predict long-term trends to facilitate managing urban environments (Loucks 1975:54). Loucks used a simple diagram to explain how the terrestrial and aquatic ecosystems can be envisioned as subsystems of the land–water interactions of a drainage basin. He attempted to follow procedures established by Frederick Smith's "Analysis of Ecosystems" (1970) and other IBP publications. Nevertheless, Loucks had no impressive conclusions from his team's efforts, explaining that their method "has a broad generality that assists development of predictive models for the movement of diverse materials in the environment" (1975:62). This may have been similar to the responses Van Dyne had received to his demands for conclusions before other grassland investigators were willing. Unfortunately, Paul Baumann et al.'s "Lake Wingra, 1837–1973: A Case History of Human Impact" (1974) apparently appeared too late for Loucks to use or cite.

The other 16 reports in the Hasler-edited volume are extremely diverse in investigated subject matter, ranging from salmon biology and migrations to drainage of Amazon tributaries.

Pierre Dansereau (1911–2011)

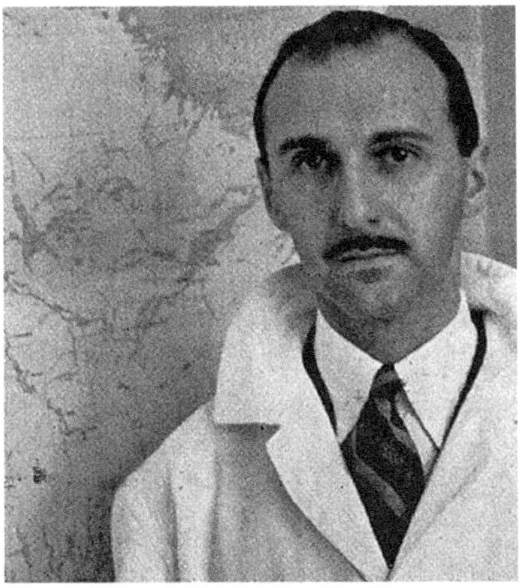

(From the Division of Archives, University of Montreal [circa 1942–1943].)

Dansereau's father was an engineer for the Quebec Department of Public Works, stationed in Montreal (Dansereau 1956 in Monk 1994:163; Dansereau 2005:15–26). The family was "upper middle-class." He attended Catholic schools that slighted science. At the University of Quebec at Montreal he earned B.A. and B.S. degrees (1931, 1936). He obtained his D.Sc. at the Université de Genève (1939) in plant taxonomy, with a dissertation on *Cistus* (Dansereau 1938, 1939, 1940, 1941a,b,c). His contact with Braun-Blanquet at Montpelier, France was about 17 days, but it had a strong influence on him, and he also learned from other European botanists during his three years in Europe (Dansereau 1956 in Monk 1994:164). He then joined the staff at the Montreal Botanical Garden, where he conducted a phyto-sociological survey of beech–maple forests of Quebec (Dansereau 1943). In 1942, he first lectured on ecology at the University of Montreal. In 1943, "I finally persuaded the Provincial Secretariat of Quebec to found a research unit which was called Service de Biogeographie" (in Monk 1994:165), with him as director until 1950. In 1946, he was a visiting professor in Brazil (Dansereau 1946; Dansereau & Segadas-Vianna 1947a,b,c). During the 1940s, he became acquainted with U.S. botanists who also influenced his development as an ecologist. That influence was strengthened by his stay at the Department of Botany, University of Michigan, where he taught a course on the vegetation of the world (1950–1955) (in Monk 1994:165–166). In 1955, he became the director of the Botanical Institute and dean of the Faculty of Science at the University of Quebec at Montreal, where he remained (except for other visiting professorships). His *Biogeography: An Ecological Perspective* (1957) drew upon his studies in North and South America and Europe. He published more than 600 academic papers (some quite brief) and is "the most authoritative and important scientific thinker to ever come out of Quebec" (Fine & DuPont 2012). His strong influence on both science and society was due to diligence, but he was also "an environmentalist with an abundance of optimism and a scientist who brought an artistic sensibility to the field."

Canada had a limnological program at the University of Manitoba and a marine project at the University of British Columbia which were "quite successful" (Coleman 2010:17). At Matador in southwestern Saskatchewan, the Canadian IBP ecologists at the University of Saskatchewan (in Saskatoon) found a prairie that they studied (Coleman 2010:18). Its principal investigators were Robert Coupland and Eldor Paul, and three others were senior scientists: Eeltje De Jong on abiotic processes, John Stewart on phosphorus and sulfur cycling, and Frank Warembourg on microbial turnover and soil respiration. Even though it was a three-hour drive from the university, the Matador project was important for training graduate students and publishing papers. Van Dyne (who flew his team to Saskatoon for a meeting, mentioned above) edited a book in 1969, *The Ecosystem Concept in Natural Resource Management*, in which the chapter on "Procedures for Study of Grassland Ecosystems" was by three ecologists at Saskatoon (Coupland, Zacharuk, & Paul 1969).

Canadian soils ecologists J. A. C. Fortescue and G. G. Marten, from the Petawawa Forest Experiment Station, Chalk River, Ontario, sought to convey both their findings and their methodology in "Micronutrients: Forest Ecology and Systems Analysis" (1970) because both subjects were equally important for ecologists learning systems analysis. They had found that their investigations had to draw upon several disciplines, both for information and for methodology. Their goal was "to discover general scientific principles (or patterns) which can explain the past, present, and future nutrient requirements of any forest" (Fortescue & Marten 1970:193–194). They constructed a diagram to indicate both sources and methods used to achieve their understanding.

Canadians from the University of Toronto initiated an Arctic IBP project at Char Lake on Cornwallis Island, directed by Frank Rigler (1928–1982) (Bocking 1997a:169–170, 2007:883, 2013). Canadians had not had a facility where computer simulations were developed such as the one at Oak Ridge, and although some students at Char Lake attempted to develop a lake ecosystem model, their effort failed. Instead, Toronto ecologists extended their previous specialized research into this new environment. The Fisheries Research Board of Canada supported their research (1968–1973) but ended funding when anticipated practical applications did not materialize. The Great Lakes Fishery Commission, jointly funded by the United States and Canada, supported a study on Lake Erie fish populations, and Ontario Hydro supported a study on the impact of heated effluent from a nuclear plant on Lake Huron fish populations. Both of these projects from the University of Toronto fishery biologists likely would have been funded without any IBP connection.

Another Canadian Arctic IBP project was at the Truelove Lowland site on the northeastern coast of Devon Island. A research camp had existed there since 1960 (Coleman 2010:18–20). This project included Eskimos as the top predator and musk ox herds as grazers. The goal was to determine the population, standing crop, and productivity (energy flow) of all trophic levels. The project director was from Ohio, Lawrence Bliss (b. 1929), in the Botany Department, University of Washington

(Bazzaz 1982; Bliss 1994). He edited a remarkably comprehensive *Truelove Lowland, Devon Island, Canada: A High Arctic Ecosystem* (1977, xxi + 714 pages) that contains 36 reports and 8 appendices by 45 authors. Bliss was also on the U.S. National Committee on the IBP.

Ten countries had Arctic IBP projects—Austria, Canada, Finland, Greenland, Ireland, Norway, Sweden, the United Kingdom, the United States, and the USSR—and by pooling their staffs and data they achieved more than any one country could do alone. O. W. Heal from the United Kingdom explained (quoted in Coleman 2010:19–20):

> The American (microbiologist) took litter samples back to the laboratory and measured the rate of respiration but the Brit (field ecologist) measured the weight loss from litter bags in the field. The modeler (mathematician) saw that respiration rate could be used to calculate weight loss for comparative purposes—but, given data on climatic conditions, it could also be used to predict decomposition at other sites.

Bliss also edited *Tundra Ecosystems: A Comparative Analysis* (1981). It was no surprise, then, that Bliss was elected ESA president from 1982–1983. The U.S. IBP ran its Arctic tundra study at Barrow, Alaska.

The IBP's formal ending in 1974 was the occasion of four tundra biome symposia, a summary of which appeared in the December 1974 issue of the *Bulletin* (pp. 24–25): (1) Climate and primary production processes, (2) Decomposition and nutrient flux, (3) Grazing behavior and population dynamics of tundra animals, and (4) Carbon flow and nutrient regulation in a tundra pond system. Details on the findings are included under each of these four headings. By the summer of 1974, the IBP had provided research funds at 60 institutions for over 700 scientists and graduate students, who produced nearly 500 publications (Mitchell et al. 1976:865).

From September 8 to 14, 1974, after the IBP officially ended, the International Association of Ecology (INTERCOL) sponsored the First International Congress of Ecology at The Hague, the Netherlands. Over 800 ecologists worldwide attended. There were 23 invited papers by 27 speakers presented at plenary sessions, which were later published, with discussions that followed, entitled *Unifying Concepts in Ecology* (van Dobben & Lowe-McConnell 1975). Eleven of these speakers were American and four were Canadian. Their papers are grouped under five headings:

- Flow of Energy and Matter between Trophic Levels
- Comparative Productivity in Ecosystems
- Diversity, Stability, and Maturity in Natural Ecosystems
- Diversity, Stability, and Maturity in Ecosystems Influenced by Human Activities
- Strategies for Management of Natural and Man-Made Ecosystems

Worldwide, the IBP thus officially ended with a unifying meeting and strong international communication among participants.

The IBP was not, of course, the only ESA activity or concern during 1960 to 1974. Robert MacArthur was one of the brightest young stars in the ESA, when he died of cancer at the age of 42. He had two degrees in mathematics before he went to Yale to obtain a Ph.D. in mathematics. However, in his second year, a fellow graduate student, Lawrence Slobodkin, influenced him to switch to ecology and he earned his Ph.D. under Hutchinson with a dissertation (1958) on niche separation versus competition in warblers (Kingsland 1985:180). MacArthur won the Mercer Award in 1959 for the publication of his dissertation, and posthumously he was awarded the Eminent Ecologist Award in 1973. Like Curtis, he surely would have become president of ESA had he lived longer.

Robert H(elmer) MacArthur (1930–1972)

(From Wilson 1973:11. With permission.)

MacArthur's early life resembled that of H. T. Odum (see below), born six years earlier. Both were sons of university professors and were significantly influenced by both their father and an older brother. Geneticist John Wood MacArthur was teaching at the University of Toronto when Robert was born, but John was a U.S. citizen, and later Robert chose to be one as well. John later taught at Marlboro College, Vermont, which is where Robert earned his B.A. degree. His older brother, John W. MacArthur, Jr., born in 1922, became a physicist and also taught at Marlboro College. Robert earned an M.S. degree in mathematics at Brown University (1953) and then applied to the University of Illinois to study under Kendeigh, but was turned down because his degrees were in mathematics. What was an impediment at Illinois was an asset at Yale, for Hutchinson was very interested in ecology students with mathematical skills. MacArthur put those skills to good use in three early papers on diversity-stability relations (1955), species abundance distributions (1957), and for a doctoral dissertation, an analysis of niche division among five closely related *Dendroica* warblers (1958), to test Georgii Gause's postulate that closely related species cannot occupy the same niche. His doctoral dissertation (Ph.D., 1958) explained the different foraging strategy of each species. The remainder of his career fulfilled the promise of its brilliant beginning, but only lasted 14 more years. His 1958 paper received the Mercer Award for 1959. He explained his concept of r- and K-selection in "Some Generalized Theorems of Natural Selection" (1962), which led to his collaborations with Joseph Connell on *The Biology of Populations* (MacArthur & Connell 1966) and with Ed Wilson on a landmark: *The Theory of Island Biogeography* (MacArthur & Wilson 1967). When MacArthur learned of his pending death from cancer, he published his expanded lectures from a course, *Geographical Ecology: Patterns in the Distribution of Species* (1972). He received the Eminent Ecologist Award posthumously in 1973 (Wilson 1973). His bibliography (Cody & Diamond 1975:13–14) shows both his outstanding academic productivity and his ability and desire to work with collaborators (Cody & Diamond 1975:vii–ix; Fretwell 1975; Kingsland 1985:see index; Wilson & Hutchinson 1989; Burgess 1996:69–70; Brown 1999; Pianka & Horn 2005; Kaspari 2008; Birkhead, Wimpenny, & Montgomerie, 2014:378–381).

A most notable book during those years was *The Theory of Island Biogeography* (1967), by Robert MacArthur, Princeton University and Edward Wilson, Harvard University. It was a monograph, not a textbook, yet its review in *Ecology* covered 2.5 pages (Preston 1968). Even so, Preston explained that his review picks up where Terrell Hamilton's review in *Science* (January 5, 1968) left off. (Hamilton thought this book "may revitalize ecology.") MacArthur and Wilson's theory is primarily about invasion, persistence, and extinction of species on islands. The serious study of species on islands goes back to Georg Steller's almost nine-month stay on Bering Island from 1741 to 1742 (Stejneger 1936:350–367; Egerton 2008b:47–50, 2012a:94–95), and included Charles Darwin's studies on the Atlantic and Pacific Islands (1832–1836) (Egerton 2010:400–423, 2012a:143–146). MacArthur and Wilson (1967:158) do cite Darwin's hypothesis on why many birds and insects on islands are flightless, but Darwin proposed no theory of island biogeography. For theory, MacArthur and Wilson began not with naturalists' observations, but rather with calculations from Lotka (1925) and Volterra (1926).

A project on the grand scale of the IBP, but with other federal government funding, was *A Tropical Rain Forest: A Study of Irradiation and Ecology at El Verde, Puerto Rico*, edited by H. Odum and R. F. Pigeon (94 authors, 114 chapters, xxiv + 1,650 pages). It is a 10-pound, gigantic volume, 3 inches thick. Its reviewer for *Ecology*, Paul Kannowski (1971), complained that many pages of the data could have been deleted, and other economies taken to reduce its bulk. However, he acknowledged that these were minor complaints, and that the book was a valuable reference work. H. Odum thought on a grand scale (not being intimidated by the book's size), and he supervised large-scale rain forest experiments, using radiation, herbicides, and huge plastic enclosures to measure physiological characteristics of this rain forest. In his review in *Science*, Joseph Connell (1971) complimented Odum's experimental originality but found his circuit diagrams difficult to understand. One of Odum's former students, Ariel Lugo, wrote two appreciative essays (1995, 2004) on the accomplishments of this "green book," in the second of which essays he quoted Odum's own retrospective summary (1995) of what was achieved.

Perhaps the best testimonial, however, came from authors unconnected to that endeavor (Hecht & Cockburn 1990:40–41):

> For three months a patch of government-owned forest in El Verde, Puerto Rico, was bombarded with gamma rays. As the forest withered under this simulation of planetary holocaust, Odum led his team of young scientists to analyze the wounded ecosystem and from this study to deduce the structure and the functions of the forest. Thus began the first comprehensive inter-disciplinary study of a tropical forest. The experience launched a cohort of tropical ecologists who, over the next generation, have had enormous influence in the North American study of tropical systems. One of the central contributions was developed by Carl Jordan who analyzed nutrient cycling and the mechanisms by which tropical forests maintain themselves on shoddy soils.

In retrospect, *A Tropical Rain Forest* was atypical of H. Odum's work, in that it was so comprehensive and with so many workers. More typical was his *Environment, Power, and Society* (1971, xi + 331 pages). Who was its audience? "Intended for the general reader, this account also attempts to introduce ecology through the energy language" (Odum 1971:vii). He had a very optimistic and flattering estimate of general readers, extremely few of whom would likely feel motivated to learn his electrical engineering language in order to understand ecology when they had a choice of less intellectually demanding popular books on ecology (Taylor 1988). It seems more likely that it was used as an introductory textbook for undergraduate courses in ecology. However, by 1971 there were a number of introductory texts, including his brother Eugene's, so which ecologists were likely to choose this one? *Science* devoted over 2 pages to its review by Egbert Leigh, Jr.; the review (1971) was a bit disconcerting, reflecting his feelings about Odum's book. On the one hand, he stated that Odum "develops a grand synthesis of the principles governing ecosystems and societies, an act of courage for which he cannot be too highly praised." On the other hand, Leigh complained about the following:

> The result is a most maddening work, which at first sight seems totally undisciplined, a chaotic mixture of the asinine, the banal, and the brilliant, with random observations, often in conflict with the available evidence, on nearly everything under the sun. Odum writes an oppressive jargon, interspersed with elaborate circuit diagrams meant to simulate energy flows, in the systems he discusses.

Leigh's reaction provides a clue about who would select Odum's book for an introductory textbook: Odum himself and his disciples.

Howard Thomas Odum (1924–2002)

(Courtesy of Mary Odum.)

H. T. Odum was as influential an ecologist as his older brother (Hagen 1992: see index; Burgess 1996:82; Brown & Hall 2004). Of her three children, his mother commented that "H. T. was the gifted one" (Craige 2001:17). H. T.'s education at the University of North Carolina (UNC) was interrupted by his service in the U.S. Air Force during World War II (he received his B.A. degree from UNC in 1947). He did his graduate school work under Hutchinson at Yale (Ph.D., 1951), and what he learned at Yale was incorporated into his brother's textbook. He was a daring theoretician who when relevant included humanity in his diagrams. He was an important contributor to systems thinking and energetics at the ecosystem level, starting with his Silver Springs paper (1957). Most of his career was spent at the University of Florida. His students were extremely devoted to him and his unique energy systems diagrams, which borrowed symbols from electrical engineering. He was a very productive author, and his influence is attested by two collections of articles by his former students and associates: Charles A. S. Hall, editor of *Maximum Power: The Ideas and Applications of H. T. Odum* (1995) and Mark T. Brown and C. A. S. Hall, editors of *Through the Macroscope: The Legacy of H. T. Odum* (2004a). Hall commented (personal communication, June 22, 2014): "HT was, in my mind and that of many others, by far the most intelligent person I ever met, an incredibly inspiring graduate mentor and very decent person. Towards the end of his life my wife and I got to know him as a friend and we greatly valued that. He was unhappy that I never signed on to emergy, but I thought his contributions to Systems thinking, energy and economics were enough for any three scientists."

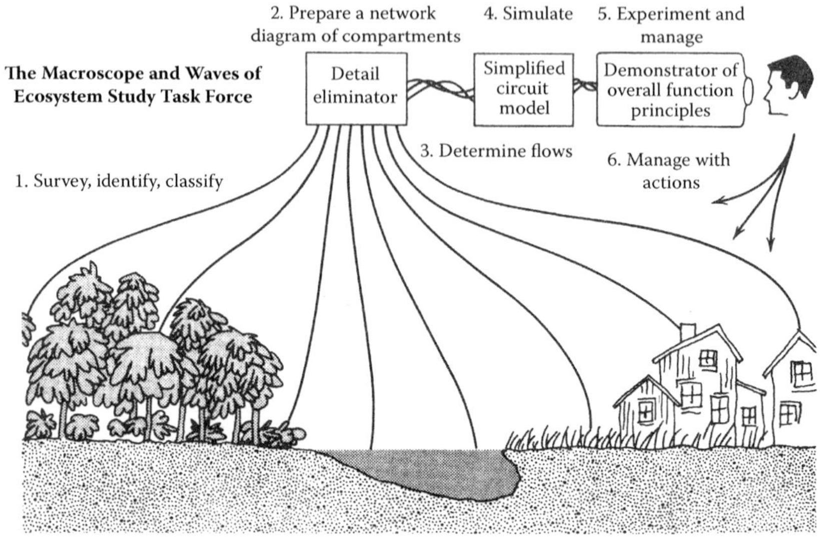

FIGURE 4.5 Macroscopic concept. (From H. Odum 1971:10. With permission.)

It was in *Environment, Power, and Society* that H. Odum introduced his "macroscope" concept (Bill Mitsch, personal communication). Viewing macroscopic systems is the opposite of microscopic analysis. With microscopes, biologists had discovered how organisms are structured and functioned. The macroscopic approach instead studies synthesis of ecosystems from their organic and inorganic components. A French ecologist, Joel de Ronsay, was impressed by the concept and wrote a book on it: *La Macroscope: Vers une Vision Globale* (1975). Ronsay's *La Macroscope* interested French limnologist Christian Lévêque (2003:87), whose discussion seemed to imply that Ronsay originated the concept. However, Ronsay's *La Macroscope* cited H. Odum's book (Jean-Marc Drouin, personal communication) (Figure 4.5).

H. Odum had a highly effective teaching style (he taught me in 1955) that infected many students with his enthusiasm for what he taught. They were willing to learn his jargon and his circuits and then teach the same to their own students. In a memorial volume published two years after his death, the editors explained that the volume pays tribute to "an individual who had a larger than life presence and impact on so many of us" (Brown, Hall, & Jorgensen 2004:1). To the uninitiated, they explained that "In the world of science that we live in there are two kinds of people: Odumites and others." Those, like Leigh, who found his jargon oppressive and his "elaborate circuit diagrams" part asinine and part brilliant, would have to find another way, as best they could.

Ruth Myrtle Patrick (1907–2013) was the daughter of a Kansas lawyer with a botany degree from Cornell, whose hobby was studying diatoms. Her father's passion was contagious, and she spent her life studying them as well (Bandler 1970; Burgess 1996:85; Cairns 2014). Although she was senior coauthor of a monograph on U.S. diatoms (Patrick & Reimer, 2 volumes, 1966–1975), she placed diatoms in a limnological context in which they served as indicators of degrees of clean/polluted waters (Patrick 2000). She first married entomology professor Charles Hodge IV, a descendant of Benjamin Franklin, and they lived in a Philadelphia suburb (Hodge was at Temple University). She became a member of the research staff at the Philadelphia Academy of Natural Sciences. In 1972, she became ESA's Eminent Ecologist (Hutchinson & Bormann 1972), and in 1975 she won the Tyler Award (Holden 1975; Miller 1975). She studied the river water from all over the United States and beyond (Patrick 2000), and her last great achievement was *Rivers of the United States* (6 volumes in 8, 1994–2003), which rivals Hutchinson's *Treatise on Limnology* as being the longest ecological treatise written by one ecologist.

There were two books that appeared in 1970 entitled *Biosphere.* Nancy Jessop's *Biosphere: A Study of Life* was actually an attractive introductory biology textbook for college courses. The other, *The Biosphere: A Scientific American Book*, was a reprint of the thematic articles of the September issue of *Scientific American*, which had an introduction by Hutchinson, followed by articles on the cycles of energy, water, oxygen, carbon, nitrogen, minerals (by Deevey), and on production processes for food, energy, and human materials. This *Scientific American* issue/book provides the biogeochemistry of the biosphere. In 1971, Paul Ehrlich, John Holdren, and Richard Holm compiled a *Scientific American* reader entitled *Man and the Ecosphere*, which does not duplicate the *Biosphere* book. The ecosphere reader takes its title from its first article, "The Ecosphere," by Lamont Cole, reprinted from the April 1958 *Scientific American.* Cole explained (1971:11):

> Probably I should apologize for using a coined word like "ecosphere," but it seems nicely to describe just what I want to discuss. It is intended to combine two concepts, the "biosphere" and the "ecosystem."

Two of the other reprinted articles in this reader contained the word "biosphere" in their titles.

We turn next to a more usual concern, the evaluation of ecology textbooks. The early spring 1974 *Ecology* issue contained reviews of an astonishing 13 general ecology textbooks, while the third edition of E. Odum's *Fundamentals of Ecology* (1971) was still in print. As a sample, I discuss only the first review, which is also the longest. (For the reviews on the other textbooks, see Hagen 2008:713–717.) Robert Ricklefs (b. 1943) wrote a text entitled *Ecology* (Chiron Press, 1973, x + 861 pages). The fact that he and his publisher thought it plausible to publish such a lengthy ecology textbook in 1973 indicates how secure ecology had become in the academic establishment. Ricklefs' *Ecology* has numerous photographs, drawings by Joel Ito, and diagrams by John Woolsey. Insect ecologist Peter Price (1974) gave it a strong endorsement in the conclusion to his 1.5-page review: "Ricklefs writes with authority on the broadest range of ecologically relevant subject matter so far published in one book. What more can we ask?"

Actually, Price had already asked a lot more of Ricklefs than his ending implies. In some instances, Ricklefs overgeneralized from vertebrate physiology/ecology to all animals. For example: "few animals feed on plants when they are small because they require a rich source of protein in their diet to support rapid growth." Price said that insects are exceptions to this claim. Ricklefs simplistically claimed that flight evolved as a strategy to escape predation, without recognizing that flight is also more energetically efficient than walking (at least in insects) and that wings also enable a species to colonize new territories and to find new sources of food when needed. Ricklefs' *Ecology*, despite its length, neglected biogeography. His chapter on social insects omitted use of the most important book on the subject, Ed Wilson's *The Insect Societies* (1971). Perhaps Australian Robert May (1989:339) had in mind Ricklefs' *Ecology* when he complained about "North American texts that encyclopaedically include everything, but which then have to be trundled around in a wheelbarrow." (In 1976 Ricklefs published a briefer version of his textbook.)

The ESA established a study committee to determine the need for a national institute of ecology, and if so, its scope. Robert Inger, at the Field Museum of Natural History in Chicago, was chairman of the study committee, and the committee enlisted the assistance of Peat, Marwick, Mitchell & Co. They jointly produced a favorable report in two volumes. Inger (1970) summarized the conclusions in the *Bulletin*, including an institutional diagram (Figure 4.6). The original name suggested was National Institute of Ecology. According to Inger's explanation, its central purpose was "the application of ecological science to both short and long term social problems," though the diagram seemed to place some emphasis on "biomes for ecosystem analysis" (1970:2). In Inger's second report (1971), the name had changed to Inter-American Institute of Ecology (IAIE), and its scope now extended from Canada to Central America. In September 1970, Edward Deevey had become co-chair of the Study Committee, and following Inger's second report Deevey (1971) proposed and answered 20 questions about IAIE.

At the October 1971 meeting of The Institute of Ecology (TIE), Arthur Hasler was asked to become its director, and he accepted (Hasler 1971):

> My acceptance was predicated on my belief that TIE offers a unique vehicle for the ecological community to perform both research and service tasks that are broader than can be accomplished by existing research institutions and universities acting alone.

He thought that "Last summer's Workshop on Global Ecological Problems at Madison offers a model which is widely replicable." That workshop, with 58 participants, mostly from throughout the United States, but with four from Canada, and one each from Mexico and Australia, was TIE's first project, supported by the NSF. The University of Wisconsin Press published TIE's workshop report on global ecological problems (Hasler 1972).

TIE sponsored six workshops in June 1973 on tropical ecology, held in Turrialba, Costa Rica. Their report, *Fragile Ecosystems*, was edited by Edward Farnworth and Frank Golley of the University of Georgia Institute of Ecology (Springer-Verlag, 1974), using the same publisher of many of the U.S. IBP reports in a volume resembling IBP reports, but the IBP was not mentioned in it. The editors' preface stated that the report was based upon worldwide correspondence with over 2,500 persons, with over 400 persons' contributing papers and ideas, and with over 100 attending the workshops. The workshop reports each list 6 to 13 participants who contributed to it. The majority of participants were Americans, with most of the others

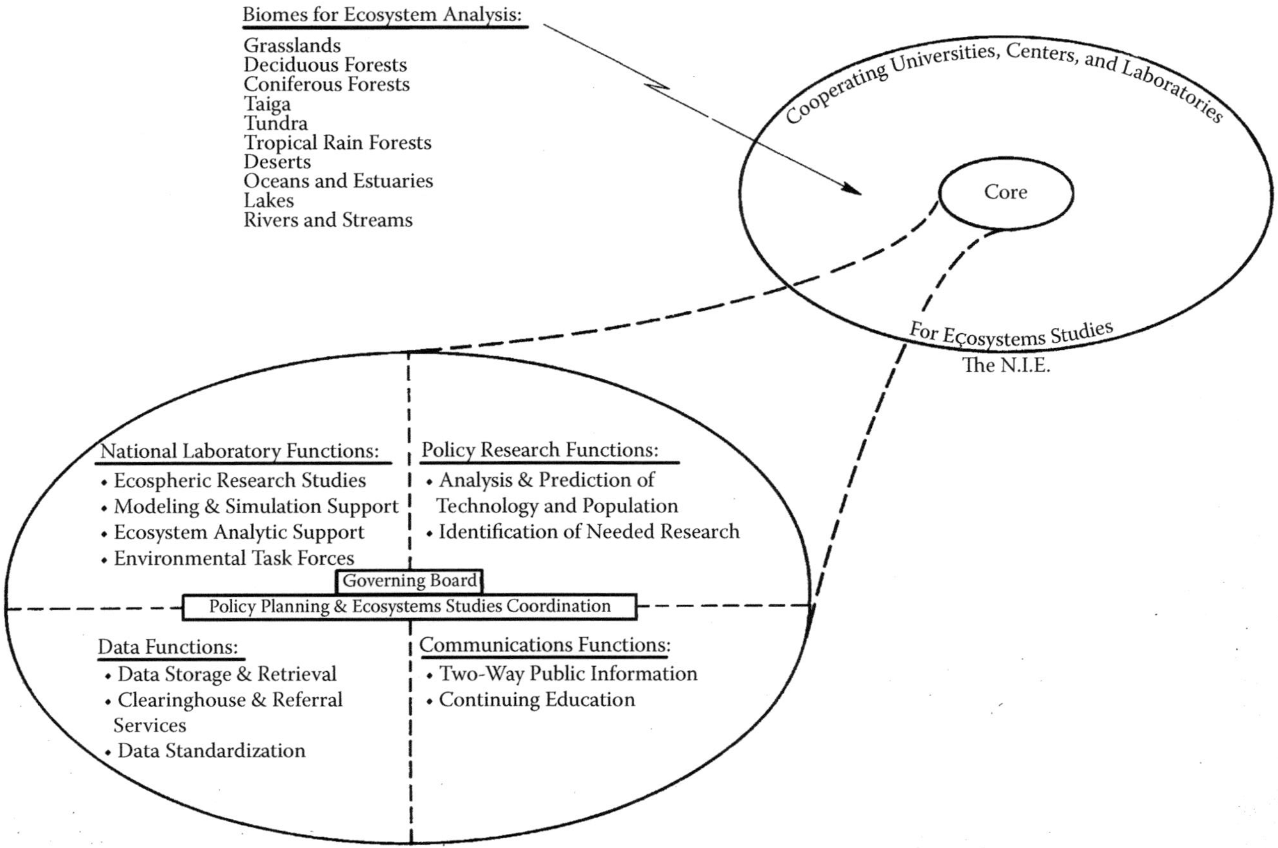

FIGURE 4.6 Proposed organizational diagram for a national institute of ecology. (From Inger 1970:15. With permission.)

coming from Latin America, but with a few from Canada and Europe. The workshops addressed these tropical topics: Population Ecology, Ecosystem Structure and Function, Ecosystem Recovery, Interaction of Man and Environments, Impacts of Regional Changes on Climates and Aquatic Systems, and Mechanisms to Support and Encourage Research and Education. In 1974, Hasler could report that there were a wide range of TIE projects underway—TIE sponsored 15 publications (1972–1974).

The saddest events in ESA history during the period of this chapter (1960–1974) were the early deaths of Curtis in 1961 and MacArthur in 1972. Since most of Curtis' publications appeared before 1960, he is discussed in Chapter 3. This chapter ends with a remembrance of MacArthur (Cody & Diamond 1975:vii):

> In November 1972 a brief but remarkable era in the development of ecology came to a tragic, premature close with the death of Robert MacArthur at the age of 42. When this era began in the 1950s, ecology was still mainly a descriptive science. It consisted of qualitative, situation-bound statements that had low predictive value, plus empirical facts and numbers that often seemed to defy generalization. Within two decades new paradigms had transformed large areas of ecology into a structured, predictive science that combined powerful quantitative theories with the recognition of widespread patterns in nature. This revolution in ecology had been due largely to the work of Robert MacArthur.

5 1975 to 1989

The International Biological Program (Concluded), the Long Term Ecological Research (LTER) Network, Ecosystems, Professional Certification, and Gender

The *official* end of the International Biological Program (IBP) in 1974 was not the end. A central data bank was never achieved (Aronova, Baker, & Oreskes 2010:209). Computers were still in the early stages of development. The IBP ecosystem models were much less developed than expected, and an anticipated development of computer models was difficult for ecologists, and data from different biomes were not comparable (Mitchell et al. 1976:865). A handbook on *Systems Analysis and Simulation in Ecology*, in four volumes, appeared too late to guide the initial modeling, though it could assist in the reporting of results (Patten 1971–1975).

Congress funneled more than $50 million to the IBP through the National Science Foundation (NSF) (Appel 2000:226–232, 261–267). At the official end of the U.S. IBP in June 1974, the NSF asked Battelle's Laboratories (Columbus, Ohio), to evaluate three of the five U.S. biome programs—Eastern Deciduous Forest, Grassland, and Tundra—and compare them with the ongoing Hubbard Brook program. The personnel who undertook this review were Ramona Meyer, an ecologist from Battelle, and two zoologists on the faculty of Ohio State University, Rodger Mitchell and Jerry Downhower.

And how did the committee evaluate the three IBP programs compared to the smaller and longer-lasting Hubbard Brook (New Hampshire) ecosystem program, that began in 1963? The evaluators examined 107 Hubbard Brook publications, about half of which were on abiotic factors. However, two synthesis volumes on Hubbard Brook ecology (Likens et al. 1977; Bormann and Likens 1979) appeared after the evaluators' study, and even later a historian of ecology wrote an account (Bocking 1997a:116–147). There were 41 scientists and 22 students involved with Hubbard Brook during 13 years. There was no attempt to model its ecosystem. Its costs were significantly lower, and it was more successful in meeting its goals than the IBP, but the IBP exerted considerably more influence upon the practice of ecology (Mitchell et al. 1976:864–865). Furthermore, the Hubbard Brook studies were at full speed when the IBP studies were getting started, so it is a somewhat unreasonable comparison. One might argue that the more numerous personnel of the IBP studies made up for their later start, but whether a fair comparison could be made still seems debatable.

The committee's comments about ecosystem modeling were as follows (Mitchell et al. 1976:862):

> [M]odels were intended to be general enough to deal with large regions, precise enough to allow for meaningful applications in management decisions, and realistic enough to add new insights to theoretical ecology. These overly ambitious goals were not achieved. Time has shown the wisdom of Levins' postulate [1966] that there has to be a trade-off between generality, precision, and reality. An increase in one aspect of a model comes at the cost of the others. The second fact which came to light is that modeling is still an art. The techniques of programming, the choice of mathematical equations—stochastic, deterministic, differential, difference—and selection of software, and the methods of condensation, building in modularity, or developing a hierarchical structure are still being explored and debated.

The committee also highlighted positive achievements, including the following (Mitchell et al. 1976:865):

> (2) It afforded support for more than 700 scientists and graduate students at 60 institutions across the United States.
> (3) It established an awareness of the advantages of team research and modeling in solving large and complex problems related to resource management.
> (4) It produced nearly 500 (as of 1974) publications that differ from nonbiome publications in the kinds of things covered.
> (11) It showed that decentralization did not inhibit the exchange of information between scientists but rather encouraged scientists to be more responsive identifying and pursuing alternative goals.

Murray Fife Buell (1905–1975)

(From ESA/History. With permission.)

Murray Fife Buell was from New Haven, Connecticut; he earned a B.A. degree from Cornell University (1930), and M.A. and Ph.D. degrees from the University of Minnesota (1934, 1935). His Minnesota advisor was a plant morphologist, but he took all of W. S. Cooper's courses. He began his career at North Carolina State University in Raleigh (1935–1946), and then settled at Rutgers University. Thereafter, he conducted his research in New Jersey. He maintained a close relationship with his students, which included a seminar for graduate students in which each student chose a prominent ecologist, living or dead, from anywhere, and researched his/her career. (Braun was the only woman chosen.) When the ecologist chosen was alive, the student wrote for information, and the responses were part of the written reports, which were eventually computerized and published (Burgess 1994). At Rutgers, he advised 39 doctoral students and arranged for the creation of the William L. Hutchinson Memorial Forest, administered by the Rutgers Department of Botany, meaning by him until his retirement in 1971. About 90 publications were based on studies there by 1975. Buell was ESA's Eminent Ecologist in 1970, and he served as both vice president and president of ESA and the Torrey Botanical Club (Pearson & Bormann 1970; Bormann & Pearson 1975; Small 1975; Burgess 1996:20–21).

Stanley Auerbach, Robert Burgess, and Robert O'Neill (1977) wrote the most authoritative response to Mitchell et al. (1976). They stated that the IBP programs were more complex than the reviewers realized, and they had not seen as many reports as they should have seen; therefore, some of their conclusions were simplistic. Not only had the NSF commissioned an evaluation of the IBP, the National Academy of Sciences did likewise, with a committee headed by plant physiologist Professor Emeritus Paul J. Kramer, which also produced a report in 1976. That report was summarized by Philip Boffey (1976:866) in *Science*:

> It simply finds that the program produced some major accomplishments, had some acute problems, met some goals, missed others, and turned in a "credible" performance, considering that it was greeted with apathy and hostility by key elements of the scientific community and had enormous difficulties meshing its goals with those of granting agencies.

Other evaluations lacked enthusiasm (Hagen 1992:177–180; Golley 1993:142, 151; Kingsland 2005:221–222). Robert McIntosh (1985:235–236) cited several published rebuttals to unfavorable reviews and counter-rebuttals, and his own comment was, "Rarely has a 'revolution' or 'paradigm change' been so grudgingly touted."

There was no expectation that all the IBP reports could be published by the June 1974 terminal date for the IBP. If the IBP achieved no new ecosystem theory, it did train far more new ecologists than ever before, with a consequential boost in the publication of ecological literature. Furthermore, "ecological theory and theorists have burgeoned in the last three decades, and very self-conscious, and sometimes self-congratulatory, groups of theoretical ecologists have come to the fore" (McIntosh 1980:197). Richard Levins and Richard Lewontin have argued in their

"Dialectics and Reductionism in Ecology" (1980) that theoretical debates between reductionists and idealists can often be recast as dialectical explorations.

If considering later IBP reports, it is appropriate to begin with the grasslands, where the U.S. IBP research began. Grassland research results enabled George Innis to edit two volumes: *New Directions in the Analysis of Ecological Systems* (1975) and *Grassland Simulation Model* (1978); and subsequently six plant ecologists and animal ecologists wrote *The True Prairie Ecosystem* (Risser et al. 1981), which acknowledged George Van Dyne's important role in organizing the IBP Grassland Biome Project. Van Dyne died in 1981, and if he had died before the volume's Acknowledgments had been written, his death presumably would have been mentioned.

Since Van Dyne's grassland IBP project was the largest in America, it seems fair to view *The True Prairie Ecosystem* as a representation of what could be achieved by large-scale ecosystem studies using modern (for that time) computers. This volume certainly provides a thorough survey of the prairie grassland ecosystem—climate, producers, consumers, decomposers (Figure 5.1)—and the answer to what could be achieved in understanding ecosystems from large-scale computerized research is in the final chapter, "Whole Systems Properties" (based on both original studies and

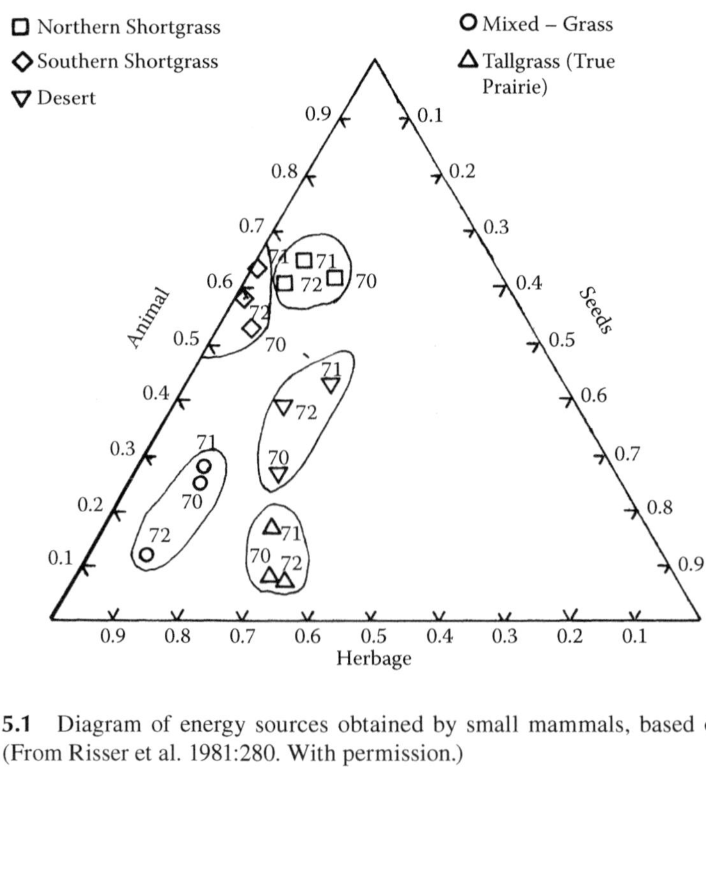

FIGURE 5.1 Diagram of energy sources obtained by small mammals, based on dietary analyses. (From Risser et al. 1981:280. With permission.)

relevant literature), and in the book's summary. Both are well done, with valuable findings, but no new theory was presented that was based upon these findings.

Orie Loucks, discussed in Chapter 4 concerning his contribution to the Lake Wingra IBP study (1975), wrote his own evaluation of the IBP in 1986, and by then he was quite enthusiastic about what the IBP had accomplished and was still accomplishing—he edited one of three IBP volumes that appeared in 1986. The productivity of IBP ecologists was by then quite impressive. Around 500 U.S. IBP scientific papers had been published by the end of 1974, with uncounted others appearing after 1974. Cambridge University Press had published 26 IBP volumes, though these included contributions from a number of countries, and counting the three IBP volumes published in 1986, the U.S. publisher Dowden, Hutchinson & Ross (later just Hutchinson & Ross) published 19 U.S. IBP volumes. However, these publications were only impressive to the extent that they significantly advanced ecological science.

The IBP assessments of the later 1970s were only evaluating four or five years of programs that lasted six to ten years (Loucks 1986:404): "Over a longer-term perspective, one can see the US IBP now as having been a scientific turning-point when research workers began to adopt a new paradigm for the study of ecosystem regulation and for biospheral management around the world." The U.S. IBP finally met the goals it sold to Congress for funding the IBP: "the emergence of a new science of ecosystem analysis" (Loucks 1986:404).

H. Odum had not been active in the IBP, and his *Systems Ecology: An Introduction* (1983, xvi + 644 pages) was not in the IBP format of volumes; it is much larger. It built upon his *Environment, Power, and Society* (1971) and had the same publisher but was much larger and broader in scope. The editors of an H. Odum memorial volume called *Systems Ecology* "a tour-de-force" that describes "the physical, kinetic, energetic, cybernetic, and mathematical underpinnings of his approach and drawing comparisons with over 50 other systems languages" (Brown & Hall 2004b:81). Odum served a term as president of the International Society for Systems Sciences. His symbolic language is also a simulation tool that could be translated into mathematical equations. One of the virtues of his systems science is that it could be used to express the results of purely ecological research, purely human economic research, or a mix of ecological and economic research (Figure 5.2). *Systems Ecology* contained illustrations of all three of these research results. Mark Brown (2004) has provided a history of H. Odum's energy systems language and simulation.

The year after H. Odum published *Systems Ecology*, Herman Shugart published *A Theory of Forest Dynamics: The Ecological Implications of Forest Succession Models* (1984, xiv + 278 pages). Shugart would not have had time to absorb any insights or lessons from *Systems Ecology*, as his own book was in the process of being written during 1983, but Shugart's bibliography did include H. Odum's *Environment, Power, and Society* (1971) which provided an introduction to Odum's circuitry models. At Oak Ridge, Shugart had participated in the Eastern Forest Biome Project of the U.S. IBP, and his book represented investigations further along the same lines. Some of his own diagrams at first glance might resemble Odum's,

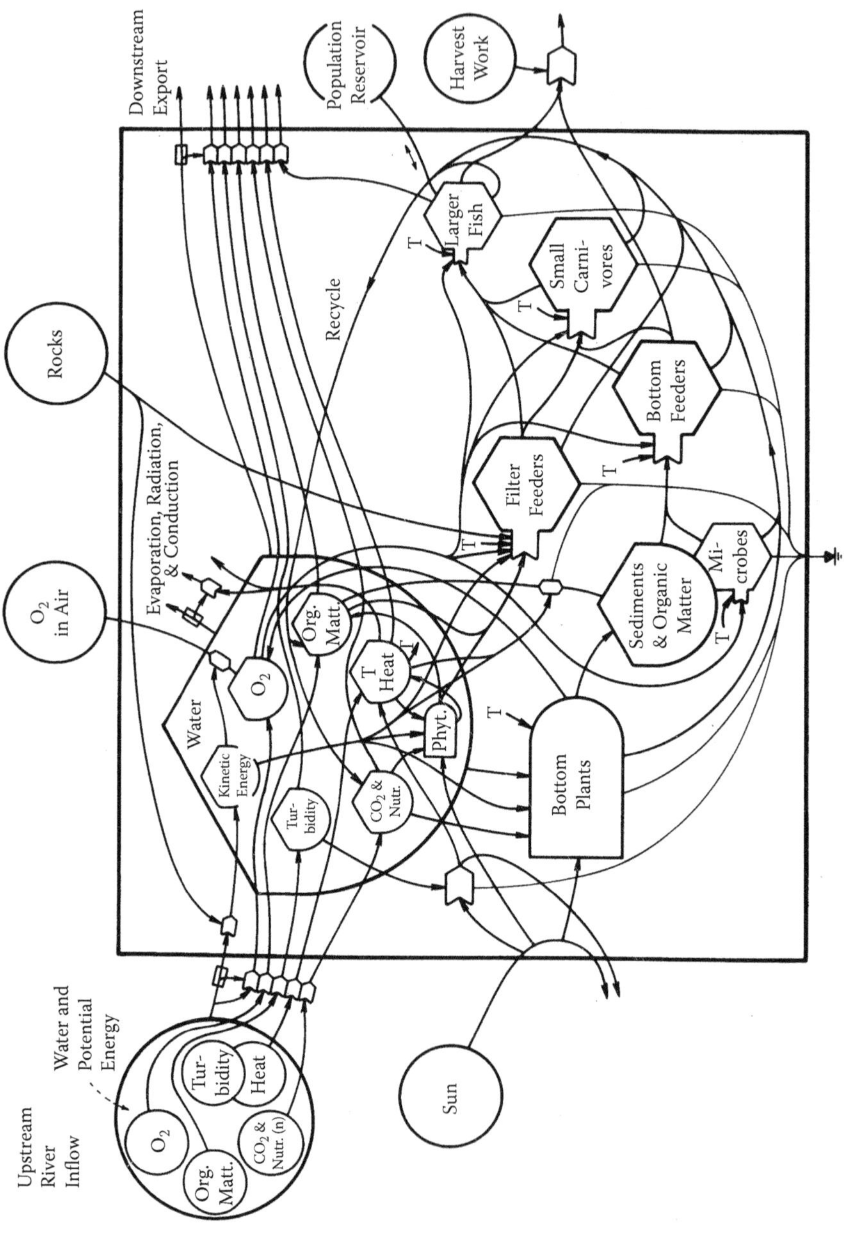

FIGURE 5.2 Generalized model for an unstratified stream section. (From Odum 1983:422. With permission.)

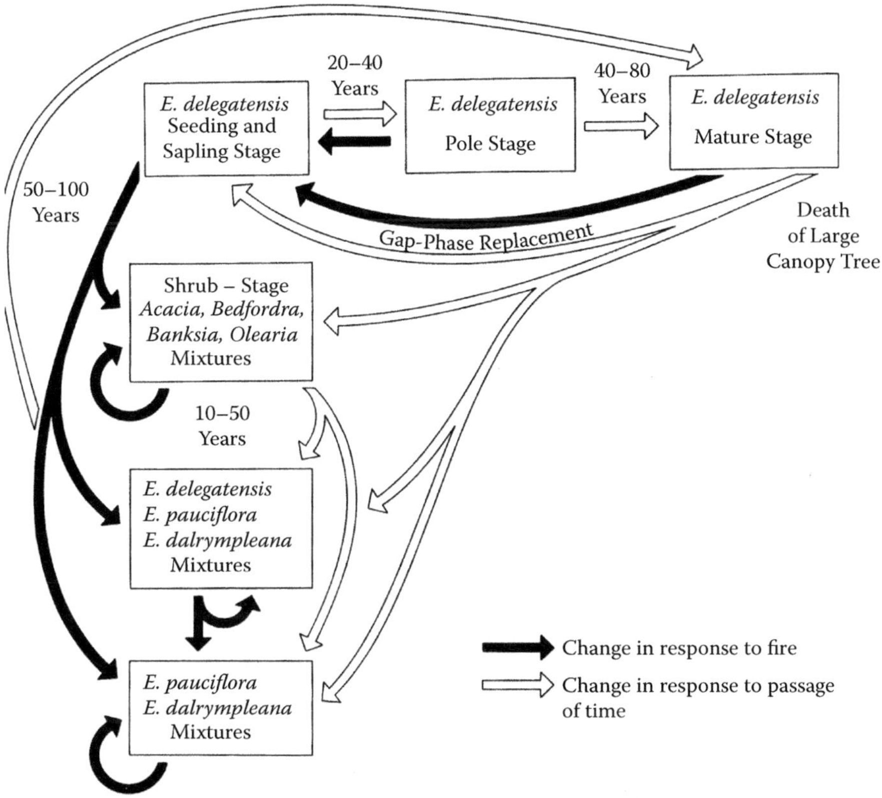

FIGURE 5.3 Simulated pattern of successional dynamics for forest types on sheltered slopes. (Originally from Shugart & Noble 1981; Shugart 1984:106. With permission.)

but they actually grew out of his previous biome work (Figure 5.3). He also did not borrow Odum's terminology.

Thomas Brock's *A Eutrophic Lake: Lake Mendota, Wisconsin* (1985) is in the same format as the IBP volumes that Springer-Verlag published and is number 55 in its ecological series. Brock acknowledged receiving NSF funds, but without stating that his funds were earmarked for the IBP. It seems fair to say that he benefitted from the U.S. IBP momentum, including computer data analysis, even though his funding was not from IBP. As Brock acknowledged in his preface, Lake Mendota is often called "the most studied lake in the world," which adds to the significance of his book. Brock was a microbiologist who was the E. B. Fred Professor of Natural Sciences at the University of Wisconsin. It is unusual for a microbiologist to tackle the entire limnology of a good-sized lake, but his scientific interests were broad, and he had a deep personal concern for a lake that he saw every day when he was on campus. Because he utilized his own research and the vast literature on scientific aspects of Lake Mendota, his book falls more or less in the same category as Risser et al.'s monograph (1981). His conclusion is in his 4-page final chapter, "Energy Flow in the Lake Mendota Ecosystem" (includes a full-page diagram and a table). His five conclusions are as follows (1985:220, slightly abridged):

1. Much more energy from primary producers passes through bacteria and detrital food than through the food chain from phytoplankton to zooplankton to fish.
2. Bacteria consume a major amount of algal production, either indirectly via algal excretion or directly.
3. Most of the organic carbon synthesized by the primary producers settles to the lake bottom either directly as algal biomass or indirectly as bacterial biomass.
4. A significant amount of the organic carbon sedimenting to the bottom of the lake is recycled because of microbial processes in the lake sediments. These processes include respiration, denitrification, sulfate reduction, and methanogenesis.
5. Most organic carbon in Lake Mendota exists in nonliving form, either dissolved in water or part of the sediments. Such organic carbon may become available to the higher trophic groups via bacterial assimilation, but probably at very low rates.

ESA planned its annual meeting in conjunction with the American Institute of Biological Sciences (AIBS) from August 26 to September 1, 1972, at the Minneapolis campus of the University of Minnesota. The program included a two-part symposium on Ethics in Ecology, moderated by Bob Burgess, on August 28th at 8 A.M. and 1 P.M. The symposium was undoubtedly in response to President Nixon, on January 1, 1970, signing into law the National Environmental Protection Act (NEPA). To prepare for that symposium, *Bulletin* editor William Niering invited two other Connecticut residents, nonacademic Frank Egler and academic Edward Deevey, to discuss ESA and ethics in the June issue of the *Bulletin*. Egler (1972) suggested that "ESA Needs a Code of Ethics and a Certification Program." With his usual blunt candor, he presented sensible arguments why both were needed. Edward Deevey, however, was not convinced (1972):

> (1) Whatever "professional ecology" may be, its qualified practitioners are also members of some other profession with standards of its own; from this it follows that (2) a legally recognizable profession of ecology does not exist. It also follows (3) that *if* there is an appropriate vehicle for further professionalization, the Ecological Society of America *as an organization* is not that vehicle.

Apparently, no consensus was reached at the symposium. Without saying whether he had attended the symposium, Bruce Welsh in the December issue of the *Bulletin* agreed with Deevey, and Egler's suggestion went nowhere in 1972.

Edward Smith Deevey, Jr. (1914–1988)

(From *ESA Bulletin* 51(4), 1970 or Livingstone 64(13), 1983. With permission.)

From Albany, New York, Deevey attended the state college in Albany for a year, then transferred to Yale University, majored in botany and graduated *summa cum laude* (1934). Linsley Pond near the campus was a convenient location for ecological studies. Hutchinson had continuing research there and obtained rides to and fro from Deevey, who decided to study the history of the lake with lake cores (Slack 2010:128–132). He failed to interest any faculty in botany in the project but did excite Hutchinson, who invited him to come over to zoology. That transfer was no problem for Deevey, whose scientific interests were quite broad. Sears and a few other Americans had already begun paleopollen investigations, but "no comprehensive study comparable to Deevey's doctoral research had previously been attempted" (Slack 2010:128). He was also the first American to study pollen from a lake bed rather than from a peat bog. He became the leading authority on paleolimnology and related studies, but he also maintained a broad research program (Burgess 1996:36; Edmondson 1997). He taught at universities in America, Canada, Denmark, and New Zealand. He served as the editor of *Ecology* and as ESA president in 1970 and became the Eminent Ecologist recipient for 1982: "Even in the select company of Eminent Ecologists, Deevey is remarkable for the diversity of fields in which he has made major contributions" (Livingstone 1983:13; Livingstone & Brenner 1989).

In 1975, Pat Halligan worried about there being too many graduate students in ecology for the number of jobs available, and in a commentary in *Ecology* suggested that graduate programs should cut back on the numbers being admitted. His solution was not met with enthusiasm, and in response, Gordon Robilliard (1977) suggested in the *Bulletin* that there were opportunities available for ecological consultants, and he offered suggestions for training ecologists to fill this niche. There were also alternatives.

In a letter published in the December 1978 issue of the *Bulletin*, Frank McCormick reminded that "The purpose of the Ecological Society of America is to stimulate and publish research in the interrelations of organisms and their environment, to facilitate an exchange of ideas among those interested in ecology and to instill ecological principles in the decision making of society at large" (p. 162). He commented that the ESA had done well in fulfilling the first two purposes and done poorly in fulfilling the third. The Standards Division of Civil Service was about to publish Minimal Qualification Standards for ecologists becoming federal employees GS-ratings 5–12, which he quoted and suggested that ESA follow with establishing a professional certification program. ESA had approved a program of self-certification that McCormick thought was inadequate. The government provided 80% of the job openings for young ecologists, and professional certification would facilitate their employment.

Nelson Beyer (1979), ecologist at the U.S. Fish and Wildlife Service, agreed with McCormick on the need for professional certification but felt that certification was not enough; ecology graduate students should have more training in applied ecology. Deane Wang argued that ESA should "take the lead in open discussion of professional ecological ethics" (1979:2). The ESA Executive Committee appointed an Advisory Certification Committee of five members—Paul Risser, Robert Burgess, Jon Ghiselin, William Laycock, and Frank McCormick—who proposed a plan for voluntary self-certification (Risser et al. 1979). They defined six levels (grades) of certified ecologists. Interested ecologists should send their notarized letter to the ESA secretary who would keep such letters on file.

At the 1979 ESA Annual Meeting in August, the ESA Council decided that ESA should poll members to see if a majority wanted a formal certification program. A detailed plan appeared in the March 1980 *Bulletin* (ESA 1980), explaining that the May 1980 ballot would include this possibility, at $40 per certification. The June *Bulletin* has the same account of the certification program (ESA 1980b) as being already in effect (so it was obviously approved in the May ballot). Is this the end of the discussion—not quite. Bruce Leon, Connecticut College, wrote "New Frontiers (Again) in Applied Ecology," explaining that the NEPA law (1970) requires environmental impact statements (EIS), but such statements were commonly inadequate (1981:13):

> Government reviewers of EIS's, rulebooks in hand, check for adequacy but not accuracy or thoroughness. The result is that many EISs are not studies but collections of data, standard formulae applied haphazardly, unsupported conjecture, and rhetoric.

Leon thought that the EIS reviewing process should be changed, requiring at least one of the reviewers of any EIS be a certified ecologist. Be that as it may, ESA finally began publishing its list of certified ecologists in March 1982 (ESA 1982).

John Neuhold became director of The Institute of Ecology (TIE) in 1975, and he reported (1975) progress on a number of projects. He also reported: "During the seven months I have been Director, we have taken a close look at our financial situation. At present, it can perhaps be best described as sound, but precarious" (Neuhold 1975:10). TIE had obtained a bank loan and had a computerized accounting system. Neuhold listed the 15 publications that TIE had published from 1972 to 1974. In 1977, TIE had some good news and some bad news (Smith 1977): TIE was exploring the possibility of establishing its headquarters at the Holcomb Research Institute of Butler University, Indianapolis, Indiana, but TIE finances were at a crisis. Smith urged ESA members to make a tax-deductible contribution. They did, and in 1978, TIE's fundraising goal of $100,000 had been met and TIE had moved to Butler University (ESA 1978). An ESA report in 1983 stated that TIE had moved to Washington, had modified its objectives, and provided information on two TIE publications of 1981 and 1982. Robert Burgess announced in the December 1984 issue of the *Bulletin* that TIE had closed operations on April 30th and that its records had been deposited in the Archives at the University of Georgia, where the ESA Archives had been stored since 1978 (Burgess 1984). Ecologists Josephine Doherty and Arthur Cooper collaborated on an excellent autopsy (1990) in the hopes of providing lessons for the future.

W(illiam) Dwight Billings (1910–1997)

(From *ESAB* 59(131), 1978 or Forman 63(12), 1982. With permission.)

Billings spent his first seven years in Washington, DC, and then his family moved to Indianapolis, Indiana, where he was mentored by two enthusiastic high school teachers (Billings 1994; Burgess 1996:15–16; Peterson 1997; Strain 1997). He earned a B.A. degree from Butler University (1933), and M.A. and Ph.D. degrees under H. J. Oosting at Duke University (1935, 1936). After two years of teaching at the University of Tennessee, he went to the University of Nevada, in Reno (1938–1952) and rose to full professor and head of the Department of Botany. He published three seminal papers on physiological ecology in 1951, 1952, and 1957. He returned to Duke University permanently in 1952 and supervised the dissertations of 52 Ph.D. students. In Nevada, he became interested in the ecology of desert vegetation and later in Arctic and Alpine vegetation. He published 125 scientific papers and a highly regarded text, *Plants and the Ecosystem* (1964, edition 3, 1978). Billings was very active in ESA affairs: he edited *Ecology* (1952–1957); he received with his graduate student Harold Mooney the Mercer Award (1962) for their study, "Comparative Physiological Ecology of Arctic and Alpine Populations of *Oxyria digyna*" (Mooney & Billings 1961); he also received the Distinguished Service Citation (1981) since his "service over nearly three decades has been a paradigm of dependability and good judgment" (Forman, Dayton, & James 1982); and was Eminent Ecologist in 1991 and he served as ESA vice president in 1960 and president from 1978 to 1979.

The vast ecological literature that was being published by 1975 is beyond the scope of this history of ESA, but reviews of some volumes in *Ecology* merit our attention. In 1975, Jack Major reviewed a symposium volume, *Oekosystemforschung* (1973), edited by H. Ellenberg. Since the 1971 symposium was sponsored by two botanical societies, the title does not mention the IBP, but presumably the investigations reported were IBP affiliated. Ellenberg ended the book with a classification of ecosystems, based only on vegetation, which Major nevertheless found interesting. All papers were in German without English summaries, but they were well done and their data and computations were easily noted. It was impractical in a review to compare these findings with similar American investigations, but Americans could themselves make the comparisons.

By 1975, any author whose book was reviewed by Frank Egler in *Ecology* would likely read the review with apprehension. We could call Egler the "Voltaire of plant ecology," though Voltaire depended more on satire and less on sarcasm than Egler did. His 1975 review was of volumes 5 and 8 of a *Handbook of Vegetation Science*, which were the first two published out of 18 planned. Egler set the tone of his review by commenting: "The list of authors in these first two volumes reads largely like a blue-blooded Social Register of the academic 'in' Jet Set." Volume 5 was the *Ordination and Classification of Communities*, edited by Robert Whittaker (1973, x + 738 pages). Egler explained that this volume was primarily on methods and procedures. "No time is lost on basic philosophic, conceptual, subjective or valuative judgments" (1973:250). Chapter 12 was by Whittaker and was introductory to the remaining "Approaches to Classifying Vegetation," and eight chapters followed on classification systems. "In many respects this book plows the same ground as Shimwell's *Description and Classification of Vegetation* [cf. *Ecology* 54 (1), 1973], without the conciseness and lack of duplication in a collection of separately authored articles." Egler, of course, wrote in *Ecology* that review of Shimwell's book he referenced.

Volume 8 was edited by R. Knapp, *Vegetation Dynamics* (1974, x + 366 pages). Egler explained that the Introduction and 27 chapters were in three languages: "English, German, and Knappic (named after the editor of this volume)" (p. 250). Knappic was more Teutonic than English and offered stumbling blocks to those for whom English is the mother tongue. Egler explained that this volume is on plant succession and that the bibliography's 1,500 titles are "overwhelmingly pro-succession." He lamented that "Some chapters are unsummarized literature surveys, or contain intermidable and almost meaningless lists of references" (p. 200). He concluded that "Vegetation Science Dynamics is in its infancy and a practical Vegetative Management is not yet born" (p. 251).

Possibly the most controversial book ever reviewed in *Ecology* was Edward Wilson's *Sociobiology: The New Synthesis* (Harvard, 1975, x + 697 pages). It was sent to an appropriate reviewer, David Barash, who would himself publish a book on the subject in 2008. Barash was enthusiastic about the book in general (Barash 1976) with some reservations about questions that needed more research. The firestorm that met the book in other journals can only be compared to the reception of Darwin's *Origin of Species* (1859), when it first appeared. The question raised in some reviews was whether Wilson's *Sociobiology* was proper science, or merely social Darwinism in a new guise (Wilson 1994:330–353). His most vocal biology critics were Richard C. Lewontin and Stephen Jay Gould. For those lacking the time or inclination to read almost 700 pages of sociobiology, Wilson wrote a 13-page summary (1976). (Don't bother looking up "sociobiology" in Simon Levin's *Princeton Guide to Ecology* [2009], where it has been chopped up into narrower, less controversial, topics.) Some three decades after Wilson published *Sociobiology*, Brian Baxter provided an excellent philosophical perspective on it in *A Darwinian Worldview: Sociobiology, Environmental Ethics and the Work of Edward O. Wilson* (2007).

Edward O(sborne) Wilson (1929–)

(From *ESAB* 75:196. With permission.)

Birmingham native Wilson graduated from the University of Alabama (B.S. 1949) and then pursued graduate studies at Harvard University (M.S. 1950; Ph.D. 1955), where he remained (Burgess 1996:114). He is a myrmecologist whose interests broadened to include evolution and ecology, and he founded island biogeography (MacArthur & Wilson, *The Theory of Island Biogeography,* 1967) and sociobiology (Wilson, *Sociobiology: The New Synthesis,* 1975), three unifying concepts for science and humanities—biophilia, biodiversity studies, consilience—and a technological advance in studying global biodiversity—The Encyclopedia of Life. Wilson's autobiography is entitled *Naturalist* (1994). Over a hundred awards he has received include the U.S. National Medal of Science, the Crafoord Prize of the Royal Swedish Academy of Sciences, the International Prize of Biology and the Cosmos Prize of Japan, two Pulitzer Prizes in nonfiction, the Nonino and Sarono Prizes of Italy, the Gold Medal of the Worldwide Fund for Nature, and the Audubon Medal of the Audubon Society. Currently, he is the Curator in Entomology and University Research Professor Emeritus.

Considerably less controversial were reviews of new textbooks. Tim Allen (1976) was enthusiastic about E. C. (Chris) Pielou's latest one, *Population and Community Ecology: Principles and Methods* (1974), only lamenting that she had omitted any discussion of the work of Curtis and Whittaker. Rather surprisingly, two new limnology textbooks appeared in 1975, by Gerald Cole and Robert Wetzel. Cole's *Textbook of Limnology* (xi + 283 pages) was an introductory text, and Wetzel's *Limnology* (xii + 743 pages) was an advanced text, requiring more background courses than Cole's. Since they were written for different student audiences, reviewer G. R. Marzolf (1976) did not face a situation of possibly recommending one and dismissing the other.

Robert May is an Australian with a Ph.D. in physics, who became a Princeton professor of ecology, and later both an Oxford ecology professor and a British lord. At Princeton he edited and contributed to *Theoretical Ecology: Principles and Applications* (1976). It was preceded by Ramón Margalef's four lectures, *Perspectives in Ecological Theory* (1968). Tim Allen also reviewed *Theoretical Ecology* for *Ecology* (1977). As a reviewer, he was not quite in the same league as Egler, but close. Overall, Allen praised *Theoretical Ecology* as "A valuable contribution and teaching tool," but he also had complaints. One was about "the overworked Lotka-Volterra competition equations." He quoted from the book on qualifications about these equations, "yet the book singlemindedly pursues the equations and the ramifications of its parts, for example their r's and K's." But Allen's complaints also extended beyond the book to ecology itself: "It suffers from a reductionism whose level is fixed by human primary perception." Perhaps May took some of Allen's comments seriously, for *Theoretical Ecology* went through two more editions. The third edition (2007) was co-edited by Angela McLean, and May is the only author in it who also contributed to the first edition. May clearly considered it important to keep the book abreast of new developments.

Herb Bormann and Gene Likens' research in the Hubbard Brook watershed was discussed in Chapter 4. Also in 1977, Likens, Bormann, and three others published

Biogeochemistry of a Forested Ecosystem, an attractive small paperback that was well adapted for plant ecology courses; and in 1979, Bormann and Likens published *Pattern and Process in a Forested Ecosystem*. Neither ecologist is a forester, but this book was given for review to R. H. Waring at the School of Forestry, Oregon State University, who (1981) praised their substantial achievement, then ended with a paragraph of topics suggested as additions to a possible later edition.

With IBP research organizations in place, and fewer than anticipated projects satisfactorily completed on time, IBP ecologists wanted to continue their research. This led to the organization of the Long Term Ecological Research (LTER) Network, also funded mainly by the NSF. Its mission was to learn to protect and manage ecosystems, their diversity, and services. James Callahan, who ran LTER for the NSF, indicated (1987) that the program formally began in 1982. There were 10 LTER cites by 1986, and five were added in 1987. The first 10 sites included the Coweeta LTER Program, which had monitoring records begun in 1934, funded by the U.S. Forest Service, and the five LTER sites added in 1987 included the Hubbard Brook Site, another Forest Service site that had begun its record-keeping and research program in 1963. A pamphlet celebrating 25 years of LTER research (2007) shows 27 sites throughout the United States. Unlike the IBP, LTER did not include Canada, but it did include two sites in Antarctica (in addition to two Arctic sites in Alaska). What kinds of research studies were pursued at these sites?

The first issue of *LTER Network News* (Spring 1987) contains the research programs of the first 10 LTER sites, and a cover photograph of a forest at the Andrews LTER site, Oregon, where a log decomposition study was underway. All the studies addressed topics that could further an understanding of that site's ecosystem. Ecosystem research allows an almost infinite number of topics to be addressed. The Central Plains Program emphasized experimentation. Coweeta began a four-year program in 1985 on effects of atmospheric deposition, in which it joined ecologists studying 15 different forest ecosystems in North Carolina, Canada, and Norway; Coweeta was the only LTER site participating in that study. Since *LTER Network News* was sent to all ESA members, it served a dual purpose, it informed network members about activities at other sites besides theirs, and it also publicized LTER research and activities to ESA members.

The second issue of the *News* (Fall 1987) contained programs for five more sites that by then had been added to the network, a biome map showing the locations of the 15 sites and an impressive 3-page bibliography of research publications produced by LTER ecologists at 13 sites. Subsequent issues of the *News* featured longer accounts of particular projects being pursued at a few sites, with photographs. Additionally, there was news of various developments and events that could interest ecologists who pursued LTER research. The third issue of *LTER Network News* (Spring 1988) included the names and addresses of participating ecologists at each site. By then, most ecologists had e-mail addresses, though a few still relied on other means of communication. Neither Bormann nor Likens was listed at Hubbard Brook, because three of their associates were running the LTER Program there. The first issue of the *News* had 8 pages, with a large photograph on the cover. The fourth issue of the *News* (Fall 1988) had information about an LTER program being established in

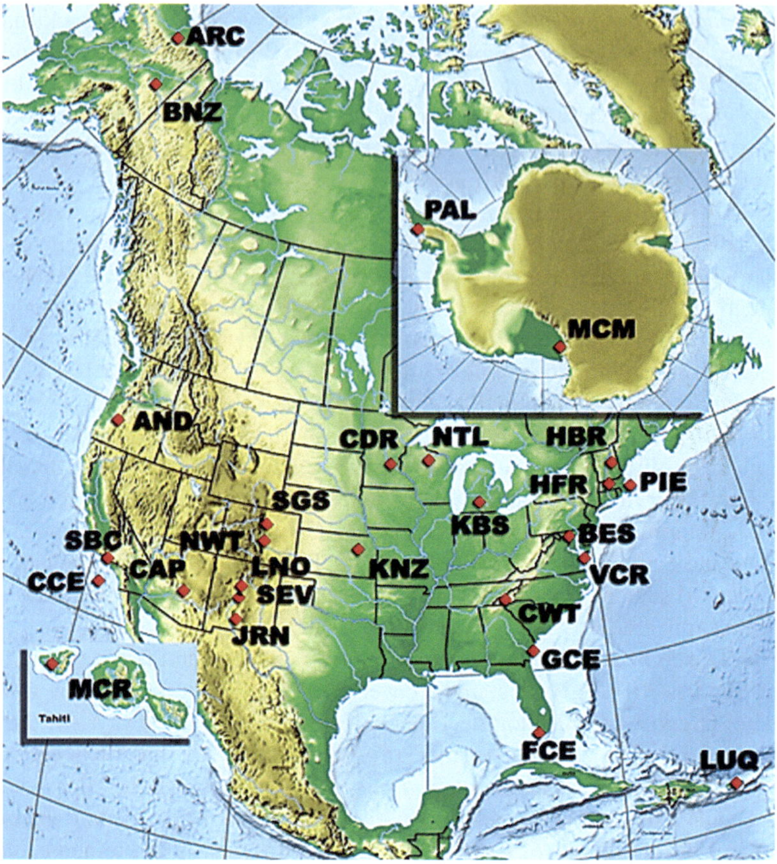

FIGURE 5.4 Map, LTER network of 27 current research sites. (From LTERnet.edu website. With permission.)

China, with initially 20 stations. Issue 6 (Fall 1989) had 12 pages, with photographs and drawings throughout. LTER was a great success, and participants shared a high morale as they conducted monitoring research and produced reports.

Ecosystem studies need not be limited to LTER sites (Figure 5.4). For example, Iowa ecologists could coalesce around the Iowa Lakeside Laboratory, where they would find that much ecosystem "homework" had already been done. This was indicated by Debby Z. Baker's bibliography (1990) of research publications from there during 80 years, and Michael J. Lannoo's *The Iowa Lakeside Laboratory* (2012) and his *Okoboji Wetlands* (1996). One or two ecologists might organize and lead such an effort. Published IBP reports could provide some guidance. There may be other places, such as marine laboratories, where the same is possible.

An important service that scientific societies provide is an opportunity for collegiality. But, human nature being what it is, the same opportunity can be used for snubs or snide remarks. On the other hand, ecologists have always argued about both fundamentals and trivia—commonly in book reviews, as discussed above. When there are hurt feelings, they surely go mostly unrecorded. It is rather unusual to read in *Science* two substantial articles by reporter Roger Lewin about bitter controversies over theory (1983a,b). Reporters do not resolve scientific controversies, but they can express the arguments of both sides clearly, and convey something of attitudes. The controversy he described was over the importance of competition theory to affect the dynamics of biotic communities.

When Levins and Lewontin (1980) proposed a dialectic approach as an alternative to some either/or ecological controversies, they had in mind Daniel Simberloff's "Succession of Paradigms in Ecology" (1980a). Simberloff welcomed their suggestion, but felt that they had misinterpreted some of his statements (1980b:81). The dialectical approach was not evident to Roger Lewin when he interviewed ecologists about competition theory. Simberloff and his colleagues at Florida State University (Strong 1980) were skeptics who believed that the competitionists lacked falsifiable theoretical claims, which Popperian philosophy demands of science. Jonathan Roughgarden, at Stanford, retorted that Simberloff and associates seem never to have published significant ecological findings, but that they had generated "a lot of bitterness" (Lewin 1983a:636). In response, Donald Strong in Florida, observed that "The MacArthurians are all in powerful positions in powerful universities," and so depended upon prestige to bolster their weak arguments. Conclusion: despite a harmonizing scientific society, collegiality can fade away during controversy.

Since Simberloff had not accepted Levins and Lewontin's suggestion of a dialectic approach to his conflict with competitionists, as seen in Lewin's reportage (1983a,b), Robert May and Jon Seger tried a somewhat different approach. They summarized some of the research of some of the competitionists, followed by praise for Simberloff and his associates (Strong et al. 1984) for their "rigorous standards in this search for community patterns" (May & Seger 1986:258). Yet, ecologists find it difficult to test inferred patterns in community structure against the "null models" that Simberloff urges, while there is a rich diversity of ideas about competitive exclusion, niche overlap, prey–predator coevolution, and more, which have stimulated much empirical research on relations within and between species (May & Seger 1986:261). They saw no need to choose one approach and banish the other. Robert McIntosh collected references on both sides of the controversy and then endorsed "Pluralism in Ecology" (1987).

Robert May in 1986, with coauthor Jon Seger, published "Ideas in Ecology," which discussed three central topics. The first topic was "Roles and Relationships," which began with comments on Gilbert White and Charles Darwin, both of whom studied plants and animals living within their environments. May was impressed by Darwin's use in *The Origin of Species* (1859) of "the image of variously shaped wedges being pushed and hammered into a barrel as a metaphor for how a set of species fit together to form a community" (May & Seger 1986:256). Next came their surprising claim that "This metaphor anticipates Hutchinson's definition (1957) of niche as an abstract hypervolume situated in a space whose axes correspond to biological and environmental variables affecting organisms studied." Reading this, I wondered why I had not noted this seminal metaphor in the *Origin* and realized its significance; so I found the passage: "The face of Nature may be compared to a yielding surface, with ten thousand sharp wedges packed close together and driven inwards by incessant blows, sometimes one wedge being struck, and then another with greater force" (Darwin 1859:67). May and Seger improved Darwin's metaphor by substituting for "The face of nature" a barrel, representing a community.

ESA's experiment in founding The Institute of Ecology had gotten off to a good start in 1970 but faded away by 1984. ESA president Arthur Cooper suggested a newer initiative in his presidential address (1982) that succeeded and prospered: opening a Washington office to give ESA a voice to educate members and influence government. In January 1983, Dr. Elliott A. Norse became the Director of Public Affairs. He was a marine biologist who had worked at the U.S. Environmental Protection Agency (EPA) and had been the staff ecologist for President Carter (ESA 1983b; Norse 2000:99–100). The June *Bulletin* provided information on the organization, objectives, operations, and budget of the Washington Office (ESA 1983c) (Figure 5.5). Norse's first report (1983) followed (3 pages later) this organizational information. He came into office during the sensational forced resignation of Ann Burford and over a dozen other EPA administrators from Reagan's administration for their harassment of EPA scientists, and her replacement by a trustworthy William D. Ruckelshaus. Norse felt certain that ESA's Washington voice could make a positive contribution to American governance.

Norse's first substantive report in the September *Bulletin* contrasted the slow pace of scientific research with legislators and government administrators' desire for prompt information. For guidance, he cited Cooper's 1982 address, "Why Doesn't Anyone Listen to Ecologists—and What Can ESA Do About It?" Norse explained that ESA's 6,700 members constituted a potential expert resource that might become available for advising government officials, and he suggested that ESA members send him information on a questionnaire that could be recorded in a "wonderful

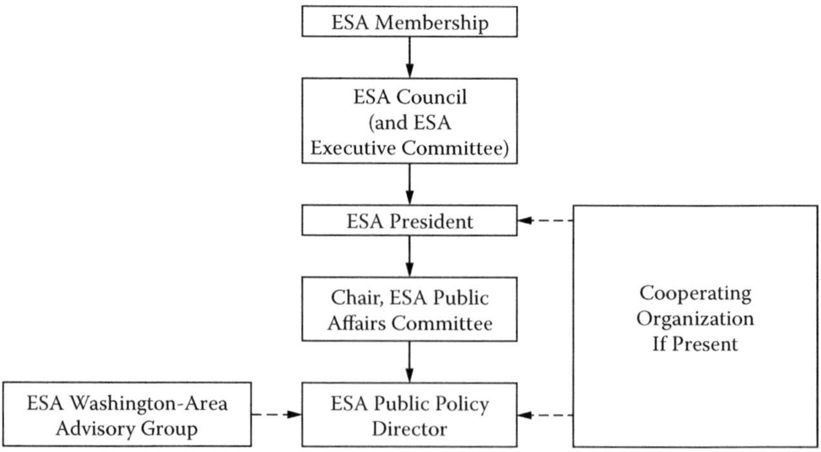

FIGURE 5.5 Administrative structure for the ESA Public Policy Director. (From *ESAB* 64(199), 1983. With permission.)

Victor 9000 microcomputer" for instant access whenever government officials wanted expert advice on a particular question. In the December issue of the *Bulletin* he discussed human impacts on the biosphere, especially potential climatic change and nuclear war—with thoughts on a nuclear war conference co-sponsored by ESA, TIE, and 29 other organizations, that he planned to attend.

In the March 1984 issue of the *Bulletin*, Norse reported that only 435 ESA members had returned his questionnaire and that he hoped to achieve a 10% response. The public had been quite interested in "The World after Nuclear War" conference (1983), which ESA and other scientific organizations had sponsored. Paul Erlich and Frank McCormick had appeared on American TV programs to discuss it, and Norse had appeared on a Canadian Broadcasting Corporation nationwide program. Norse cited four articles on the conference already published, and Herbert Grover (1984) reviewed the conference in the *Bulletin* following Norse's report. In the June issue of the *Bulletin*, Norse again discussed the Ecological Information Network being compiled for expert advice for government, but in the following two *Bulletins* he reverted to further discussion of nuclear war aftermath. Additionally, in the December issue of the *Bulletin*, his "Report of the Public Policy Director" explained what he had accomplished in the 18 months of operations since January 1983.

The status of UNESCO's Biosphere Reserves was what Norse discussed in March 1985. Countries could attach this designation to bureaucratically differently classified lands. There were 226 reserves designated, 40 of which were in the United States. The United States approved of this system but was in the process of withdrawing from UNESCO for political reasons. In the December issue of the *Bulletin*, he explained the various kinds of projects the Public Affairs Office undertook, devoting a good bit of his discussion to the Ecological Information Network, which had proved to be quite popular and had attracted $20,000 in funds from several sources. In March 1986, Norse discussed "Biological Diversity and Tropical Agroecology." Congress was reauthorizing the Endangered Species Act and was preparing to allocate $10 million to the Agency for International Development for protection of biodiversity worldwide. There were various nonlegislative activities that Norse also described, including how to make agriculture more sustainable.

U.S. presidential candidate Ronald Reagan in 1980 "strongly criticized incumbents who presided over budget deficits and promised to stop deficit spending if elected" (Norse 1986:186), yet under his administration the national debt had doubled from $715 billion in 1980 to $1,510 billion in 1985. This motivated Congress to pass the Balanced Budget and Emergency Deficit Control Act (Graham-Rudman-Hollings Act) in 1985. If Congress and Reagan could not agree on a budget for 1986, this act mandated automatic cuts in federal spending, with deeper cuts in 1987 likely. Such cuts would affect research funding for the NSF, the U.S. Fish and Wildlife Service, and other agencies. A positive note in the September issue of the *Bulletin* was the fact that over 860 ecologists had sent in their information on an Ecological Information Network questionnaire, which was reprinted for others. By 1986, Norse had an assistant, Amy Matthews, and her name began appearing with, or without, his name on information from Washington (Matthews & Norse 1986).

American Society of Naturalists	1868
American Fisheries Society	1870
American Ornithologists' Union	1884
Botanical Society of America	1894
Society of American Foresters	1905
Agronomy Society	1907
British Ecological Society	1913
Ecological Society of America	1914
Limnology Society of America	1936
Wildlife Society	1937
Society for the Study of Evolution	1946
Range Management Society	1947
Association for Tropical Biology	1962
INTECOL	1967
Animal Behaviour Society	1971*
International Society of Ecological Modelling	1980
International Society of Chemical Ecology	1984
Society for Conservation Biology	1986
International Society for Behavioral Ecology	1986

* First independent meeting.

American Naturalist	1868
American Midland Naturalist	1909
Journal of Ecology	1913
Ecology	1920
Ecological Monographs	1931
Journal of Animal Ecology	1932
Oikos	1949
Limnology and Oceanography	1956
Researches on Population Biology	1962
Journal of Applied Ecology	1964
Oecologia	1968
Theoretical Population Biology	1970
Agriculture, Ecosystems, and Environment	1974
Ecological Modelling	1975
Marine Ecology, Progress Series	1982
Biogeochemistry	1984
Journal of Tropical Ecology	1985
Evolutionary Ecology	1987
Conservation Biology	1987
Functional Ecology	1987

FIGURE 5.6 A list of the founding dates of ecologically oriented biological societies and journals. (From Root 1987:541, 539. With permission.)

Every decade, ESA activities seem to expand, and accompanying that expansion has been an an increase in ecological publications, in both titles and number of pages. Richard Root, in his past president address (August 1987), put this increase in pages of ecological literature published per year in historical perspective by reminding ecologists of the dates of the founding of ecologically oriented scientific societies and the founding dates of ecologically oriented journals (Figure 5.6). He also provided three charts that indicated the increase in pages published in journals and the increase in the number of abstracts published before annual meetings (Figure 5.7).

Since male scientists commonly participate in raising their children, Jonathan Cole and Harriet Zuckerman argued in *Scientific American* (February 1987) that the reason female scientists published only half as much as males was not explainable by the burden of childrearing—married women with and without children published at the same rate. Craig Loehle brought this situation to the ESA *Bulletin* (1987) and offered three answers for the lesser productivity of female ecologists: (1) women received less research funds, laboratory space, and graduate student

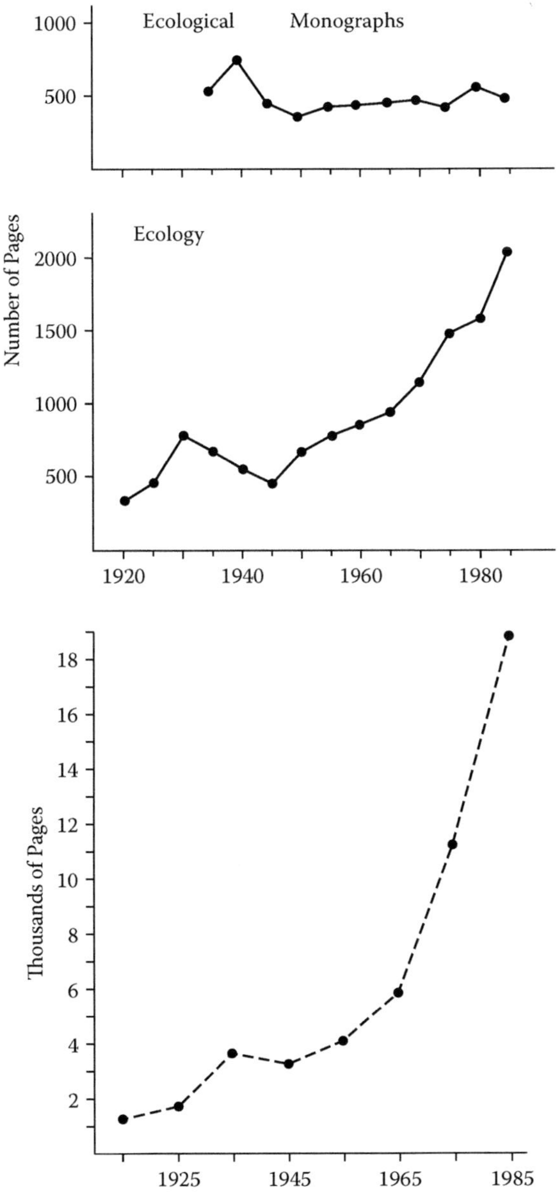

FIGURE 5.7 Two charts on the increase in journal pages in ecology journals over time, and one chart on the increase in abstracts published before the annual ESA meetings. (From Root 1987:538, 540. With permission.)

assistants; (2) women often get heavier teaching loads than men; and (3) women were less aggressive than men and spent more time than men "going over the data, redoing experiments, and reworking manuscripts." Andrew Sih and Kisa Nishikawa responded with a survey, "Do Men and Women Really Differ in Publication Rates and Contentiousness?" (1988). They found that (1) women in ecology do not publish at a lower rate than men, (2) women do not publish less in major journals, but (3) women do publish critiques less often than men, though they had doubts about Loehle's hypothesis that it is because women are less aggressive than men. They thought editors or reviewers might discriminate against women, and that women had not yet reached the later stages in their careers when writing critiques was most likely.

Jessica Gurevitch learned (1988) that Karen Blumer had surveyed (1984) the participation of women speakers at the 1983 ESA Annual Meeting, and Gurevitch decided to compare her own survey on the 1987 ESA Annual Meeting with Blumer's. She found the following (1988:158):

> (1) Women are invited to present the results of their research in symposia at the annual ESA meeting less frequently than they offer contributed papers on their work; (2) women are far less likely to be invited to participate in a symposium if the symposium organizers are exclusively male; and (3) women are being invited to participate in symposia slightly more frequently in 1987 than in 1983, but the increase is associated only with sessions in which women are part of the invitation process; the number of such sessions rose dramatically over this period.

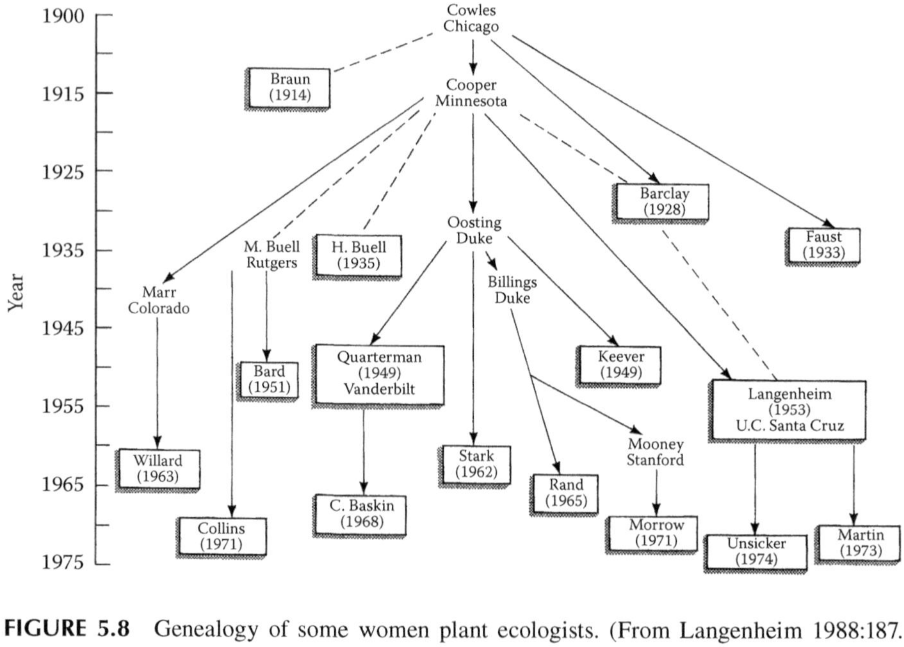

FIGURE 5.8 Genealogy of some women plant ecologists. (From Langenheim 1988:187. With permission.)

Jean H. Langenheim was president of ESA from 1987 to 1988, and her outgoing president address was on "The Path and Progress of American Women Ecologists" (1988). It would be difficult to summarize her lengthy talk, but she included very clear charts that conveyed much of her message. The covers of the December 1988 *Bulletin*, in conjunction with her talk, reproduced 16 photographs of prominent ESA women (Figures 5.8 and 5.9). Her talk, however, did not "dispose" of the issue of women ecologists. Diana Freckman became chair of an ESA Committee on Women and Minorities, and she wrote a letter in the *Bulletin* (1989) stating that women and minority students should be recruited for ecology professions and explained that the committee was conveying this message to academic administrators and other scientific societies and sponsors of workshops and symposia (Figure 5.10).

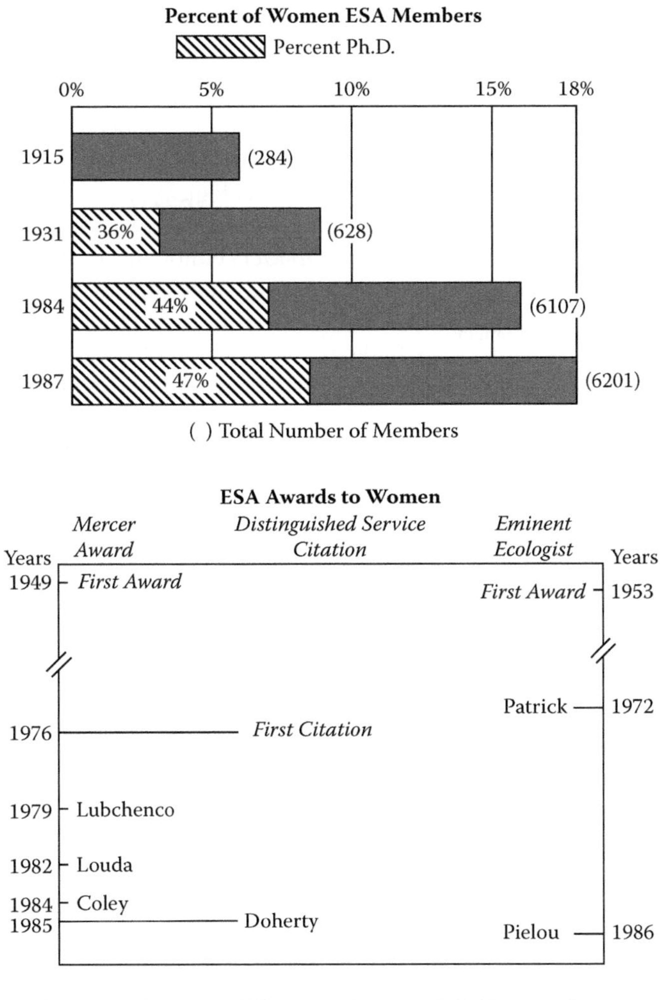

FIGURE 5.9 Percent of women ESA members and ESA awards to women. (From Langenheim 1988:191, 193. With permission.)

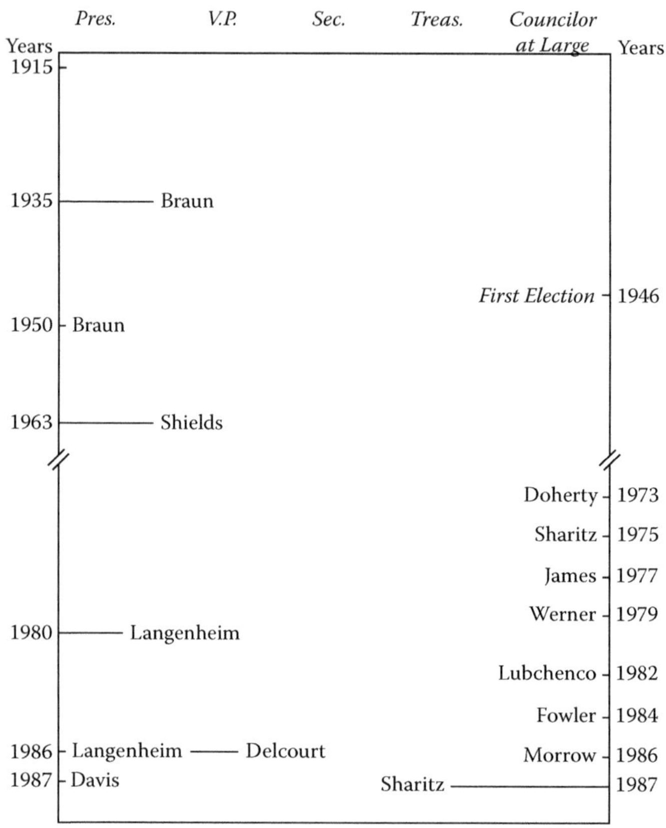

FIGURE 5.10 ESA women officers. (From Langenheim 1988:192. With permission.)

Robert Harding Whittaker (1920–1980)

(From Westman 1982:11. With permission.)

From Kansas, Whittaker, son of a professor of zoology, earned a B.A. degree from Washburn College (now Washburn Municipal University, Topeka, Kansas, 1942), then spent 3.5 years in the U.S. Air Force as a meteorologist (Whittaker 1994:617). Afterward he studied animal ecology under Shelford and Kendeigh at the University of Illinois but was also influenced by University of Illinois's skeptical ecologist Arthur G. Vestal (1888–1964), whom Whittaker viewed as a plant ecologist, though Vestal had also studied animals. Whittaker studied plant and animal communities in the Great Smoky Mountains for his doctoral dissertation. He began that research as an orthodox Clementsian, but ended up in the Vestal-Gleason camp, independently from Curtis, who was doing similar gradient analysis. He earned a Ph.D. in 1948 and then joined the Department of Zoology, Washington State University (1948–1951), where he studied copepod communities in saline water in semi-arid southeastern Washington, before moving over to the U.S. Hanford Laboratory at Richmond (1951–1954), where he studied movement and concentration of radiophosphorus in aquarium and pond ecosystems (Whittaker 1994:618–619). He then taught and/or conducted research in three other institutions before settling at Cornell University in 1968. Whittaker's "most significant contributions to ecology lie in the development of theories and methods of gradient analysis (Whittaker 1967). Through the verification of the individualistic hypothesis with field data from many regions, and the subsequent development and dissemination of methods for studying species distributions along continua, he helped replace the Clementsian paradigm with a Gleasonian one" (Whittaker 1958; Westman 1982; Westman & Peet 1982:98; Hagen 1992:31–32; Burgess 1996:113).

A very important function of ESA is presenting awards to worthy members. Awards boost morale, enhance opportunities for highly desirable and highly competitive employment, and inspire members to do their best in their science and for ESA. The awards also enhance the prestige of ESA. By 1988, there were six awards to individuals, plus a corporate award, and two travel funds. Award requirements were as follows (ESA 1988):

EMINENT ECOLOGIST: To an ecologist for continuing, significant contributions to the science of ecology. Nominees may be from any country and need not be ESA members. Recipients receive free, lifetime, active membership in ESA.

GEORGE MERCER: For an outstanding ecological research paper published by a researcher under 40 years old. Papers by two authors are eligible, but both authors must be under 40. Author or authors may be from any country and need not be ESA members.

ROBERT H. MacARTHUR: For an active and prominent ecologist in mid-career. Nominee may be from any country and need not be an ESA member. Recipient is invited to address members at an annual meeting.

W. S. COOPER: Honors an outstanding contributor to geobotany and physiographic ecology, the fields in which Cooper worked.

DISTINGUISHED SERVICE CITATION: Is given to recognize dedicated and important service to ESA.

HONORARY MEMBERSHIP in ESA: Is open to an ecologist who has made exceptional scientific contributions and whose principal residence and site of research are outside the United States and Canada. Up to three awards may be presented in any year.

CORPORATE: Recognizes a corporation, business, division, program, or individual for accomplishments incorporating sound ecological concepts, knowledge, and practices into planning and operations.

ROBERT WHITTAKER TRAVEL FUND: Promotes active collaboration and exchange of ideas between foreign and U.S. ecologists, to defray costs of travel to USA.

FORREST SHREVE SONORAN DESERT RESEARCH FUND: Supports research by young investigators in the Sonoran Desert, for no more than $750.

Other awards were added at various future times. Some recipients of these awards are noted in the ecoSketches.

Beyond the practicalities (or impracticalities) of Washington politics were various aspects of theoretical ecology. Charles Hall and D. L. DeAngelis asked, "Models in Ecology: Paradigms Found or Paradigms Lost?" (1985). They cited skeptics Daniel Simberloff (see above), Robert Peters (1980a,b), and Glenn Suter (1981) as having expressed doubts in print, which some others shared about theoretical ecology. Despite lingering questions about models, they nevertheless felt that "ecology cannot proceed without models" (Hall & DeAngelis 1985:340). Therefore, they thought it helpful to distinguish between analytic and simulation modeling. Analytic modelers study populations and communities and are influenced by Nicholson, Hutchinson, and MacArthur. Simulation modelers study energy, biomass and bioelements, and the effects of abiotic factors on populations, and are influenced by Andrewartha and Birch, the Odum brothers, and Kenneth Watt. "Both approaches probably are necessary, and there will be problems for which one is more appropriate than the other" (Hall & DeAngelis 1985:345). Modelers in both groups have little appreciation and communication with the other group. They discussed the strengths and weaknesses of both approaches and concluded that "models have delivered far less than was once promised and far less than it seems that they might" (1985:345). Modelers had begun 25 years earlier when ecologists imagined nature was simpler than it had since been discovered to be.

Other issues that provoked extended discussions were more attitudinal than theoretical. Two examples were creationist ecology and whether ESA journals should publish articles from apartheid South Africa. In a September 1985 commentary, 16 ecologists at the University of Wisconsin–Madison (Schneider et al. 1985) protested the publication of three articles in *Ecology* in 1984 on research conducted in South Africa. (Four signers were limnologists and two signers were botanists who were not ESA members.) In the December issue of the *Bulletin*, Norse (1985) responded that he had discussed the Wisconsin commentary with other ecologists who had "exceptional international (including African) experience, sensitivity, and political acumen" before voicing opposition. He acknowledged that publishing South African research tended to confer legitimacy on the apartheid government, but he had four counterarguments that seemed more compelling: (1) ecologists did not know the attitude of those authors toward apartheid, (2) "if ESA bans the work of South African scientists because their government's policies are repugnant, then what other nations should join our list of pariahs?" (3) South Africa's biota is one of the richest in the world, and (4) "Bishop Tutu recently stated that immediate withdrawal of invested funds would harm the South Africans that well-intentioned outsiders seek to help." The *Bulletin* published six more letters responding to this dilemma in the January 1986 issue.

This chapter ends with mention of two Jubilee Symposia volumes to celebrate the 75th anniversary of the British Ecological Society (not counting Sheail's history of the BES [1987] mentioned in my introduction): first, *Ecological Concepts: The Contribution of Ecology to an Understanding of the Natural World* (1989, viii + 385 pages), edited by J. M. Cherrett; four of its eleven chapters were written by ESA members, including Robert May, who was then at Princeton. The first chapter is Cherrett's Introduction to the 10 following chapters: Ecosystems: Fluxes of Matter and Energy; Food Webs; Ecological Niche; Diversity and Stability; Predator–Prey and Host-Pathogen Interactions; Population Regulation in Animals; Competition; Life-History Strategies; Optimization; and Levels of Organization. These were topics of great interest on both sides of the Atlantic, and by including U.S. and Canadian authors of the chapters, *Ecological Concepts* illustrated the capacity of the two societies to work together when there was reason to do so. The second Jubilee volume was *Toward a More Exact Ecology*, edited by Peter Grubb and John Whittaker (1989, x + 468 pages). It features 18 chapters concerning ecological research techniques, with one Canadian and six American authors.

6 1990 to 2004
New Journals, the Sustainable Biosphere Initiative (SBI), Strategies for Education in Ecology, Development, and Sustainability (SEEDS), and More

In 1990, ESA celebrated its 75th anniversary at the August annual meeting, held at the Snowbird Resort in Little Cottonwood Canyon in the midst of the National Forests in the Wasatch Mountains, Utah. Three retrospective symposia held there are reviewed in the March 1991 *Bulletin*. In unusual detail (4.5 pages), Earl Werner reviewed the symposium which Richard Root and Donald Shure had organized, "Changing Perspectives on Some Long-Standing Problems in Ecology." One speaker was James Brown, and Werner stated that this symposium "prompted" Brown and Leslie Real to edit their *Foundations of Ecology: Classic Papers with Commentaries*. In their preface, however, they indicate that they had begun work on it after Real taught a course on ecology classics in 1987, but that the Snowbird Symposia had "given impetus" to their project (Real & Brown 1991:xiii). The other two symposia reviewed in March were "Long-Term Dynamics of Vegetational Patterns in Mountainous Regions of North and South America," organized by Jane Beiswenger and Stephen Jackson (Baker 1991) and "Equilibrium and Non-Equilibrium Paradigms," organized by S. T. A. Pickett and Margaret Bryan Davis (Murdoch 1991).

When animal ecologist Nelson Hairston, Sr. published in the September 1991 issue of the *Bulletin* "The Literature Glut: Causes and Consequences: Reflections of a Dinosaur," many overworked ecologists must have eagerly sought his answer. He estimated that there were 1,500 ecological and closely related papers published a year, which reduced ecologists to reading many titles, some abstracts, and only a small portion of papers in their entirety. This led him to suggest that review articles help bridge the gap between what one wants to know and what one can actually read. The remainder of his article was about the hazard of depending too heavily upon the accuracy of review articles, exemplified by how his own research papers had been misunderstood or misevaluated by Andrew Sih et al., "Predation, Competition, and Prey Communities: A Review of Field Experiments" (1985).

Bulletin editor Solomon invited Sih to respond to Hairston's article. Sih (1991) concluded that Hairston had singled him and his colleagues out because they had reported that the experiments that Hairston, Smith, and Slobodkin had published had not substantiated the conclusions (1960). Hairston and Sih's 1991 papers together only cover 8 pages, but both papers carry the reader through very complex arguments, which only the cognoscenti were apt to fully understand. Jeannette Zamon had turned eagerly to Hairston's article only to find that he had an ax to grind, and so she wrote "The Literature Glut: Some General Advice to Ecologists (A Response to Hairston)" (1992), in which she advised both new graduate students and their advisors on how to deal with the question that Hairston had raised, but barely answered.

Nelson G(eorge) Hairston, Sr. (1917–2008)

(From Wilbur & Gill 2008:277. With permission.)

North Carolinian Hairston grew up on his family's plantation on the banks of the Yadkin River. He earned B.A. and M.A. degrees (1937, 1939) at the University of North Carolina, then went to Northwestern University to study under Orlando Park. His studies at Northwestern were interrupted by his military service during World War II, when he worked on malarial therapies in the South Pacific. Upon completion of his doctorate (1948), he joined the Zoology Department at the University of Michigan, where he and animal ecology flourished: "Nelson was the senior member of the dynamic trio of ecologists at Michigan with Fred Smith and Larry Slobodkin, who together spawned and cultivated for two decades a vibrant and productive cohort of population and community ecologists" (Wilbur & Douglas 2008:266–267). While the three of them were together in Michigan, they wrote their landmark "Community Structure, Population Control, and Competition" (Hairston, Smith, & Slobodkin 1960). "As an elegant example of synthesis, HSS's note is hard to beat both for brevity and impact. It proposed an ingenious resolution to a long-standing debate by reconciling opposing viewpoints; like most such resolutions, it also launched a whole new set of debates" (Kingsolver & Paine 1991:312). The dynamic trio did not remain at Michigan, however; Frederick Smith went to Harvard and Slobodkin to Stony Brook (New York) in 1969, and Hairston returned to the University of North Carolina in 1974 as William R. Kenan Professor of Zoology. "His modern approach to vertebrate natural history, as presented in his 1991 textbook *Vertebrate Zoology: An Experimental Field Approach*, was dedicated to 'the 520 students who, over thirteen years, made it a joy to teach vertebrate field zoology'" (Wilbur & Douglas 2008:268). He received ESA's Eminent Ecologist Award in 1991.

Teaching ecology is a persistent concern of ecologists, and discussions on it appeared from time to time in the *Bulletin*. In the June 1990 issue, Paul Keddy asked, "Is Mutualism Really Irrelevant to Ecology?" No one would claim that mutualism is irrelevant, but Keddy compared pages devoted to it to pages on competition and predation in introductory textbooks and found it was implicitly treated as inconsequential. A. H. Cherif made the same observation in 1990 in *American Biology Teacher*. Judith Bronstein (1991) pointed out that this neglect had been earlier noted by Risch and Boucher (1978) and Boucher (1985) without effecting a change. In December 1991, the *Bulletin* began running a regular feature on teaching entitled "Ecology 101," edited by Harold Ornes. It proved to be a popular addition to the *Bulletin*.

ESA had three extremely successful journals by 1990, all of which had gradually increased in length: the *Bulletin* (begun 1917), *Ecology* (begun 1920), and *Ecological Monographs* (begun 1931). Why have more? *Ecological Applications* began in 1991 because by then ecology seemed sophisticated enough to be applied to ecosystem management. In fact, there had been a good number of articles in the three earlier journals relating to ecosystem management, but by 1991, ESA officers felt that such articles needed their own journal. The June 1995 issue of the *Bulletin* announced ESA's newest journal, *Conservation Ecology*, a free online journal, and the September *Bulletin* announced that it would begin in early 1996. However, that venture soon broke away and became a paper journal of the Conservation Ecology Society, which overlaps to some extent with the Society for Conservation Biology and the Society for Ecological Restoration.

In 2003 a fifth journal, *Frontiers in Ecology* began, which publishes 10 times a year and is sent to all ESA members in a paper format but also appears online. The need for *Ecological Applications* seems obvious, but why *Frontiers*? In the year 2000, the Visions for the Future of ESA Publications Committee recommended such a journal to publish synthesis articles on all aspects of ecology. A secondary goal was to provide current information to policy makers, resource managers, and other decision makers (Silver 2003). Start-up funds were obtained from the Andrew W. Mellon Foundation and the David and Lucille Packard Foundation. *Frontiers*' "Write Back" feature mostly received responses to articles, but also some more general letters.

These letters reduced the number of published letters to the *Bulletin*'s editor. In the August 2004 issue of *Frontiers*, a new "Pathways to Scientific Teaching" began appearing, which rivals "Ecology 101" in the *Bulletin*. Although this chapter mainly draws upon ESA's *Bulletin*, Chapter 7 mainly draws upon *Frontiers*.

In 1988, National Academy of Sciences president Frank Press challenged the sciences to prioritize their goals in the face of rising needs of society and diminishing means to satisfy those needs. At the 1988 ESA Annual Meeting under the leadership of ESA president Hal Mooney, ESA appointed a committee under Jane Lubchenco to define such priorities for American ecologists (Mooney & Levin 1991). The 16-member committee included at least three future ESA presidents. Its 41-page report appeared in *Ecology*, "The Sustainable Biosphere Initiative: An Ecological Research Agenda" (Lubchenco et al. 1991). Here was democracy at work in American science: the president of a respected federally supported organization made a simple request of a large nongovernmental organization, that responded by appointing a large ad hoc committee that produced and published a lengthy report three years later. Press sent an appreciative letter to Lubchenco published in the September 1991 issue of the *Bulletin* thanking ESA for its prompt response, explaining that it and a similar report from astronomy were the first two submitted and would be used as a model for other science reports and could be submitted to a congressional committee illustrating science's willingness to respond to an important request.

The committee report (Lubchenco et al. 1991) identified three research priorities for assisting in the management of the earth's resources:

1. Global change: Its causes and consequences in climate, soil, and water.
2. Biological diversity and habitat diversity: Conservation of rare and declining species.
3. Sustainable ecological systems: Restoration of damaged systems, study pests, pathogens, and diseases.

To fulfill these priorities, the committee had three research recommendations:

1. Examine ways ecological complexity controls global processes.
2. Study importance of biological diversity in controlling ecological processes and the role processes play in shaping patterns of diversity at different scales of time and space.
3. Establish a major integrated program of research on sustainability of ecological systems, to understand ecological processes in natural and human-dominated ecosystems to prescribe restoration and management strategies for sustainability.

How would these plans be implemented? The committee listed eight steps to take in the coming year:

1. Organize workshops on SBI and global change.
2. Organize workshops on biodiversity.
3. Organize workshops on sustainable ecological systems and processes.

4. Publish nontechnical education documents for students and the public on the importance of ecology and ecological research.
5. Develop short- and long-term strategies to enhance ecological knowledge among students and the public.
6. Explore how ecologists can become more responsive to critical problems.
7. Meet with leading foreign ecologists to evaluate SBI and begin construction of an operational framework for international cooperation.
8. Develop a framework to coordinate and obtain funds for SBI.

If all of this was accomplished, what incentive would ESA members have to plan future research along suggested lines? Perhaps grant applications that conformed to these expectations would get preferential consideration from federal agencies.

Jane Lubchenco (1947–)

(From Niering et al. 1980:37 or Paine 1992:163. With permission.)

From Denver, Lubchenco entered Colorado College when the Ford Foundation funded an experimental Ford Independent Study Program for 20 highly qualified students (Richard G. Beidleman, personal communication). She successfully completed the program and received a B.A. degree in 1969, an M.S. from the University of Washington in 1971, and a Ph.D. from Harvard University in 1975. From 1975 to 1977, she was the first woman assistant professor in Harvard's Department of Biology (Langenheim 1996:38). In 1971 she married Bruce Menge, who had obtained a Ph.D. at the University of Washington, also in ecology, in 1970. He taught at the University of Massachusetts, Boston (1971–1976). They negotiated with Oregon State University possibly the first split tenure-track position in America, with each holding a half-time assistant professorship—a convenient arrangement while raising their two sons (Lubchenco & Menge 1993). In 1979, they won the Mercer Award (Niering, Marks, & Weigl 1980). "Lubchenco's research has been directed toward linking ecological patterns and processes occurring at different scales and different levels of organization in rocky intertidal communities and near shore ocean ecosystems" (Langenheim 1996:39). She was the fourth woman to be elected president of ESA (1992–1993, vice president 1988–1989; Paine 1992), and was also elected as a Fellow in the American Academy of Arts and Sciences (1993), National Academy of Sciences (1996), the American Philosophical Society (1998), the Royal Society (2004), a John D. and Catherine T. MacArthur Fellow (1993), and president of American Association for the Advancement of Science (AAAS) (for 1997). In 2009, President Obama appointed her as Administrator of the National Oceanic and Atmospheric Administration (NOAA) and Undersecretary of Commerce for Oceans and Atmosphere, where she remained until February 2013, when she returned to Oregon State University.

Jane Lubchenco and Paul Risser, on behalf of the Sustainable Biosphere Initiative (SBI), applied for a grant to open an SBI office in Washington and received an interagency grant from the National Science Foundation (NSF), National Aeronautics and Space Administration (NASA), U.S. Environmental Protection Agency (EPA), U.S. Forest Service, and the U.S. Fish and Wildlife Service. The grant allowed SBI to hire three full-time staff, located in an office adjacent to EPA's offices (Dennis 1992:219). The elaborate plan that had been explained in *Ecology* bore fruit. The Institute of Ecology had also had good initial funding but had fizzled by 1984. Would this be different? Only time would tell. Lubchenco was president of AAAS in 1997, and her presidential address was on "Entering the Century of the Environment: A New Social Contract for Science" (1998), which sustained the earlier commitment of both the ESA and federal agencies and served as a progress report. Later, her "Reflections on the Sustainable Biosphere Initiative: SBI + 20 Celebration ESA Annual Meeting" (2012) presented persuasive arguments that SBI was a success.

In 1990, the Public Affairs Committee decided that ESA should have a greater understanding of its membership, and in 1991 ESA obtained funds to enable the Public Affairs Office to conduct and analyze a survey of its members. That office's report, *Profiles of Ecologists: Results of a Survey of the Membership of the Ecological Society of America* (Holland et al. 1992), first appeared as a separate publication and was then serialized in three issues of the *Bulletin* (Lawrence, Holland, & Morrin 1993; Morrin, Holland, & Lawrence 1993). To what extent, if any, the SBI program influenced this curiosity about ESA members is not stated. The report did, among many other things, document the increasing number of women in the ESA (Figure 6.1).

The granting agencies were not the only respondents to the SBI report. Ecologists fully appreciated its importance and began reacting to it. The original 16 authors of the 41-page report in *Ecology* added three more ecologists to their team and boiled that report down to 7.5 pages in the March 1992 issue of the *Bulletin* (Huntley et al. 1992). That summary was followed by another explanatory article and a commentary in the

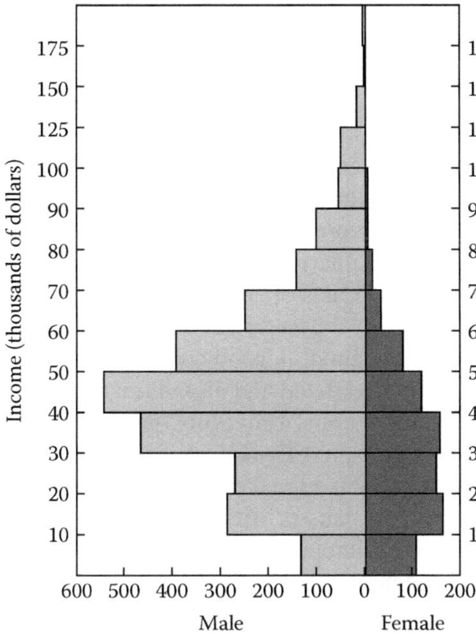

FIGURE 6.1 Male and female pyramid of respondents, showing increased membership of women. (From Holland 1992:8; from Lawrence et al. 1993:24. With permission.)

same March issue of the *Bulletin*, responding to the long version. That explanatory article, by Anthony Janetos (1992), at the National Aeronautics and Space Administration, discussed SBI and federal agencies. With participant diligence, he thought SBI prospects seemed good. The commentary was by six ecology students in the Botany Department at the University of Wyoming (Brewer et al. 1992). Their four concerns were about their possible futures within SBI. They also felt that SBI goals would not be realized unless there were effective means of conveying the findings and conclusions to the public.

Ecologists criticizing ecology is a persistent feature of the ecological literature. No one reasonably expects ecology to be as precise and theoretical as physics or chemistry, but could it not be a bit more precise and theoretical like physiology or genetics? Robert H. Peters (1946–1996) from McGill University published *A Critique for Ecology* (1991), which drew substantial fire and some praise from fellow ecologists (Kramer & Power 1997). His most general complaint was that "Academic ecology poses unanswerable questions" (Peters 1991:13). To illustrate this claim he quoted two such questions from seven prominent textbooks, such as, "What are the major attributes of the community?" John Lawton, at the NEERC Centre for Population Biology in England reviewed Peters' *Critique* unfavorably in *Nature* (1992): "It is a disappointing, uninspiring, negative book." In reaction to his review, Paul Keddy from the University of Ottawa, wrote "Thoughts on a review of a critique for ecology" (1992:234). He acknowledged that there were "irritating parts of [Peters'] analysis," but he thought that all ecologists knew of problems "in the way we conduct ecological research." Keddy suggested that how one reacts to

Peters' provocation may depend upon one's personality. One's reaction might depend upon how one would have reacted to Jesus chasing money changers from the temple: approving, disapproving, or ambivalent.

Three reactions to Peters' *Critique* in 1993 apparently appeared without any mutual influences. Arthur Shapiro (1993) of the University of California–Davis agreed with Peters' diagnosis of ecology and recommended his *Critique* to his graduate students to read, but when it came to a cure, Shapiro noted a lack of mechanisms in Peters' prescriptions. Furthermore, "Peters would run all the big intellectual questions out of ecology and focus on narrow (and hence relatively easy) ones; he would ask not 'Why are there so many kinds of animals?' but 'How many turnip thrips are there per acre in Manitoba in July?' The former question may be formally unanswerable, but it is much more fun. I cannot in good conscience tell my students that ecology is not science unless it is trivialized and all the fun taken out of it" (1993:178).

Samuel Scheiner, Northern Illinois University, responded to Keddy's comments with some "Additional Thoughts on *A Critique for Ecology*" (1993). He complained that Peters' "only goal of science is to make predictions." Peters denied that we can ever know "anything about true causes." At a practical level, it would be foolish to introduce a predator of a pest without modeling the expected population dynamics. And, "Because [Peters' epistemology] denies any theoretical basis, it is impossible to extrapolate the results from one system to another, and even interpolation is suspect." Peters, Scheiner thought, presented a very empty view of ecology.

Ecology's reviewer, Naomi Cappuccino, Université du Québec à Montréal, found Peters' book "engaging (and occasionally enraging)" (1993:1907), and she decided that all ecologists should read it. He "outlines the symptoms of the malaise, describes the numerous pitfalls in the way we construct ecological theories, and prescribes 'predictive ecology' for what ails our science." She thought he was extreme in dismissing tautologies, such as natural selection. His diagnoses were provocative, but his cures were insufficient. Most ecologists thought ecology had deficiencies, but determining what they were and what to do about them were questions beyond consensus.

Carita Bergman and Ian Adams (1993), zoologists at the University of Guelph, took Cappuccino's advice and read Peters' *Critique* and then tested some of his complaints. They devised a questionnaire of 13 questions and submitted them to faculty in four groups (a total of 80 professors): physics, chemistry, nonecological biology, and ecology. They presented their findings in the *Bulletin*, with eight graphs and two tables. They concluded that ecology does differ from other sciences, but that the differences did not make it a "soft" science. Bergman and Adams addressed process issues among Peters' complaints, such as competiveness among scientists, but not his intellectual complaints, such as whether ecologists studied unanswerable questions. But even at the level of their investigation, Robert Tamarin, Chairman of the Biology Department, Boston University, complained that their letter "seem[s] to exemplify many of the problems that Peters discusses" (1994:194).

Peters' comprehensive *Critique* grabbed attention but also raised resistance. A less polemical critique would have been to simply say, "here is a new approach to ecology." This was the approach taken by Timothy Allen and Thomas Hoekstra in *Toward a Unified Ecology* (1992), and by Simon Levin in his MacArthur lecture, "The Problem of Pattern and Scale in Ecology" (1992). These two theoretical programs have much in common. Pattern and scale in ecology were problems that both these works addressed. Two contemporary reviews of Allen and Hoekstra's book are by Jurek Kolasa (1993) and Frank Egerton (1995). Since my own review consists of uncritical praise, I will discuss Kolasa's. He thought their book "provokes, illuminates, and twists perspective" (1993:1605). They attack the hierarchy of ecological systems: "biosphere, biome, landscape, ecosystem, community, population, organism, cell" (Allen & Hoekstra 1992:6), but in doing so, the authors "become entangled in terminological and empirical diversity and contradictory evidence." However, having explained their problems as he saw them, Kolasa nevertheless concludes the following: "These deficiencies will not deprive you of the excitement the book is likely to generate."

From the beginning, ESA members have urged their colleagues to become involved in preserving nature. That impetus was behind the production of the *Naturalist's Guide to the Americas* (Shelford & Shreve 1926), and discomfort with politicizing ESA via conservation activities was behind the founding of the Ecologist's Union in 1946, which later changed its name to the Nature Conservancy and became the largest environmental organization in the world (Dexter 1978). This politicization issue, however, can reappear in different forms. The December 1992 issue of the *Bulletin* carried a letter by Cindy Sagers from the Department of Biology at the University of Utah, asking ESA to cancel its planned annual meeting in Utah, on the grounds that Utah ranked 48 in its dollars spent per child in public schools. Utah was the only state that had never appointed a female to any major administrative role within a state university, it passed the most restrictive laws in America on reproductive options, and it executed people (1992).

Six letters responding to Sagers' request appeared in the June 1993 issue of the *Bulletin*, one supportive—by Anna Korn—and five against. One of the unfavorable letters was from Vincent Tepedino and Staci England, of Logan, Utah; the other four were from men who did not reside in Utah. The December issue of the *Bulletin* included letters from two ESA members—Raymond Pierotti and Richard Brewer—supporting Sager. The March 1994 issue of the *Bulletin* included a long letter from Brian Meurer from the Department of Zoology at Brigham Young University, Provo, Utah, disagreeing with Sagers. His tactic was to discuss the Mormons, their church, and the state in positive tones while acknowledging they were all conservative. The 1995 meeting was held again (as in 1990) at the Snowbird Resort, Utah.

In December 1993, the *Bulletin* included a separate supplement, the *ESA Handbook, 1993–1994* (38 pages). It consisted of a list of ESA officers, the constitution and bylaws, code of ethics, bylaws of the 11 sections, and five chapters. There is no indication as to whether *Bulletin* editor Allen Solomon or some other official initiated this *Handbook*. What seems noteworthy is that it may be the second ESA handbook, the first appearing in March 1917. That first one included the first directory of ESA members. The September 1993 issue of the *Bulletin* had also included a supplement—a directory.

Since the 1940s, "ecosystem" had been one of the fundamental concepts and units of ecology. Yet, some ecologists never tired of reexamining it. "Ecosystem: A Powerful Concept and Paradigm for Ecology," by Thomas Burns (1992), Oak Ridge National Laboratory, seemed to sell something virtually everyone had already bought. However, Stan Rowe from British Columbia, and Burton Barnes from the University of Michigan, wrote a critical commentary in the March 1994 issue of the *Bulletin* on

"Geo-ecosystems and Bio-ecosystems." They thought that "niche," "succession," "biodiversity," and "landscape" were fuzzy terms, and that "ecosystem" was as well, if used without clear definition. One should specify whether one discussed a geo-ecosystem or bio-ecosystem. A geo-ecosystem "resembles a giant terrarium or aquarium with a particular developmental history at a particular location. It is a volumetric, layered, site-specific object—such as a lake, a particular landform-based forest, or a more complex tract of land–water terrain—into and out of which mobile organisms come and go." A bio-ecosystem is abstract and heuristic, "a conceptual device relatively elastic in its space/time dimensions," which adds "an 'abiotic' dimension to studies of individual, population/community and with the same spatial/structural concreteness (once boundaries have been set) as an individual organism" (Rowe & Barnes 1994:40).

Another theoretical foray came from three ecologists at Northern Illinois University, Samuel Scheiner, André Hudson, and Mark VanderMeulen: "An Epistemology for Ecology" (1993). They attacked the claim that ecology has no "laws." They undertook to demonstrate (in 4.5 pages) that "ecology has a very strong and rich theoretical basis" (1993:17). They began with the ecosystem concept. They explain that an ecosystem is composed of two sub-entities—biotic community and abiotic resources—and that these sub-entities are made up of further sub-entities. The sub-entities are linked by two processes: the flow of energy and the cycling of materials. They next discussed Lindeman's paper (1942b) and the first two laws of thermodynamics. They also discussed other entities and linking processes and then summarized their thinking in a table that illustrates their epistemological framework for ecology (Figure 6.2).

Entity	**Theory**	**Process**	**Sub-Entities**
Biome (= Landscape)	Climax	Succession	Ecosystems Meta-populations
Ecosystem	Energy pyramid (Second Law of Thermodynamics)	Energy flow	Community Niches
	First Law of Thermodynamics	Cycling of matter	
Community (= Food web)	Predator–prey theory	Predation	Guilds
Guild	Competitive exclusion principle Theory of natural selection	Competition	Populations
Niche	Competitive exclusion	Life histories Predation	Individuals Resources
	Malthusian principle Theory of natural selection		
Meta-population	Island biogeography theory	Migration Extinction	Populations
Population	Malthusian principle Theory of natural selection Population dynamics theory	Life histories	Individuals

FIGURE 6.2 An epistemological framework of ecology, showing the entities, the sub-entities that constitute them, the processes that link the sub-entities, and the theories that define the relationships. (From Scheiner, Hudson, & VanderMeulen 1993:19. With permission.)

Whether or not Scheiner, Hudson, and VanderMeulen's approach solved every ecologist's theoretical confusion is not discussed here. Rather, their foray illustrates a common theoretical uncertainty among ecologists, the opportunity that the *Bulletin* provides to express one's insights, and the willingness of ESA members to accept that opportunity. Whatever merits this epistemology might have for ecology did not concern Keddy (1993), who cited it in passing but took aim at Scheiner's earlier critique of Keddy's 1992 essay on emotional influences on one's judgment. Scheiner, however, was no more willing to let Keddy have the last word than Keddy had been to let Scheiner have it. Scheiner therefore explained "Why Ecologists Should Care about Philosophy: A reply to Keddy's Reply" (1994).

This is not a business history of ESA, but some business is too important to ignore. In the June 1993 *Bulletin*, the ESA Executive Committee, led by president Jane Lubchenco, published a letter to members on "A Proposal to Establish ESA Headquarters in Washington." There was already a Washington Public Affairs Office, but Bruce Patten had run the Business Office from Tempe, Arizona, for five three-year terms (ending August 1994), and the proposal was to unite both offices and hire an executive director over both. The plan described in their letter would be discussed at the business meeting in Madison, Wisconsin, August 3rd, open to all members. There were expenses in establishing this united office and hiring an executive director, and the nine members of the executive committee contributed $1,800 toward this expense and invited others to contribute to this fund. It was a well-planned proposal and received unanimous support at the Madison meeting. *Bulletin* editor, Solomon, reported substantial progress toward obtaining that expanded Washington office in the December 1993 issue. ESA members know that the plan was successfully implemented, as seen in the existing Washington office. Ecologists were adept at handling practical matters; theory was the domain of conflict.

Paul R(alph) Ehrlich (1932–)

(Courtesy of Ilka Hartman, 1974.)

Philadelphian Ehrlich earned a B.A. degree from the University of Pennsylvania (1953), and M.A. and Ph.D. degrees from the University of Kansas (1955, 1957), specializing in butterfly ecology. In 1959, he joined the Stanford University faculty and became Bing Professor of Population Studies in 1977. Ehrlich and E. O. Wilson are the foremost living ecologists. Ehrlich's online bibliography covers 61 pages (October 2013), and he is author or coauthor of 43 books. His fame began when he published the first edition of his third book, *The Population Bomb* (1968). In it he described many aspects of environmental degradation and pollution, including the "death" of Lake Erie (Ehrlich 1968:62). He predicted dire consequences, including mass starvation if steps were not taken to curb human population growth. Business Administration Professor Julian L. Simon (1932–1998), University of Maryland, was skeptical. Ehrlich's final chapter was entitled: "What if I Am Wrong?" in which he acknowledged that scientists can make predictions that are later falsified (Ehrlich 1968:197–198). His reply was essentially that it is better to be safe than sorry—the world would be better off if it followed his suggestions anyway. In 1980, Simon offered Ehrlich a wager: Ehrlich could select five commodity metals, and Simon would bet $1,000 that their prices would all decline by 1990. Simon would argue in his forthcoming *The Ultimate Resource* (1981) that new ideas and technologies would resolve resource challenges as the world population increased (Sabin 2013:4). Ehrlich, John Harte, and John Holdren accepted, lost the wager, and paid up.

In 1995, Simon complained in a commentary in the *San Francisco Chronicle* that Ehrlich and other "doomsaying environmentalists" were still preaching their incorrect ideas (Sabin 2013:200), that every measure of the environment would improve, and challenged Ehrlich to another bet. Ehrlich and climatologist Stephen Schneider then countered in a commentary in the *San Francisco Chronicle* with a $1,000 per trend wager of his own that 15 environmental quality indicators would all be worse in a decade (Sabin 2013:201). Simon declined; he was willing to wager on commodities but not on environmental factors. In 1996, he did bet David B. South from the School of Forestry at Auburn University, $1,000 that the price of timber would decrease in five years. In 1997, Simon saw that he would lose this time and paid up (Burgess 1996:38–39; Ehrlich 2000). Ehrlich's message remains unchanged. One of Ehrlich's prominent students was Michael Soulé (b. 1936), founder of conservation biology, who like Ehrlich, has been so busy teaching, researching, and publishing (Hannibal 2012) that he has not found time to be conspicuously active in ESA.

The balance of nature is a familiar ecological concept discussed occasionally in the ecological literature. It is the title of Stuart Pimm's *The Balance of Nature? Ecological Issues in the Conservation of Species and Communities* (1991) and John Kricher's *The Balance of Nature: Ecology's Enduring Myth* (2009) and is the main theme in Mark Sagoff, "Biodiversity and the Culture of Ecology" (1993), though Sagoff does not use that phrase, which may be why he overlooked Frank Egerton, "Changing Concepts of the Balance of Nature" (1973), which might have been helpful to his discussion. Sagoff's title is vague, and the phrase on which he focused was "the great chain of being," which he obtained from Arthur Lovejoy's history of that concept (1936). He did state that "A controversy … rages between those who believe that nature must exhibit a 'balance' or 'order' and those who argue that it is all chaos and contingency…" but that is not quite stating "the balance of nature." Sagoff was at the Institute for Philosophy and Public Policy at the University of Maryland, so presumably he was familiar with the philosophies of Plato and Aristotle. Yet, he argued that it is the "ecologists who followed the path of Platonic rationalism … who transformed, however inadvertently, an essentially religious cosmology into a basis for engineering" and that "Naturalists such as S. J. Gould and E. O. Wilson stand firmly within an Aristotelian tradition of empiricism, yet they turn this tradition into a spiritual quest" (Sagoff 1993:380). Sagoff's goal appears to have been merely to explore a situation without suggesting any changes. However, Warren Platts (1994) challenged Sagoff's interpretation of the perspectives of Robert MacArthur, Edward Wilson, and Stephen Jay Gould.

This narrative discusses only a few topics that filled the pages of the *Bulletin*, yet one may still gain insights into how meaningful such discussions were to ESA members. To cope with the increasing volume of such contributions, editor Solomon announced in the March 1994 issue that the size of the *Bulletin* was being increased again (done previously in 1970). The larger format increased the page size by 40%. The earliest *Bulletin* format was one column per page, the second format was two columns per page, and this third format was three columns per page. Another announcement was that Dr. Brian D. Keller had become ESA's first executive director on April 1st. He was an ecologist with prior administrative experience. That enlargement of the scope of the Washington operation coincided with Marge Holland's departure from the Public Affairs Office (PAO) for a position with the EPA in North Carolina. This prompted her to publish a history of the decade of PAO activities and accomplishments (Holland 1994, eight photographs).

The ESA continued to present an important array of awards annually to deserving ecologists, all of whom are featured in *Bulletin* accounts. However, the March 1994 issue carried a full-page account (but spread on 2 pages) of the 1993 John and Alice Tyler Award for achievement in environmental science that confers great benefit upon mankind, to ESA's dynamic duo, Herbert Bormann and Gene Likens. Established by the Tylers in 1972, their prize gave $200,000 and a gold medal to the recipients. The account of this award provided a fresh account of the achievements by ESA's most famous and illustrious research collaborators, focused upon the Hubbard Brook Ecosystem Study. By 1993, they had drawn 125 other scientists into their study, leading to 61 doctoral dissertations, and the ongoing results provided a model for other ecosystem studies worldwide.

Another notice in that same March 1994 *Bulletin* announced a proposal for the American Meteorological Society, American Geophysical Union, Oceanographic Society, and ESA to plan a new electronic journal tentatively called *Earth Systems Interactions*. The notice was for interested persons to get involved. Possibly, no one at first knew exactly where this proposal would lead. This project became a science education support system. Of interest here is that it was an opportunity to explore a venue that ESA would use in 2001 to dispense with its paper *Bulletin* in favor of an online-only version—just seven years after establishing its attractive largest paper version.

"What do ecologists do?" must be asked frequently when ecologists attend parties or other events with members of the public. However, ecologist Peter Stiling, University of South Florida, Tampa, thought it a question worth documenting (1994). The study produced by the Public Affairs Office (Holland 1992, serialized in the *Bulletin*, as Lawrence et al. 1993) possibly aroused his curiosity, and it certainly provided useful information. However, Stiling was more interested in the subjects and methods of their research than in where ecologists were educated and where they worked. Most of his own research for this study came from three journals—*Ecology, Oecologia,* and *Oikos* (1987–1991)—and the NSF provided the research funds. Nine tables that he provided indicated the frequency of taxa studied, number of species studied, trophic level studied, habitat types used in research, primary methods used, comparison of methods used with the taxa studied, concepts studied, comparison of research methods with the level of organization, and comparison of the taxa studied with the level of organization. One of Stiling's conclusions was as follows: "Because there are more studies on invertebrates and vertebrates (over 1600) than on plants (1076), population ecology remains better represented in ecological journals than community or ecosystems ecology" (Stiling 1994:120).

Apparently, there will always be new things to say about ecosystems. Roger Blew of the Environmental Science and Research Foundation, Idaho Falls, Idaho, brought a managerial perspective to his discussion (1996). He began by reminding readers of three approaches to ecosystem studies: (1) with organism as focus and ecosystem components as resources; (2) ecosystem as a set of processes and focus on flow of matter and energy, with organisms as vessels for matter and energy; and (3) geographic places. All are valid, but the General Accounting Office (1994) considered only the geographic approach. Rowe and Barnes (1994, discussed above) distinguished "bio-ecosystems" and "geo-ecosystems" and traced the latter to Tansley (1935:299), where he stated that we cannot separate organisms "from their spatial environment…" Unfortunately, Blew explained, Rowe and Barnes misquoted Tansley, who actually stated "from their special environment." Blew thought that "Tansley was trying to tell us that an ecosystem is characterized not by the biotic and abiotic components themselves, but by the interactions among these components." Blew's concern was that "we not allow the definition of 'ecosystem' to become so restrictive nor so diluted that it loses relevance to either science or management" (Blew 1996:171). The January 1997 *Bulletin* carried Rowe's apology for the error in quoting Tansley, which Blew noted, and in that same issue, Victor Marin, Universidad de Chile, added his own perspective on ecosystems to the debate. As did John Fauth in the October 1994 *Bulletin*.

In the summer of 1991, the Fourth Cary Conference, at the Institute of Ecosystem Studies, Millbrook, New York, assembled 68 scholars from throughout the United States and 4 foreign countries to discuss *Humans as Components of Ecosystems: The Ecology of Subtle Human Effects and Populated Areas*. In 1993, editors Mark McDonnell and Steward Pickett published the 21 papers and 4 conclusions from different perspectives presented at Millbrook, under the above title. Subtle human effects "covers a wide variety of often inconspicuous or unexpected interactions of humans with ecosystems" (1993:3), which McDonnell and Pickett itemized. Only a spotlight on those effects would enable society to appreciate their significance and take remedial measures. The volume ends with eight summary conclusions that the editors distilled, number 6 being: "Ecology needs new, integrated models and theories in order to effectively address both basic questions and environmental problems in human dominated ecosystems" (Pickett & McDonnell 1993:316). This was not an ESA-sponsored symposium, though a great majority of participants were ESA members.

When the U.S. Government developed a plan for "Integrating the Human Dimension in Ecoregion/Ecosystem Studies" (Sheifer 1996) and published the plan in the ESA *Bulletin*, one might expect that *Humans as Components of Ecosystems* would be an important resource, cited in Isobel Sheifer's bibliography. But no, she mainly cited government reports. She did cite Paul Stern, "A Second Environmental Science: Human-Environment Interactions" (1993), which appeared in *Science* in the same year as McDonnell and Pickett's volume. Stern can be excused for not citing McDonnell and Pickett, since he could not have cited it unless he had prior knowledge of it before he published, which is not a reasonable expectation. Sheifer's conclusions resemble Pickett and McDonnell's.

The July 1996 issue of the *Bulletin* had two supplements, both thicker than the *Bulletin*: the annual meeting program (191 numbered and 9 unnumbered pages) and meeting abstracts (503 pages). This was not unique to that year but is an example of the ongoing response to a steadily increasing membership. *Bulletin* editor Solomon noted in a 1997 editorial that ESA published a few hundred pages a year for its first 70 years, but in 1996 it printed 4,519 pages. A growing membership provided increasing numbers of occasions for pages published.

Judy Meyer's past president's address was "Beyond Gloom and Doom: Ecology for the Future" (1996). She became an ecologist after reading *The Subversive Science: Essays toward an Ecology of Man*, collected by Paul Shepard and Daniel McKinley (1969). A quarter-century later, she focused upon what that subversive science had accomplished and what its prospects were for the future. She discussed five areas of accomplishments: monitoring, conservation, ecosystem management, adaptive management (dams), and restoration and rehabilitation. Myer proposed that ESA establish "a new award for ecological research in the public interest" (1996:205). She concluded by suggesting that ecologists adopt a slogan from the computer revolution: "The best way to predict the future is to make it happen." Bolstering her message was a report from Anthony Janetos on NASA's Mission to Planet Earth, "Six Years of Public Affairs Activities by ESA: An Evolving Agenda" (1996), a progress report on SBI and other initiatives. However, Janetos' report also carried a warning that the last congressional session had witnessed "an extraordinary attempt to roll back not only substantial progress on environmental quality and regulations, but also to curtail support for the underlying science…" (Janetos 1996:229).

In April 1997, James Callahan began editing a new regular feature in ESA's *Bulletin*, "Focus on Field Stations." Callahan was on an NSF panel that reviewed proposals for improving field stations. This new feature was a convenient way to update Robert Hiatt's *World Directory of Hydrobiological and Fisheries Institutions* (1963), but also to add terrestrial stations. The April *Bulletin* featured the Oregon Institute of Marine Biology (Shapiro, Butler, et al. 1997), part of the University of Oregon. Research and teaching began in the Coos Bay area in 1924, with tents for laboratories and dormitories, and in 1931 the Army Corps of Engineers deeded to the university part of the Coos Head Military Reservation (built by the Civilian

Conservation Corps), with buildings that could be used for classrooms, laboratories, and dormitories. During World War II, the Army reoccupied the site, and it was returned to the University of Oregon afterward. In 1966, the University of Oregon began extensive repairs to the buildings, and the institute became occupied year round. New buildings were built in 1985. The October *Bulletin* focused on the Baruch Marine Field Laboratory, near Georgetown, South Carolina, part of the University of South Carolina (Fletcher & Allen 1997).

Boris Zeide (1937–2012) was born in Moscow and received an M.Sc. degree in 1959 from the Moscow College of Forestry and worked as a forester from 1959 to 1960. He received a Ph.D. from that college in 1970, lectured at the Hebrew University of Jerusalem (1974–1976), taught at Harvard and Rutgers universities (1976–1980), then settled at the University of Arkansas School of Forestry in 1980. In the ESA *Bulletin*, he raised the question of "Affirmative Action or Parasitism?" (1998a) concerning birch-fungus and fir-fungus relationships which Suzanne Simard et al. (1997) had discussed as mutualistic and which David Read (1997) had summarized. Zeide thought their interpretations were biased. Thomas Rooney, Department of Botany, University of Wisconsin–Madison argued (1998) that, instead, it was Zeide who was biased—against mutualistic interpretations.

If so, that was not Zeide's only bias: "Calculating the value of ecosystem services is the cutting edge of ecological research at present. Alta Vista already contains 396,802 documents on 'ecosystems + services'" (Zeide 1998b:215). At this point he could have cited as a convenient example Gretchen Daily's edited *Nature's Services: Societal Dependence on Natural Ecosystems* (1997), but he gave no indication of having seen it. He claimed that authors focus upon positive ecosystem services and overlook negative ones: diseases and pest competition. He thought "This omission is a manifestation of the currently fashionable ideology, which may be called biodiversity worship. It is best expressed by its chief advocate, Edward Wilson [1992:32]" whom he quoted claiming biodiversity is priceless. Anantha Prasad's response to Zeide (1999) explained why ecologists had emphasized the positive aspects of ecosystem services.

What others called his bias, Zeide viewed as merely the skepticism of a realist. In "Another Look at Leopold's Land Ethic," he attacked the chief god of environmentalism: "He himself apparently felt no contradiction between his call for a land ethic to respect wild animals and his lifelong pleasure in hunting" (1998c:13). However, Zeide's main concern was Leopold's claim: "If the land mechanism as a whole is good, then every part is good, whether we understand it or not [Leopold 1966:176–177]" (Zeide 1998c:15). Zeide viewed this as uncritical defense of ecosystems: "When the American chestnut (*Castanea dentata*), a key species of the eastern forests, disappeared, oaks and hickories filled the gaps, and today's ecosystems are as healthy as before" (1998c:14). Zeide, therefore, complained that "The very term *ecosystem* is a misnomer. A system means a set of coordinated parts that form a whole, such as an animal body or an engine. The precision of a body differs in kind from what would be better called an 'ecomedley'" (1998c:15). The editor of the *Journal of Forestry* invited a Leopold scholar, Baird Callicott, to respond to Zeide's critique (1998); he complained that Zeide's critique of Leopold was biased and logically flawed, that Zeide believed in a zero-sum world, whereas Leopold believed in win–win solutions in which forests can be logged to maintain prosperity while protecting our ecological heritage.

Zeide resumed his attack on uncritical support for biodiversity in "Nature's Surprises" (2004), in which he challenged the ideas of Leopold, Rachel Carson, and Edward Wilson: "along with our food and fluffy birds, biodiversity includes chiggers, ticks, SARS viruses, and other pests" (2004:120). One cannot say for sure what influence Zeide's critiques exerted, but when four international conservation agencies joined to produce an elegant coffee-table book, *The Wealth of Nature: Ecosystem Services, Biodiversity, and Human Well-Being* (Mittermeier 2009), their book addressed Zeide's concerns: Chapter 6 discussed protection against natural hazards, Chapter 8 was on pest control, Chapter 9 was on waste regulation, and Chapter 10 on disease regulation.

Following Rooney's critique of Zeide was a critique by Diane Srivastava and Jennifer Ruesink (1998), Centre for Biodiversity Research, University of British Columbia, of an appeal, also in the January issue of the *Bulletin*, by John Blair (1998), Division of Biology, Kansas State University, for ecologists to sign a letter to world leaders expressing alarm over the growth of human population and environmental degradation. They argued that "Population Growth and Environmental Destruction: It Isn't that Simple!" Blair was appealing to a presumed consensus among ecologists that increasing human population leads to increased consumption of resources and environmental degradation. He had thought it unnecessary to provide data to support this concern. However, Srivastava and Ruesink did think they needed data and a counterargument to question Blair's assumption. They claimed that "we too are alarmed by present and projected rates of environmental destruction, and we welcome any actions to reduce such degradation" (Srivastava & Ruesink 1998:219). They were disturbed that Blair's letter contained no recognition that per capita effects are as important as population growth in causing environmental degradation. Their data indicated that per capita resource use was much higher in the northern than in the southern hemisphere: per capita impact for CO_2 from 1960 to 1988 was 65 in the North and 54 in the South, but the growth of population was greater in the South. They argued further that the way to reduce population growth was not by government incentives (poor results), but by fighting poverty, increasing education, and improving the status of women.

David Pimentel and Walter Dodds at Kansas State University in Manhattan, strongly supported Blair's appeal for signatures on the letter to world leaders (1999). They also agreed with Srivastava and Ruesink's argument that the situation could be seen in a more complex perspective than the letter seemed to indicate, but felt that it did not negate the value of the letter's appeal for controls of population and environmental destruction. They presented the case in more detail than Blair had and urged ecologists to sign on.

In 1998, ESA established the Leopold Leadership Program (leopoldleadership.stanford.edu/history). Jane Lubchenco persuaded two other former ESA presidents, Harold Mooney and Paul Risser, to join in calling for such a program. She recalls the following (personal communication, August 9, 2014):

> …need for more mid-career ecologists to be more adept at communicating ecological science with non-scientists. In response ESA constituted an ad-hoc committee, the Spring Green Committee (after the annual meeting in Madison when a number of

SBA leaders visited Spring Green, Wisc.) to explore what a training program might look like. The Packard Foundation provided funds. The result was the Aldo Leopold Leadership Program which was designed to be independent of but serve the ESA community, and provides intense leadership and communications training to 20 mid-career environmental scientists a year. The program has been a stellar success and has enabled ecologists to be much more effective and engaged in connecting their science to policy, management, business and public understanding.

In 2005, the program moved to the Stanford Woods Institute for the Environment. The co-leaders in 2014 were Pam Matson and Steve Palumbi.

Two British zoologists, Daniel Haydon and Alum Lloyd, published in the July 1999 issue of the *Bulletin* "On the Origins of the Lotka-Volterra Equations." The point of their commentary was to emphasize that although Lotka and Volterra's equations have the same mathematical structure, the assumptions of their derivations were different, as are the interpretation of parameters of their formulations. "Popular recognition of these alternate formulations would, on occasion, save theoreticians from the charge of fanciful modelling, and empiricists from measuring the wrong parameters" (Haydon & Lloyd 1999:206). Haydon and Lloyd cite Leslie Real and Simon Levin's essay on "The Role of Theory in the Rise of Modern Ecology" (1991), which, in turn, cites Sharon Kingsland's *Modeling Nature: Episodes in the History of Population Ecology* (1985), but these British zoologists seem not to have consulted Kingsland's book to see what she wrote about Lotka and Volterra.

Frank B(enjamin) Golley (1930–2006)

(From Pomeroy et al. 2007:16. With permission.)

Chicagoan Golley earned a B.S. degree in agriculture from Purdue University (1952), an M.S. in wildlife management from Washington State University

(1954), and a Ph.D. in zoology from Michigan State University (1958). He went to the University of North Carolina to teach in January 1958, but Eugene Odum persuaded him to come to the Savannah River Plant in September. In 1961, he became the first Director of the Savannah River Radiation Ecology Laboratory. Golley embraced the ecosystem ecology of the Odum brothers and began a long productive collaboration in research and institution development at the University of Georgia (Golley 2001; Pomeroy et al. 2007:15). In 1967, he moved to the Athens campus as executive director of the new Institute of Ecology. In 1979, during a leave of absence, he became director of the Division of Environmental Biology at the NSF, returning to Georgia in 1984. Golley was the author or editor of 17 books and more than 170 articles or book chapters. He was the founder and first editor of *Landscape Ecology* (volume 1, 1987). He served as ESA president (1976–1977) and also served terms as president of the International Association of Ecology and the International Society of Tropical Ecology. Although he retired from the Institute of Ecology in 2000, he remained active there afterward.

In 1982 and 1995, the National Research Council published reputational ratings of doctoral programs for 41 fields. Michael Brett and associates (1999) correlated a reputation for the top 50% of U.S. doctoral programs in ecology, evolution, and behavior (EEB), which was 63 programs. They determined the top 64 of the EEB journals, which they ranked according to the publication frequency of faculty in these 63 programs, and they also ranked the journal-article productivity of the faculty of these 63 institutions (Figures 6.3 and 6.4). They found that the "best predictor of a program's reputational rating was its research productivity in scholarly journals" (Brett et al. 1999:255). Their two figures of frequency of journal usage and faculty productivity at the top institutions have a general interest beyond their particular purpose.

Robert ("Mac") McIntosh, plant ecologist at the University of Notre Dame, a former student of John Curtis at the University of Wisconsin–Madison, shared Curtis' admiration of H. A. Gleason and Gleason's individualistic concept of plant associations. McIntosh's major contribution to the history of ecology was *The Background of Ecology: Concept and Theory* (1985), which can be characterized as a bibliographic guide, because he commonly presents the literature on both sides of past ecological controversies without pronouncing winners and losers. In April 1999, he participated in a celebration of the centennial of Henry Cowles' publication of his studies on the ecological succession of plants at the Indiana dunes, and McIntosh published his talk from that celebration in the December 1999 issue of the *Bulletin*: "The Succession of Succession: A Lexical Chronology." Henry David Thoreau first used the term "succession" in its ecological meaning in his essay, "The Succession of Forest Trees" (1860, 1988:72–92). Cowles used the terms "progression" and "development" in "The Ecological Relations of the Vegetation on the Sand Dunes of Lake Michigan" (1899), but in "The Physiographic Ecology of Chicago and Vicinity" he wrote: "There must be, then, an order of succession of plant societies, just as there is an order of succession of topographic forms in the changing landscape" (1901:79b). McIntosh discussed the elaboration of the succession theory from Cowles to the 1990s.

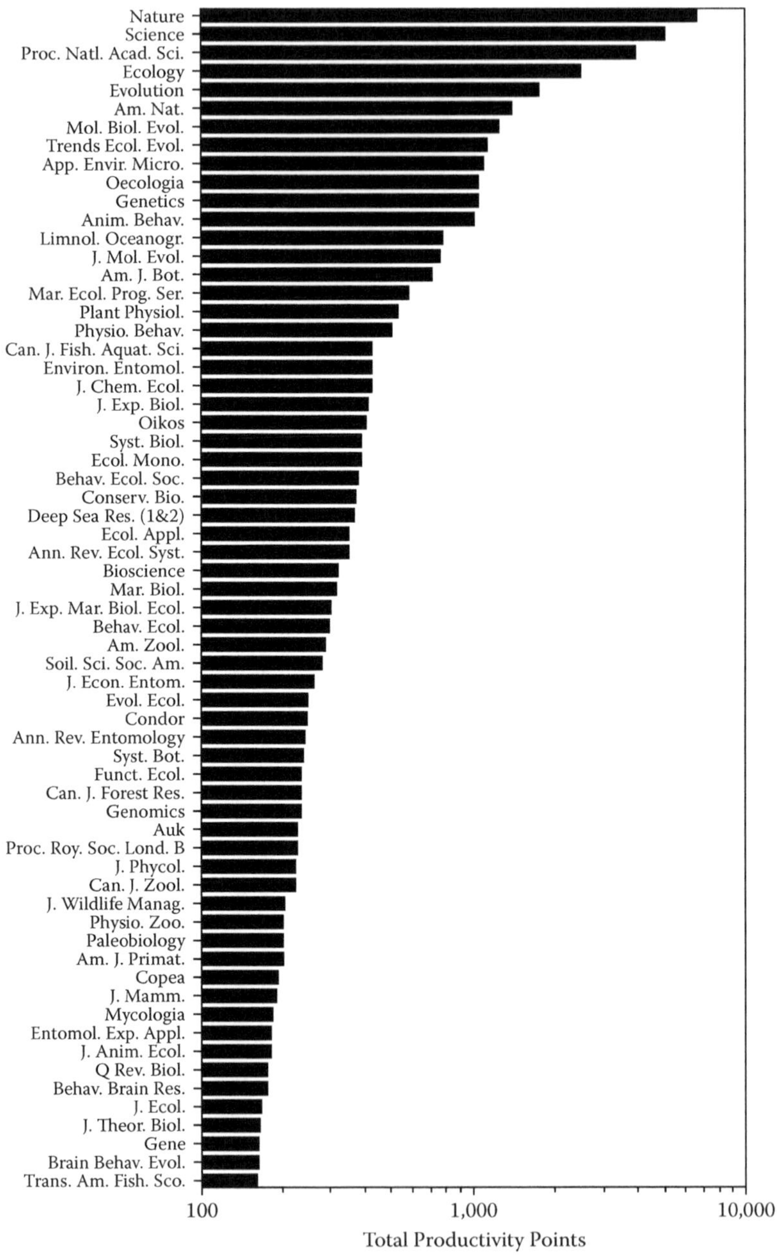

FIGURE 6.3 The top 64 EEB journals that carry the most articles by EEB faculty from the top-ranked EEB institutions. (From Brett et al. 1999:251. With permission.)

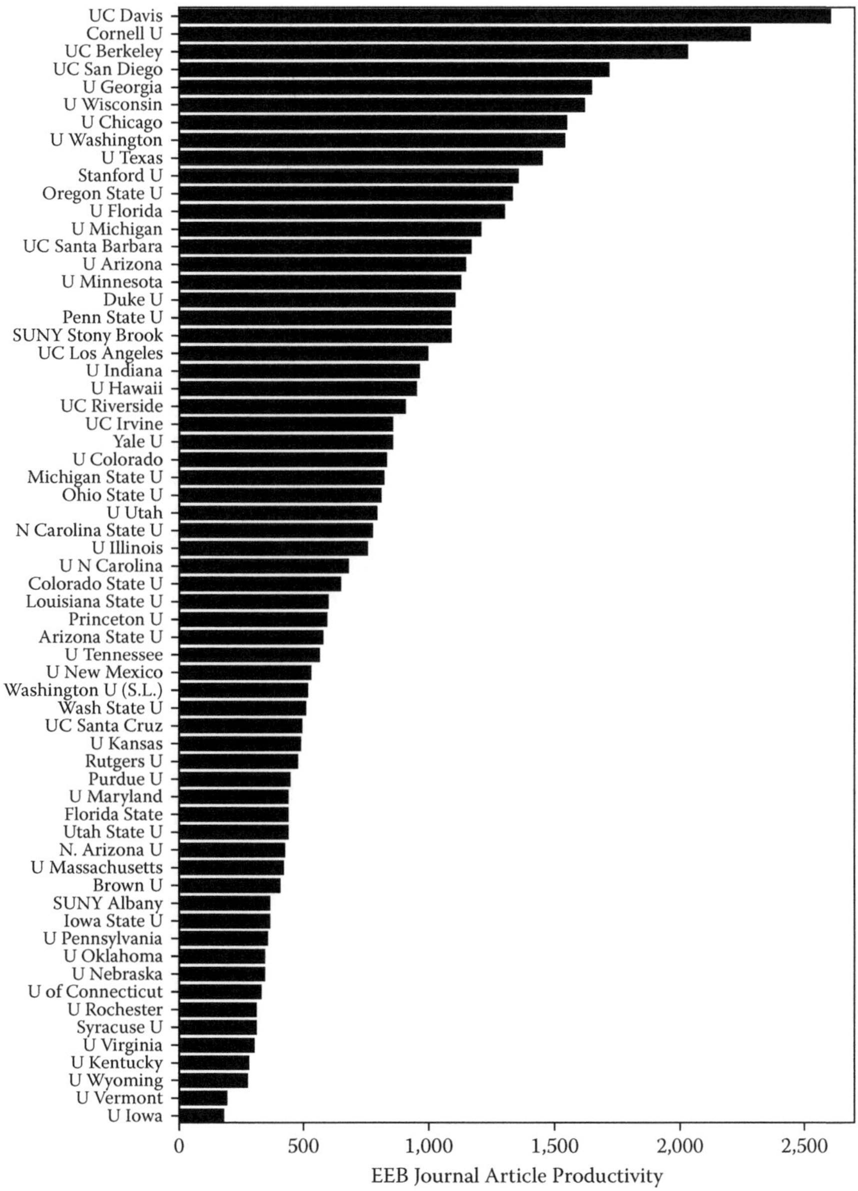

FIGURE 6.4 A ranking of EEB institutions according to publications in the top EEB journals. (From Brett et al. 1999:252. With permission.)

The July 1999 issue of the *Bulletin* announced that in the year 2000, *Ecology* would appear monthly, *Ecological Applications* would appear six times a year, and journals would publish about 25% more pages than before, all of which would reduce the delay in publishing accepted manuscripts. For the year 2000, ESA began publishing its annual report as a separate (paper) supplement to the *Bulletin*, and in that report it provided some statistics: total membership 7,807, domestic 6,465, foreign 1,342; membership professional affiliation—academic 48%, government 11%, consulting 5%, nonprofit 3% (the missing 33% of members apparently did not identify a professional affiliation); subscriptions—*Bulletin* total 8,007 (members 7,807, institutions 200), *Ecology* total 6,074 (members 4,058, institutions 2,009, other 7), *Ecological Applications* total 3,633 (members 2,357, institutions 1,271, other 5), and *Ecological Monographs* total 2,961 (members 1,641, institutions 1,315, other 5).

Larry Slobodkin published an interesting 4-page commentary in the July 2000 issue of the *Bulletin*, "Proclaiming a New Ecological Subdiscipline," which is "societal ecology." One might say that he discussed a persistent, if not perennial, theme among ecologists: the need of society to be guided by ecological understanding and insights in the uses and modifications of the Earth's environments and resources. The problem being that only occasionally does society beat a path to ecology's door and seek such advice (as Art Cooper had complained in the *Bulletin* in 1982). Therefore, ecologists must not wait for such opportunities to present themselves, but must seek opportunities to inform political and other decision makers of ecology's practical uses.

In contrast to Slobodkin's solo essay, eight authors from seven countries wrote "Biodiversity and Ecosystem Function: An Issue in Ecology" (Wardle et al. 2000), complaining about the one-sided perspective in ESA's fourth *Issues in Ecology* (1999) online publication. It "adequately considers only one of the two major schools of thought on the whole diversity-function issue, selectively ignores a large subset of the published literature, and states opinions as facts" (Wardle et al. 2000:235). David Wardle and coauthors thought it unprecedented for ESA to endorse one school of thought on an issue and ignore the other school. Strangely, Wardle and coauthors do not cite in their bibliography the fourth *Issues in Ecology* publication, though they cite its title and mention that it can be downloaded from the ESA website. *Issues in Ecology*, number 4, on biodiversity and ecosystem function, had 12 authors, including some who became presidents of ESA (Naeem et al. 1999), 12 pages, and illustrations. Wardle et al.'s complaint elicited two responses that followed their commentary.

David Tilman, editor-in-chief of *Issues in Ecology*, explained that "Each report represents a consensus reached by a panel of scientists who have skills relevant to the topic being addressed" (2000:240). And that *Issues in Ecology* reports are not ESA position papers. A completed report is sent to eight members of an editorial board for comment, which is more of a review than any article received by ESA's journals. Tilman welcomed skepticism about the content of any ESA publication but felt that the editorial procedures for *Issues in Ecology* were exceedingly fair. Shahid Naeem, senior author of the fourth *Issues in Ecology*, wrote a 5.5-page response to Wardle et al.'s 5-page complaint. Naeem's response is wide-ranging, but his main point was as follows (2000:242):

> Wardle et al. describe a debate in which one party subscribes to a "population biology" perspective, while another subscribes to an "ecosystem property" perspective, suggesting that such perspectives are mutually exclusive. This description is incorrect.

From the standpoint of ESA handling internal disputes, this is an exemplary illustration of fairness. Both sides aired their cases as they wished, and other ESA members could read both cases and the fourth *Issues in Ecology* and draw their own conclusions. Around the same time that this debate was occurring in the *Bulletin*, a workshop was held on the topic (when and where unspecified), and a report on it appeared in the January 2001 issue of the *Bulletin* (Alvo & Ponomarenko 2001).

That debate on biodiversity and ecosystem function was theoretical, but Ariel Lugo et al. reminded *Bulletin* readers that "Managed Ecosystems Deserve Greater Attention," for "conservation of biodiversity and the sustained production of goods and services that we depend upon" (2001:92). Management can fail, because of "insufficient understanding and/or poor applications." Population growth and increased economic activity require greater control of biodiversity. Collaboration between managers and ecologists was needed for managerial success, and knowledgeable management could also improve understanding of ecosystems.

ESA initiated its professional certification program in 1981, and in October 2000, the chair of the Board of Professional Certification, Virginia Dale, and other board members, wrote a guest editorial, "Let's Make It Work." They explained that in 1982 there were 72 persons certified, 154 certified by 1983, but only 346 by 1999, which was only 5% of ESA's over 7,500 members. Academic ecologists probably saw the program as irrelevant and the $125 fee (by 2000) as burdensome. They argued that only 40% of trained ecologists would get jobs in academia, and so professors needed to train students for other professional opportunities in which certification would be very helpful. The certification process had been streamlined at the August 2000 meeting of ESA's Board of Governors, and the Board of Professional Certification urged members to support this important program. The August meeting of Board of Governors also approved the revised version of ESA's Code of Professional Ethics, which was also published in the October issue of the *Bulletin*.

On March 29, 2000, a research vessel carrying ecologists from the University of California, Davis, capsized during a storm in the Sea of Cortez off Baja California, Mexico, and ecologist Gary Allen Polis, postgraduate researcher Michael D. Rose, and three students from Kyoto University, Japan, drowned (Strong & Hastings 2001). Four other scientists and students from the University of California, Davis, swam to an island and survived. Polis' happy, outgoing personality made him popular on *National Geographic Explorer* films and PBS documentaries. He published *The Biology of Scorpions* (1990), *The Ecology of Desert Communities* (1991), co-edited *Food Webs: Integration of Patterns and Dynamics* (Polis & Winemiller 1996), and coauthored a children's book, *Scorpion Man: Exploring the World of Scorpions* (1994). Polis had served as president of the American Society of Naturalists, and if he had lived more than 53 years, he perhaps would have served as an ESA officer.

Arthur W(ells) Cooper (1931–)

(From ESA *Bulletin* 81:172. With permission.)

Art Cooper, from Washington, DC, earned B.A. and M.A. degrees (1953, 1955) from Colgate University, then studied with Professor Stanley Cain at the University of Michigan for a Ph.D. (1958) with a dissertation on the relationship between plant forms and microclimates (Seneca 1980; Stearns 1985). In 1958, he joined the botany faculty of North Carolina State University (NCSU). As a student in his classes during 1959 to 1960, I can attest to his enthusiasm, energy, and humor, which made him an outstanding teacher, researcher, and administrator. He took a leave of absence from 1971 to 1976 to serve as Assistant Secretary for Resource Management in the North Carolina Department of Natural and Economic Resources. In 1976, he returned to NCSU, but in the Department of Forestry, and became department head in 1979. His activities in ESA led to his election as vice president in 1975 and president from 1980 to 1981, and he received the Distinguished Service Award in 1984. Yet, his activities as a professor, administrator, and ESA activist were not the whole of his career. He was also active in other professional societies and on other committees, where his articulate judgment was highly valued.

In April 2001, David Inouye initiated a new feature in the *Bulletin*, asking authors who had published prize-winning articles at least 10 years before, to comment on the preparation and publication of their articles. The first article he chose was by Jane Lubchenco and Bruce Menge, "Community Development and Persistence in a Low Rocky Intertidal Zone" (*Ecological Monographs* 48, 1978, 67–94), which won the Mercer Award in 1979. Their brief retrospective essay mentioned no personal information, but most members of ESA would have known that they were a married couple, he had received a Ph.D. from the University of Washington in 1970 and she had received an M.S. degree from there in 1971 (her Ph.D. was from Harvard in 1975). In their research, she focused upon herbivore–plant interactions and he upon predator–prey interactions. Their findings confirmed their hypothesis that the "community structure at wavier rocky headlands was an indirect consequence of physical stress on food web structure along gradients of wave exposure" (Menge & Lubchencho 2001:124).

Their nice story was followed by a conflict: Professor Fakhri Al Bazzaz from Harvard University, lamented heated arguments among ecologists over ecosystem biodiversity versus productivity. The news report in *Science* identified the combatants as David Tilman and John Lawton (FRS) versus Michael Huston and Philip Grime (FRS). Instead of calling experiments "irrelevant" and "politically motivated," Bazzaz suggested calling experiments "incorrectly designed" or "incorrectly discussed" (2001:136). Bazzaz listed seven issues that must be considered and four questions to be answered to resolve the conflict through scientific investigation. Bazzaz's lament was not a new issue, but it shows that previous attempts to resolve it were unsuccessful.

Joan Roughgarden from Stanford University, received a sabbatical to study at the National Center for Ecological Analysis and Synthesis, Santa Barbara, California. Her investigation resulted in a "Guide to Diplomatic Relations with Economists" (2001). The logic of her perspective was that ecologists might be more effective in influencing ecosystem decisions if they could work with and communicate effectively with economists in planning human uses of ecosystems. Her advice appeared in the January *Bulletin*, and two responses came in April. Professor Emeritus John Cantlon from the Department of Botany at Michigan State University, reminded readers that he had been ESA president from 1965 to 1966; he had collaborated with Professor Emeritus Herman Koenig from the Department of Electrical Engineering at Michigan State University, publishing articles in *Ecological Economics* and the *Journal of Industrial Ecology* relating to Roughgarden's theme. Cantlon (2001) complemented her for "sage advice" on expanding the dialog between ecologists and economists. His comment was advice on a narrow, technical point concerning a proposed national Board of Technological Risk Advisors setting Pigovian tax rates.

The other commentary, by John Gowdy from the Department of Economics at Renselaer Polytechnic Institute, Troy, New York, and Karin Limburg from SUNY College of Environmental Science and Forestry, Syracuse, New York, also complimented Roughgarden on her essay, then warned that there are two schools of economics: neoclassical economists who subscribe to general equilibrium theory, and ecological economists who take a more eclectic view. Their commentary was about the environmental fallacies of the former (2001):

> Policy recommendations of neo-classical economists, whether about global warming, biodiversity protection, acid rain, or whatever, invariably attempt to reconstruct the "ideal" market (by "getting the price right") rather than dealing directly with the problem at hand.

Gowdy and Limburg claimed that three Nobel laureates in economics (named) were well aware of the crisis in economic theory, but most economists were not. In other words, Roughgarden's respectfulness toward economists needed a serious qualification.

The July 2001 issue of the *Bulletin* announced the publication by Oxford University Press of three volumes reporting on 20 years of multidisciplinary teams of researchers at three Long Term Ecological Research (LTER) sites: *The Structure and Function of an Alpine Ecosystem: Niwot Ridge, Colorado*, edited by William D. Bowman et al.; *Standard Soil Methods for Long-Term Ecological Research*, edited by G. Philip Robertson et al.; and the Konza Prairie site's *Grassland Dynamics*, edited by Alan K. Knapp et al. These volumes showed the ecological community the rewards of careful monitoring of sites for two decades and were challenges to researchers at other LTER sites to produce comparable volumes on their similar research.

In 1999, the ESA Governing Board appointed an Ad Hoc Task Force on Membership to determine "needs and membership status of professional ecologists in North America (United States, Canada, and Mexico) and to propose future member services targeted at maintaining a diverse and stable membership base for the Society" (Chazdon 2001:181) (Figure 6.5). The committee took surveys and proposed "several ways for the structure of ESA to become more open and more inclusive of the full diversity of professional ecologists entering the field." The task force found: "Membership in ESA is stable, but recruitment of new members and members active in Sections is low," "Journals have been and still are ESA's lifeline," main interests of members was in community and population ecology, applied and government ecologists do not find ESA membership as beneficial as do ecologists doing basic research," "One-third of ESA members do not attend Annual Meetings, do not subscribe to any journal, and are uninformed about ESA Headquarters activities," most members are interested in regional meetings but do not belong to regional chapters, member priorities are "related to employment status and professional needs and rewards." The task force made 10 recommendations, including "Make the ESA a truly American society that fully embraces members from Canada, Mexico, and other Latin American and Caribbean countries" and translate all abstracts into Spanish, French, and Portuguese on the ESA website.

The second award-winning article that Inouye chose for an author's retrospective was Orie Loucks, "Ordinating Forest Communities by Means of Environmental Scalars and Phytosociological Indices" (*Ecological Monographs* 32, 1962, 137–166), which won the Mercer Award in 1964. A Canadian, Loucks had received B.Sc.

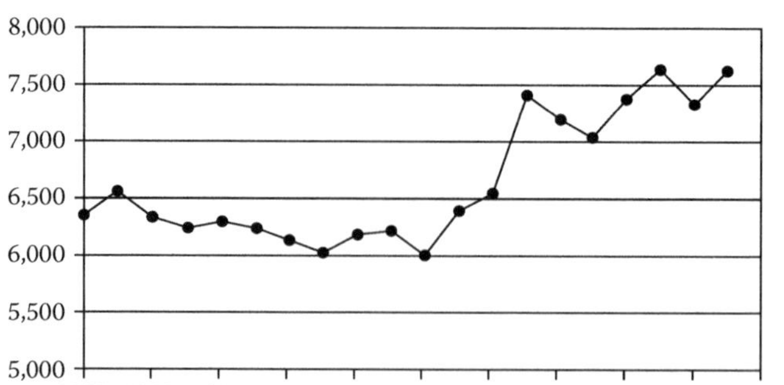

FIGURE 6.5 ESA membership, 1980 to 1999. (From Chazdon 2001:183. With permission.)

and M.Sc. degrees from the University of Toronto before going to the University of Wisconsin–Madison to earn a Ph.D. (1960) under plant ecologist John Curtis. Loucks' prize-winning paper was based upon his doctoral dissertation. His research had been conducted at the Gaspé highlands of northwestern New Brunswick, Canada. His focus had been on "what factors differentiated the distribution of sugar maple in this landscape (occurring in nearly pure stands on the hilltops) from the conifers on locally rich soils in the valleys" (Loucks 2001:187). Within a decade, the community-based ordination method he had used was replaced by ecosystem simulation modeling.

By 1995, Loucks, at Miami University, Oxford, Ohio, was collaborating with Jianguo Wu from Arizona State University West, Phoenix, Arizona, in moving "From Balance of Nature to Hierarchical Patch Dynamics: A Paradigm Shift in Ecology." The bibliography to their joint article (Wu & Loucks 1995:463–466) shows the paper trail of articles that led both to this collaboration. Loucks' retrospective essay (2001) was looking back almost four decades, and seeing that his research career had changed directions a few times since then.

A third retrospective essay by a Mercer Award winner in 2001 was by Svata Louda, concerning her "Distribution Ecology: Variation in Plant Recruitment in Relation to Insect Seed Predation" (*Ecological Monographs* 52 (1982a) 25–41). That article was one of three she published based upon her doctoral dissertation (Louda 1982a,b, 1983). In 1971, while a graduate student at the University of California, Santa Barbara, she wondered: "Could insect herbivory be a significant factor influencing the population density of some native plants?" (2001:229). One species she studied, *Hazardia squarrosa*, grew from the coastal mountains to the shore in San Diego County but its density declined as it approached the shore. That fact was explained by a gradual change in its environmental conditions. However, she noticed that insect herbivory increased as it approached the shore. She used an insecticide to control insects in one plot and compared that plot with a control plot and found *H. squarrosa* denser in the sprayed plot.

In October 2001, Dan Fiscus from the University of Maryland Center for Environmental Science Appalachian Laboratory, Frostburg, began a three-part discussion of "The Ecosystematic Life Hypothesis." Part I included an introduction and definitions. He began with a simple question while working on a master's degree: how to determine the ecological health of soil. He decided that he would define its health relative to human values of food and fiber production and sustainability. He worked with "free-living (i.e., not plant-parasitic) nematode communities due to their abundance, functional diversity, and significant roles in decomposition and nutrient cycling processes" (2001:248). He subjected nematodes to tillage, fertilizer, and pesticides and found that some genera suffered while other genera were stimulated. Fiscus ultimately wanted to know: "What is life?"

That led to his ecosystemic life hypothesis: "an autotroph functionally integrated with a heterotroph—is the original, general and self-perpetrating unit of life" (Fiscus 2001:249). He viewed cells and organisms as" subunits of life that cannot persist in isolation." He hypothesized that "the iteration of simpler ecosystemic processes may have preceded and generated much more complex cells and organisms." The details of this concept he developed from ideas from Lotka, Odum, Ulanowicz, and others (Figure 6.6). His bibliography cited R. E. Ulanowicz, *Ecology, the Ascendant Perspective* (1997). Lotka undoubtedly referred to his 1925 book, and reference to Odum without a first name could mean Eugene Odum, who he mentioned on p. 250. However, in Part II (Fiscus 2002a:94–96), it became clear that he meant H. Odum. In Part II, he also linked concepts from Lotka (1925), H. Odum (1970), Ulanowicz (1997), and R. Rosen, *Life Itself* (1991). Fiscus summarized (2001:96):

> I see Rosen's metabolism-repair model as dynamically analogous to Lotka's coupled transformers, Odum's origin via production and consumption, and Ulanowicz's growth and development of ecosystems. The ecosystemic life hypothesis exists at the place of congruence of these concepts.

However, he realized he needed to explain "The Hypothesis and Its Implications," which he did in Part III (2002b). He envisioned

> *two functional processes*: a molecular string *composer* (proto-autotroph) and a molecular string *decomposer* (proto-heterotroph). The composer function is to harness energy (solar, chemical, or thermal) to form chemical bonds and build molecular strings. The decomposer function is to harness energy in chemical bonds made by the composer, and to break molecular strings into component parts. The composer also stores energy in molecules, and the decomposer also dissipates some of this stored energy.

Fiscus concluded, therefore, that "The ecosystem is thus a plausible intermediate dynamical realm between physics/chemistry and cellular/organismic life" (2002b:147). He then explained how to falsify his hypothesis.

Theory and introspection are clearly two favorite intellectual preoccupations among ecologists. After Fiscus' discussion of theory came an introspection by Robert Swihart, John Dunning, and Peter Wieser, who asked (2002:149): "Has ecology stagnated as a scientific discipline?" They cite a commentary by R. J. O'Connor, "Why Ecology Lags behind Biology" (*The Scientist* 14(20), 2000, 35), who thought slow ecological progress was because ecologists asked "what" questions rather than "how" or "why" questions. Is this just a rehash of Peters (1991)? They gave no indication of having discovered his book of complaints. "O'Connor (2000) claimed that ecology lags behind successful life sciences because ecologists have remained fixated on descriptions of patterns." Swihart et al. argue that the ultimate goal of science is to understand the processes underlying observed patterns. To test the progress of ecology, they examined a random 20 articles published every decade in *Ecology* and rated them on a scale of 0 to 5, 0 being purely descriptive and 5 being purely deductive process theory and constructed a graph of their findings. They urged ecologists to improve the training of ecological students (Figure 6.7).

Six papers on ecological paradigms, with a brief introduction, appeared in *Ecology* in June 2002 (Figure 6.8). They attest to ecologists taking seriously Thomas Kuhn's account of scientific paradigms and revolutions in *The Structure of Scientific Revolutions* (edition 2, 1970). Many ecologists clearly want ways to measure theoretical progress, and these papers illustrate that quest. Michael Graham and Paul Dayton's paper was "On the Evolution of Ecological Ideas: Paradigms and Scientific Progress" (2002). They cited Robert MacArthur's broken-stick model (1957) as

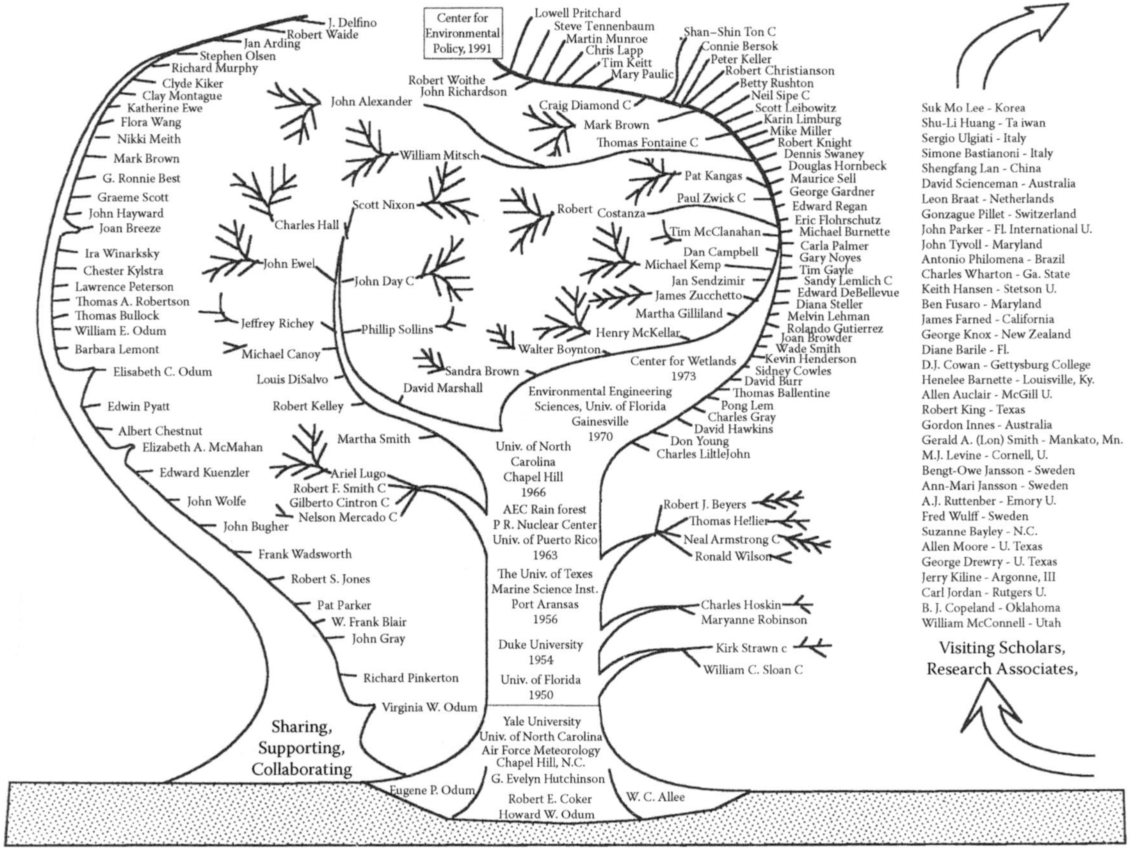

FIGURE 6.6 H. T. Odum's students and research associates. (From Hall 1995: Frontispiece. With permission.)

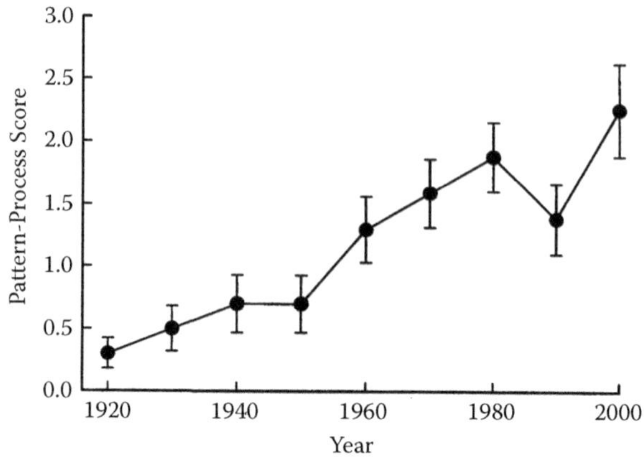

FIGURE 6.7 Mean pattern-process scores for 20 articles published in *Ecology* in every decade, from 1920 to 2000. (From Swihart et al. 2002:151. With permission.)

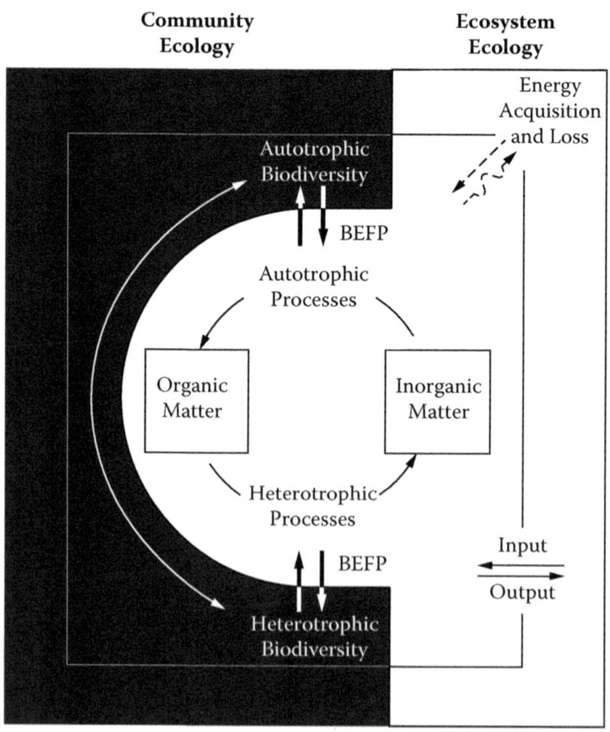

FIGURE 6.8 Synthesizing community and ecosystem ecology through the biodiversity-ecosystem function paradigm. (From Naeem 2002:1547. With permission.)

an example of a concept that quickly gained supporters, only to be demolished by conflicting empirical data (Preston 1962). Unfortunately, there are few such examples of simple ecological progress, so Graham and Dayton turned to other kinds of ecological progress, such as the growth in the number of ecological periodicals and pages. Robert Paine (2002:1557), author of the sixth paper, concluded that "because ecology in the broadest sense is a pluralistic, multiple-causation discipline, revolutions are unlikely." Paradigms, yes, revolutions, no.

The *Bulletin* appeared in both print and online in 2001 and 2002, and after 2002 only online. The members were no longer sent the paper annual meeting program and abstracts. Ed Johnson (2005), who succeeded Allen Solomon as *Bulletin* editor, praised Solomon for initiating that transition, because it allowed the *Bulletin* to be significantly more flexible. The new paper journal, *Frontiers in Ecology and the Environment*, launched in 2003, appearing 10 times a year and sent to all members, carries ecological and environmental news, but not ESA news, which remains the domain of the *Bulletin*. The main focus of *Frontiers* is synthesis articles on ecological or environmental topics, with generous maps, diagrams, and photographs—in color, as needed.

The organizational ferment during 1996, discussed above, had included the modest establishment of a new program: Strategies for Education in Ecology, Development, and Sustainability (SEEDS), which was aimed at providing college students, especially minority students, an introduction to ESA. By 2003, SEEDS was successful enough to be described to the membership (Anon. 2003). ESA encouraged organization of campus ecology chapters, which received ESA grants each semester to conduct ecological activities, including field trips, travel funds for attending ESA meetings, and career development. SEEDS's motto is "Diverse People for a Diverse Science." Its continuing development can be followed in its online newsletter and at its website (esa.org/seeds).

Sociology graduate student Sara Grineski from Arizona State University, Tempe, published "Toward an Integrative Urban Ecology: Voices from the IGERTs" in the December 2003 issue of the *Bulletin*. Her motivation was an NSF decision to fund two Integrative Graduate Education Research and Training programs at Arizona State and at the University of Washington, Seattle. She surveyed participants in the two programs and found different perspectives among ecologists and sociologists about the scope of urban ecology. Ecologists started with the environment and fit humans into it, whereas sociologists saw urban ecology as a subset of human ecology and considered forces of production and consumption in a city and their influence on the environment. She suggested that "Blending these perspectives could prove key to theoretical innovation and scientific discovery" (Grineski 2003:200). Her essay also reminded ESA members that after ecologists lost interest in human ecology, sociologists had taken it up. *Urban Habitat* is an online journal (urbanhabitat.org) that began publication in December 2003. Urban ecology, the domain of sociologists, was not the same as human ecology, the domain of anthropologists.

In the January 2004 issue of the *Bulletin*, the Hubbard Brook Ecosystem Study leadership announced a new governance structure. Since this was "one of the longest running ecosystem research projects in the world" (Groffman et al. 2004:5) and was part of the Long Term Ecological Research (LTER) network, the reasons for reorganization had broad implications. As they entered the third decade of LTER, they addressed these questions:

> Who will maintain the integrity of the long-term experiments and data at the site? How will new scientists be brought into the program? How will new leaders be developed? How will a flow of new ideas and visions be maintained?

They thought that other LTER site leaders would also want to face these questions, and that those leaders might find some help in the way Groffman et al. answered them.

"Studies of resource partitioning and niche had their golden era in the 1970s to the 1980s, but fell in disgrace after being convicted of circular reasoning and not considering alternative hypotheses," so wrote André Cunha and Marcus Vierira (2004:25), Laboratório de Vertebrados, Departmento de Ecologia, Universidade Federal de Rio de Janeiro, Brazil. However, niche theory remained important. They briefly reviewed niche literature from Grinnell (1917) to Pianka (1999) and argued that available space is the most important niche factor and also the easiest to measure.

This chapter ends with Nancy Stamp's contribution to Ecology 101: "Misconceptions about Plant-Herbivore Interactions, Especially Plant Defenses" (2004). She discovered that there is widespread public ignorance about the fact that spices are actually toxins which plants manufacture as defenses against insect herbivores. This fact is ecologically significant but also has practical significance, since spices are often perceived to have medicinal values. If used in large doses or with certain other medications, spices can actually endanger human health. (An example of Boris Zeide's "Nature's Surprises.") In 2005, Stamp continued her lessons with "The Problem with the Messages of Plant–Herbivore Interactions in Ecology Textbooks."

7 2005 to 2015

A Sustainable Biosphere and the National Ecological Observatory Network (NEON)

Although this chapter covers only about 10 years, compared with 15 years for the other six chapters, documentation for this chapter from the *Bulletin* and *Frontiers* is exceedingly abundant—even when the subject of this chapter is restricted to a sustainable biosphere—since *Frontiers* contains 568 pages per year. The biosphere is a theoretical concept that has not been viewed as a political threat to any interest groups. Global warming is a theoretical concept that has been viewed as a political threat by air polluters who do not want to have their enterprises or profits threatened by laws to inhibit global warming. The ESA documentation of global warming and its consequences are within the scope of this history, and the political struggle over it is briefly mentioned and documented. Applied ecology achieved greater prominence during this period than before.

The University of Vienna Professor of Geology Eduard Suess (1831–1914) coined the term "biosphere" in his book *Die Entstehung der Alpen* (Vienna, 1875:159), and explained it in his more important *Das Antlitzder Erde* (3 volumes, Vienna, 1883–1909). Suess credited Lamarck and Darwin with the concept of the unity of life, and he thought the biosphere concept was implicit in their work. Suess only briefly explained in general terms his biosphere concept. The relevant passage is quoted from the French edition of this three-volume work by Jean-Marc Drouin (1993:84). Vernadsky (1998:91) acknowledged Suess' "biosphere" term. Vladimir Vernadsky (1863–1945) was the son of a liberal Moscow University professor and became one himself (Fedoseyev 1976; Grinevald 1987, 1998; Bailes 1990). He traveled and studied in Europe and also traveled in North America. Vernadsky has been credited with being first to recognize life as an important geological force, though that credit belongs to Lamarck (1964:85–106). However, Vernadsky did establish the science of biogeochemistry. His *Biosphere* is now in English (1997), with extensive annotations. His annotator, Mark McMenamin, correctly noted (Vernadsky 1997:91, note 172) that the biosphere concept is implicit in Lamarck's *Hydrogéologie* (1802, English 1964). One might assume, therefore, that Suess' reference to Lamarck was to *Hydrogéologie*, but that is unlikely. Lamarck published only a few copies of *Hydrogéologie* at his own expense, and it seems unlikely that Suess ever saw a copy, but maybe read Lamarck's biological works.

In 1896, the Swedish chemist–physicist Svante Arrhenius (1859–1927) argued that the ice ages were likely influenced by the concentration of CO_2 in the atmosphere, with a condensed version of his discussion translated into English, also in 1896 (reprinted in Snelders 1970:302; Weart 2008:5–8; McKibben 2012:19–30). Because CO_2 is a small percentage of atmospheric gases and the ocean absorbs CO_2, few, if any scientists paid attention, until 1938, when the English steam technology engineer Guy S. Callendar (1898–1964) compiled average temperatures for England and found that average temperatures had risen slightly since the 1890s (1938), and he defended Arrhenius' hypothesis (reprinted in Fleming 2007:65–87; Weart 2008:2–3; McKibben 2012:33–37). Callendar provided interesting but inconclusive evidence. Chemist Hans Suess announced in 1955 that the ancient carbon 14 isotope in the atmosphere came from burning fossil coal and oil (Weart 2008:25–26). Oceanographer Roger Revelle (1909–1991), director of the Scripps Institution of Oceanography, immediately hired Suess to assist him in studying carbon in the ocean (Weart 2008:26–28). Rather quickly, they investigated, then published "Carbon Dioxide Exchange between Atmosphere and Ocean and the Question of an Increase of Atmospheric CO_2 during the Past Decades" (Revelle & Suess 1957, reprinted in McKibben 2012:39–42). Their measurements and interpretation were not definitive either, and their paper was widely ignored, though Gilbert Plass was saying the same thing at the same time (1956a,b, 1959). Revelle was determined to obtain definitive data. And so, as Al Gore (1992:4–5) tells it, "Roger Revelle had, through sheer persistence, convinced the world scientific community to include as part of the International Geophysical Year (1957–1958) his plan for regularly sampling CO_2 concentrations in the atmosphere." Revelle's colleague Charles David Keeling (1928–2005) actually took measurements from the top of the Mauna Loa volcano in Hawaii, from April 1958 for the rest of his life. In 1964, Harvard created an endowed chair for Revelle, and in 1965, Revelle informed the President's Science Advisory Committee about increasing CO_2 and its effect on global warming. President Lyndon Johnson explained this in a message to Congress (Oreskes & Conway 2010:170–171).

The first to alert the public of the dangers of global warming was author Bill McKibben, *The End of Nature* (1989), and in 2008 he added activism to his writing and speaking by founding 350.org (McKibben 2013). Gore had learned of global warming in an undergraduate course he had taken at Harvard under Revelle, and Gore's *An Inconvenient Truth* film and book (2006:38–41) featured "A scientific hero: Roger Revelle." In 1992, conservative columnist George Will in the *Washington Post* claimed that Gore had exaggerated Revelle's concern over global warming, but Revelle's widow and children wrote a letter stating that it was Will, not Gore, who had exaggerated Revelle's conclusions (Ehrlich & Ehrlich 1996:193–194) (Figure 7.1).

James Edward Hansen (b. 1941) is an Iowan who obtained three degrees in physical sciences from the University of Iowa, and after obtaining a Ph.D. in 1967, joined the Goddard Institute for Space Studies. He became interested in anthropogenic changes in the earth's atmosphere. In 1981, Hansen et al. from Goddard published a paper on global temperatures from 1880 to 1980: "Climate Impact of Increasing

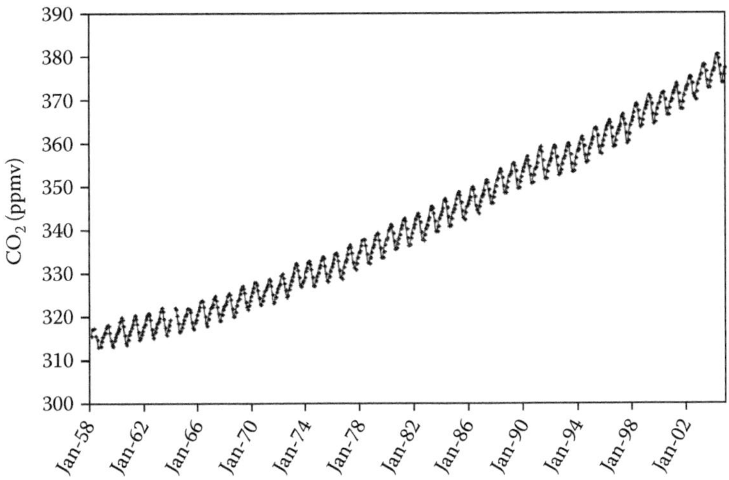

FIGURE 7.1 CO_2 concentrations in the atmosphere at Mauna Loa Observatory, Hawaii, which Keeling began measuring in April 1958. This chart (for the years 1958–2004) is from Lovett et al. (2007:256); versions of it also appear in Al Gore's books (1992, 2006) and in McKibben's *Global Warming Reader* (2012:44–45).

Atmospheric Carbon Dioxide." In August 1988, he testified before a U.S. House committeee that "The greenhouse effect is real, it is coming soon, and it will have major effects on all peoples" (quoted from Oreskes & Conway 2010:128; Hansen's testimony is reprinted in McKibben 2012:47–53). In 1989, when Senator Gore invited him to testify before a U.S. Senate committee, the Bush White House tried to censor his claims to make them less positive (Gore 1992:176; Bowen 2008). In 2006, Gore consulted with him on his *An Inconvenient Truth* project (film and book). Hansen advised him, but felt that Gore, as a Democratic politician, could never convince right-wing skeptics of global warming (Pooley 2010:253). Hansen also wrote his own book with the same message: *Storms of My Grandchildren: The Truth about the Coming Climate Catastrophe and Our Last Chance to Save Humanity* (2009). For his persistent publicizing of the dangers of global warming and the science supporting his conclusions, the Nation Institute and Fertel Foundation awarded Hansen the Ridenhour Prize in 2013 (Romm & Hansen 2013).

In the United States, there are still determined global warming skeptics (Jaques, Dunlap, & Freeman 2008; McCright & Dunlap 2010). Physician–novelist Michael Crichton published *State of Fear* (2004), which expressed his skepticism of global warming (excerpted in McKibben 2012:193–207); Joseph Mascaro criticized Crichton's novel in a *Frontiers* editorial, "State of Ignorance," but in 2005 not all academics were global warming believers. In a June *Frontiers* letter, Peter Geddes (2005a), Foundation for Research on Economics and the Environment, Bozeman, Montana, warned of "The Dangers of Scientific Consensus," with his example being global warming: "We should resist being stampeded into public policies with huge costs and few, if any, benefits." In her "Finishing Lines" column, in the May issue of *Frontiers*, Katherine Ellison may have given some support to Geddes' argument by reporting that new evidence indicated that the earth was dimming due to pollution mixing with clouds, which could diminish sun ray penetration. However, she also reported that predicted increases in the global temperatures of oceans were being confirmed—a confused picture.

Three letters in the August *Frontiers* issue attacked Geddes' letter. Val H. Smith, from the Department of Ecology and Evolutionary Biology at the University of Kansas, accused Geddes of using "selected snippets from the history of science to call into question our scientific knowledge of global warming, and to raise doubts about the need for immediate action" (Smith 2005:469-470). Michael Marsh, Seattle, Washington, did not think that Geddes' letter should have been published in *Frontiers*, because it cited no scientific evidence supporting his argument. Bob Kinzie, Department of Zoology, University of Hawaii, Honolulu, argued that "It is simply disingenuous to posit that as long as there are any dissenting opinions the question is still open…" (Smith, Marsh, & Kinzie 2005:302).

Geddes did not fall silent. In the November *Frontiers* issue (2005b:469) he claimed: "Whether climate change is real, however caused, and potentially dangerous is not the issue I was raising. Rather, the important policy question is: In a world of limited resources, how do we deal responsibly with climate change?" None of his skeptics had detected this question in his previous letter. Well, try this one: "Might we increase the well-being of the world's poorest more by eradicating malaria, provide access to clean drinking water, reducing infant mortality, increasing female literacy, and increasing access to primary education?" And furthermore, the Climate Stewardship Act was defeated in Congress because it is difficult to persuade politicians to make sacrifices now for the benefit of future generations. Smith responded that both of Geddes' letters were recycled from his newspaper columns, and that both letters used red herrings to make his points. There is no indication whether this particular Peter Geddes had trained as a scientist, but Naomi Oreskes and Eric M. Conway found enough scientists of similar ilk to fill a book: *Merchants of Doubt: How a Handful of Scientists Obscured the Truth on Issues from Tobacco Smoke to Global Warming* (2010); Geddes' name is not indexed.

Since the year 2000, numerous books for the public have been published on global warming, some by ecologists (Johansen 2010). Mostly these books take a global view, but Joseph Di Mento and Pamela Doughman edited a collection of reports on *Climate Change: What It Means for Us, Our Children, and Our Grandchildren* (2007), though their paper with the same title is the only one of eight papers in the book with that theme. Ecologist Amy Seidl brought the story even closer to home in *Early Spring: An Ecologist and Her Children Wake to a Warming World* (2009), in which she describes the actual changes attributable to global warming for plants, animals, and people in her surroundings and beyond. In a sequel, *Finding Higher Ground: Adaptation in the Age of Warming* (2011), she announces that global warming is here and we should find the best means of adapting, and that plant and animal species will either adapt or disappear.

The February 2005 *Frontiers* issue had seven articles on sustainability, introduced by an editorial, "Ecology and the Transition to Sustainability," by ESA president

Jerry Melillo, president-elect Nancy Grimm, and past president William Schlesinger (Melillo 2005). They reminded readers that ESA had begun its Sustainable Biosphere Initiative in 1991 and had success in stimulating new research and in advancing ecological knowledge in the service of society. However, achieving sustainability in a changing world, with new pressures coming from increasing human populations that change the world as it draws upon natural resources, is a very complex challenge.

Margaret Palmer and 19 coauthors set out a plan to ensure "Ecological Science and Sustainability for the 21st Century" (2005). Since most oceanic fish stocks are overfished, Rebecca Goldburg and Rosamond Naylor discussed "Future Seascapes, Fishing, and Fish Farming" (2005). Most fish raised on oceanic fish farms are predatory, and such farms merely change the fishing pressure to other, mainly herbivorous, species. Fish farms also can cause problems if domestic fish escape and mate with wild stocks, and even worse is aquacultural pollution from metabolic waste.

Philip Robertson and Scott Swinton addressed "Reconciling Agricultural Productivity and Environmental Integrity" (2005). Their goal of managing agriculture for multiple services in addition to food and fiber required "integrative research both ecological and socioeconomic as well as policy innovation and public education." Supplementing their concerns was a discussion by Laura Myerson et al., "Aggregate Measures of Ecosystem Services" (2005). They recommended quantification studies of ecosystem services and dissemination of this information to both decision makers and the general public for the sake of responsible management of natural resources.

Kim Cuddington from Ohio University, and Beatrix Beisner from the University of Quebec at Montreal, were two "bright-eyed postdocs" who organized a symposium for the Tucson 2002 ESA Annual Meeting on ecological paradigms and theory change. Later, they collected and edited the 16 revised papers by 21 authors from international locations into *Ecological Paradigms Lost: Routes of Theory Change* (2005a). By then, the dozen articles on "Conceptual Issues in Ecology" that had appeared in *Synthese* (1980), including Simberloff's "A Succession of Paradigms," were ancient history. *Paradigms Lost* is another collection of historical perspectives in quest of the holy grail of ecology: unified theory. Both the 1980 and 2005 papers clarified the thinking of participating ecologists and others. However, Cuddington and Beisner's conclusion for the volume was merely: "the answer to the question of how theory changes in ecology seems to be that it changes in a variety of ways" (2005b). These 16 papers and that conclusion confirmed McIntosh's "Pluralism in Ecology" (1987).

Should hunting be an ethical issue for ecologists? Talks from a forum on "The Ethics of Hunting" appeared in the September 2005 issue of *Frontiers*, consisting of an introduction by a philosopher, Roger King from the University of Maine and five ecologists discussing hunting within the sphere of their work (King et al. 2005). King presented an evenhanded account of both pros and cons: (1) animals have rights, and the pleasure of hunting does not justify ending those rights, yet in some cases human needs outweigh animal rights; (2) hunting can harm a species or biotic community, yet managing species can require population control; and (3) hunting represents a dominating attitude toward nature, yet hunting can protect against animal overpopulation, and predation is part of nature.

Animal behaviorist Patrick Bateson of Cambridge University, evaluated hunting deer with dogs for the English government and concluded that using dogs was not the most humane method (2005). Heather E. Evans and Christine M. Wolf, "The Bushmeat Crisis Task Force in Africa" (2005), reported that natives hunting for food is not controversial, but they also hunt to sell bushmeat in cities for profit, which is controversial. Fred Nelson of the Sand County Foundation Community Based Conservation Network, also reported on a similar situation in Africa (2005). Ted Kerasote, Kelly, Wyoming, defended his elk hunting as consuming less energy than buying food from stores (2005). Scott Yaich, Director of the Ducks Unlimited Conservation Program, pointed out that hunters have played a crucial role in wildlife conservation, and that the purchase of hunting and fishing licenses provides 65% of the budget of state wildlife conservation programs (2005). If there were ecologists who defended animal rights per se, they were not part of this forum.

"Although there has been scientific consensus on the primary drivers of anthropogenic environmental degradation for well over a decade, little progress has been made in determining the precise relationship between drivers and impacts" (Dietz, Rosa, & York 2007:13). Three who decided to bridge this gap were ecologist Thomas Dietz of Michigan State University, and sociologists Eugene Rosa of Washington State University and Richard York of the University of Oregon. They felt that "This gap constitutes a significant barrier to identifying the policies that have the most potential for reducing human impact on the environment, projecting future impacts, and estimating the level of effort needed to reduce adverse effects on the environment" (2007:13). Data to determine an ecological footprint (EF) per country and future projects were already available, which could be used in a STIRPAT model (stochastic impacts by regression on population, affluence, and technology). Their main finding was "that population size and affluence are the principal drivers of anthropogenic environmental stressors, while other widely postulated drivers (e.g., urbanization, economic structure, age distribution) have little effect" (Dietz et al. 2007:13). However, affluence could be mitigated as a factor if affluent countries invested in significant pollution controls. They developed a map indicating the extent of national ecological footprints, showing that the countries with the most extensive and intensive ecological footprints are the United States and Russia. However, the map was somewhat misleading, because a chart showing the footprints of the 20 countries with the heaviest footprints in 2001 and their projected increase by 2015 indicated that the heaviest footprint was the United States and the second was China, and that in 2015, both were expected to maintain their positions, while Russia was only seventh in magnitude and was the only country projected to have less of a footprint in 2015 than in 2001, as it is probably the only country losing population.

The Long Term Ecological Research Program (LTER) had begun in 1980 at six sites, four of which had been International Biological Program (IBP) sites. In 25 years it grew to include 26 sites (including one in Puerto Rico and two in Antarctica), with connections to an international network, and LTER published a celebratory pamphlet that explained its research goals. It is still going strong, and it is fair to ask the questions posed concerning the IBP: has it been worth the trouble and expense? David Coleman (2010) has provided a history of both the IBP and LTER, including a sketch on the leadership contributions of Tom Callahan at the National Science Foundation (NSF), who supervised, first the IBP, and then LTER (Coleman 2010:101–104). Coleman's 13-page account of major LTER research findings was illustrative but not exhaustive (2010:108–121). He based his discussion on scientific papers that ecologists at different LTER sites had published.

Coleman's LTER summary discussed the following: Biodiversity/Productivity Relationships, Patterns of Invasive Species in Ecosystems, Ecosystem Variance, Implications of Long-Term Climate Trends, Effects of Large-Scale Climate Drivers, Watershed Ecosystem Analysis, and Integrative Long-Term Experiments. The NSF was quite interested in intersite synthesis; an example that Coleman discussed is *Long-Term Dynamics of Lakes in the Landscape* (Magnuson, Kratz, & Benson 2006). Coleman ended with a comparison of the achievements of the IBP and LTER: "Because of the wider time spans covered by the LTER volumes in most cases, the caliber and extent of synthesis has been considerably higher" (Coleman 2010:130). An example of expanding LTER research to an international scope was an international meeting that the NSF funded, which Coleman arranged for October 7 to 15, 1988, on "Dynamics of Tropical Soil Organic Matter," held in Maui, Hawaii. He had to convince NSF officials that this was not a vacation junket, and the volume assembled from the conference was tangible proof (Coleman, Oades, & Uehara 1989).

Environmental biologists at the NSF decided to hold a contest in 1995 to choose a location for a center for analysis and synthesis. It was established in 1996 at the University of California, Santa Barbara (Coleman 2010:142–144). It was named the National Center for Ecological Analyses and Syntheses (NCEAS), and an announcement about its establishment appeared in *Science* (Baskin 1997). Both senior scientists on sabbatical and postdoctoral ecologists could apply to participate in the Center for a research period.

The National Ecological Observatory Network (NEON) is another ecological monitoring program (Coleman 2010:153–164). Bruce Hayden developed the plan for NEON after he became the head of the Biological Sciences Division of the NSF. It involved establishing a network of environmental sensors, coordinated nationwide. The NSF arranged workshops around the United States from 2000 to 2002, which Scott Collins chaired, with about 1,300 participants. Later, a NEON committee of 14 ecologists, David Tilman, chairman, developed six major environmental challenges which NEON should address (Tilman et al. 2004:24–25): Biodiversity, species composition, and ecosystem functioning; Ecological aspects of biogeochemical cycle; Ecological implications of climate change; Ecology and evolution of infectious diseases; Invasive species; and Land use and habitat alteration. In the September 2005 issue of *Frontiers*, past ESA president Jerry Melillo, president Nancy Grimm, and president-elect Alan Covich published an editorial, "NEON: Lighting the Way Forward," explaining that NEON would "document and forecast changes in ecology, at both regional and continental scales, caused by climate and land use change" (Melillo, Gimm, & Covich 2005:351), and that ESA members should write their congressional representatives to support it.

G. David Tilman (1949–)

(From the College of Ecology, Evolution, and Behavior—University of Minnesota.)

Illinoian Tilman earned a B.S. degree (1971) in zoology and a Ph.D. (1976) in ecology from the University of Michigan (Davis 2004). He joined the University of Minnesota faculty in 1975, and in 2002 he became Regents' Professor and Distinguished McKnight University Professor in the Department of Ecology, Evolution and Behavior. Tilman is a Fellow of AAAS and American Academy of Arts and Sciences and a member of the National Academy of Sciences. He edited *Limnology and Oceanography* (1978–1981) and was editor-in-chief of ESA's *Issues in Ecology* from its beginning in the spring of 1997 through issue 11 in 2003. In 1989, he won ESA's William S. Cooper Award for his second book (1998), *Plant Strategies and the Dynamics and Structure of Plant Communities* (Peet et al. 1990). In 1996, he won the Robert H. MacArthur Award for his focus "on broad questions related to biodiversity, including interspecific competition, succession, community composition, and ecosystem functioning" (Wiegert et al. 1997). In 2008, he won the International Prize for Biology presented by the Japan Society for the Promotion of Science. In 2010, he won the Heineken Prize for Environmental Science, awarded by the Royal Netherlands Academy of Arts and Sciences.

A progress report by David Schimel, William Hargrove, Forrest Hoffman, and James McMahon appeared as a guest editorial in *Frontiers* (March 2007). Schimel was in charge of its Boulder, Colorado, headquarters. A website (neoninc.org) explained that "NEON, Inc. is an independent 501 (c) (3) corporation created to manage large-scale ecological observing systems and experiments on behalf of the scientific community. NEON, Inc. has been tasked with planning and managing the NEON project." The editorial explained that there were 16 domains within the

48 contiguous states, one each for Puerto Rico and Hawaii, and two in Alaska. At the 2008 ESA Annual Meeting, a hand-out map showed locations for the 20 permanent stations, and there was a "Town Hall Meeting" about NEON at that meeting. (A map and a list of core sites is also in Coleman 2010:156–157.) A newer editorial in *Frontiers* by Michael Keller, Schimel, Hargrove, and Hoffman (2008) explained that "NEON differs from other environmental monitoring networks because, by design, it integrates processes, interactions, and responses." Each domain has four packages of equipment to measure different kinds of environmental variables, and measurements are made at core sites, relocatable sites, and mobile sites. There is also a Stream Experimental and Observational Network (STREON) that manipulates nutrients in small streams and a Global Change Experiment (GCE) that manipulates CO_2 concentrations and terrestrial temperatures. A cyberinfrastructure (CI) organizes data from the four science packages into information for forecasting change. Information from these processes will be available to professional societies, government agencies, educational institutions, and community organizations. The expectation is to conduct NEON for 30 years, so that young scientists and staff can be recruited.

There are multitudes of ecological research groups and organizations, and two with an international scope deserve notice: FLUXNET and the Terrestrial Ecosystem Response to Atmospheric and Climatic Change (TERACC). FLUXNET coordinates over 140 micrometeorological sites worldwide that measure CO_2, water vapor, and energy between the biosphere and atmosphere. Coleman (2010:152) summarized some of its findings and assessed prospects of its continuing success as being very bright. TERACC was established in 2001, includes 135 sites in 25 countries, and uses experiments and models to understand ecosystems and global change (Chapin et al. 2008; Coleman 2010:175).

Lawrence B(asil) Slobodkin (1928–2009)

(From Colwell & Futuyma 2011:19. With permission.)

"A child of the Bronx, New York City, Slobodkin was strongly influenced by the artistic, intellectual, cultural, and political milieu in which he developed; his mother was a writer and his father a noted sculptor" (Colwell & Futuyma 2011:19). That cultural environment included the Bronx Zoo and the Bronx High School of Science (Slobodkin 2009:330). At age nine, he kept a menagerie: canary, rabbit, tortoise, horned toad, newts, baby turtles, and tropical fish. He also had a child's microscope with which he observed rotifers and ciliates. At age 11, he decided to become a protozoologist. When almost 16, he went to a Christian college, Bethany, in West Virginia, which he enjoyed for its scenery and his opportunity for independent study. He earned a B.S. degree there (1947), then went to Yale, where he earned a Ph.D. under Hutchinson at the age of 23 (1951). "His doctoral research, a detailed study of the role of age structure in the growth of experimental populations of the microcrustacean *Daphina* epitomized his approach—a quantitative experimental test of a mathematical theory that was intended to apply broadly" (Colwell & Futuyma 2011:19). He worked two years for the U.S. Fish and Wildlife Service on the cause of the red tide. "My Fish and Wildlife Service supervisors were furious at my claiming to have solved the 'Mystery of the Red Tide.' They had an institutional commitment to working on the 'Mystery of the Red Tide,' but not to its solution" (Slobodkin 2009:335). In 1953, he went to the Zoology Department, University of Michigan as a research associate and stayed to become a full professor. He published his first book there, *Growth and Regulation of Animal Populations* (1961), drawing upon his extensive experiments with hydra. In 1968 he settled at the State University of New York, Stony Brook (now Stony Brook University) as chairman of the Program (now Department) of Ecology and Evolution. He became ESA's Eminent Ecologist for 2005. Like his own mentor, Hutchinson, he was an encouraging, successful mentor to graduate students at both Ann Arbor and Stony Brook (Slobodkin 2009:347–349; Colwell & Futuyma 2011:21–22). Students he had mentored and some colleagues wrote a festschrift in an issue of *Evolutionary Ecology Research* (2009). His impressive bibliography is published by Colwell and Futuyma (2011:23–32). He was the Eminent Ecologist of ESA in 2005.

The ESA has included Canada and the United States from the beginning; both countries have a common language and rather similar traditions and societies. The ESA welcomes Latin American members; however, their traditions, language, and societies have been less similar to those of Canada and the United States, and ecologists have emerged in Latin America more slowly than among their English-speaking northern neighbors. Puerto Rico is a special case, since it became an American possession after the Spanish–American War of 1898. That fact encouraged Americans to take an interest in its biological resources. The New York Botanical Garden sent a collector there in 1899, assisted by the U.S. Armed Forces (Kingsland 2005:79). English-language ecological publications began to appear by 1911 and are mentioned in Herminio Lugo's report on Puerto Rico's ecological studies (1981). Ismael

Vélez first taught an ecology course there in 1935, at the Polytechnic Institute (now the Inter-American University).

In the June 2006 issue of *Frontiers*, 10 authors collaborated on a report, "The Evolution of Ecology in Mexico: Facing Challenges and Preparing for the Future" (Martínez et al. 2006). They drew upon José Sarukhán's "Mexico" (1981), and their report includes information on ecology in other Latin American countries. Ecology began in Mexico with the establishment of the National Commission for Study of Dioscoreas in 1960, supported by the Forestry Ministry and pharmacological industry that used yams to synthesize cortisone and corticosteroids. This Commission studied yam ecology and habitat in tropical forests and yam regenerative capacity in forests undergoing secondary succession. These studies provided the first long-term funding for population and community ecology research. The Commission also provided funds to train students in ecology.

Other Mexican institutes followed: the National Institute for Investigation of Biotic Resources, the Institute for Ecology, the Center of Ecology at the National University, and five biological stations, all with regular budgets. Subsequently, other institutions and graduate programs increased to 18, more than in any other Latin American country (Figure 7.2). Despite regular budgets, limited funding is a major obstacle to further development of ecology in Mexico and other Latin American countries. In 1980, Mexico had only five Ph.D. ecologists; in 2006 it had about 350, and adds about 10 more each year. Latin America has only one scientist per 5,000 inhabitants (except for Cuba, which has more), which is 10 times lower than in industrialized countries. Cuba emphasizes applied ecology, focused on agronomy, fisheries, biomedicine, and biotechnology. Mexico now has a Consejo National de Ciencia y Technologia and a Sistema National de Investigadores, which provide funding to encourage scientists to publish papers in international peer-reviewed journals and to maintain high standards in graduate programs and also award scholarships to promising students. In 2006, Latin America published only 5.5% of the world's ecological papers.

Martínez and nine coauthors were correct about the history of formal ecology in Mexico, but if one asks about the history of Mexican natural history and nature conservation, Mexican ecology can be seen as having deeper roots. Professor Enrique Beltrán (1903–1994) received a Ph.D. in protozoology from Columbia University in 1952, taught at the Universidad de Mexico, headed the Department of Protozoology at Mexico's Institute of Health and Tropical Diseases, and also served as Secretario Perpetuo de la Sociedad Mexicana de Historia Natural (and later became its Presidente Honorario). He had an interest in the history of biology and of natural history, and he sent me, during some years, 20 reprints of his articles on these topics (1948–1974), and I responded with my reprints on the history of ecology. One of Beltrán's articles (1955) is on La Sociedad Mexicana de Historia Natural conferring honorary membership upon Paul Sears, who went to Mexico City for the occasion. Another (Beltrán 1956) is a 20-year history of that society. His role in Mexican nature conservation is even more impressive—one of the three leaders of his time—as detailed by Lane Simonian's history, including over a page of Beltrán's relevant publications (1995:132–140, 279–281). It seems relevant to also mention here American Starker Leopold's "Vegetation Zones of Mexico" (*Ecology* 31, 1950, 507–518).

Panel 3. Scientific Societies in Latin America Fostering Interactions among Ecologists in the Region

At least eleven professional botanical, zoological, biological, and ecological societies and associations based in Latin America foster interactions between local ecologists and colleagues from related disciplines (see below). The Red Latinoamericana de Botánica (www.uchile.cl/rlb/) plays an important role in promoting student and teacher exchanges among countries in the region. Latin American ecologists also belong to international societies such as the Ecological Society of America, Association for Tropical Biology and Conservation, Society for Conservation Biology, British Ecological Society, and Society for Ecological Restoration. In addition, Latin American societies produce 11 journals with national/regional distribution. However, only five of these (*Revista Chilena de Historia Natural*, *Acta Zoológica Mexicana*, *Acta Botánica Mexicana*, and *Revista Mexicana de la Biodiversidad*) are internationally indexed. Papers by Latin American ecologists can also be found in regional science journals, such as *Interciencia*, and the journals of the aforementioned societies and other professional organizations. In September 2005, the Mexican Society of Ecology was founded. Currently, it has nearly 200 members.

Largest Scientific Societies Including Ecologists Based in Latin America

Organization	Year Founded	Members/Meeting Attendance	Web Page
Sociedad Biológica de Chile	1928	500/na	www.biologiachile.cl/
Sociedad Botánica de México	1941	1,200/2,000	www.socbot.org.mx/
Sociedade Botânica do Brasil	1950	na/na	www.botanica.org.br/
Asociación Latinoamericana de Botánica	1986	na/1,250	www.botanica-alb.org/
Sociedad Mesoamericana para la Biología y la Conservación	1996	768/na	www.socmesoamericana.org/ev.php

na = data not available.

FIGURE 7.2 Panel 3 on Scientific Societies in Latin America. (From Martínez et al. 2006:266. With permission.)

In the early 1980s, an environmental justice (EJ) movement arose, inspired by civil rights and feminist movements (Middendorf & Nilon 2007:161). Initially, it challenged the placement of waste dumps in minority communities. This new movement demanded that "affected communities should be apprised of environmental issues affecting them," so that remediation may occur. For the 2000 ESA meeting, George Middendorf from Howard University, discussed ESA's relationship to EJ and three others (named) discussed aspects of EJ. At the 2002 ESA Annual Meeting, Middendorf and 13 others discussed aspects of EJ; and in 2006, Middendorf, Charlie Nilon from the University of Missouri–Columbia, and Leanne Jablonski from the Marianist Environmental Education Center, Dayton, Ohio, organized an ESA symposium on EJ, and the *Bulletin* published the nine symposium papers by 20 authors (88 [2007] 160–203). Topics discussed included EJ and ecological theory, EJ and tropical conservation, exploitation of natural resources and people who extract them, restoring justice/restoring ecosystems, and developing an ESA agenda on EJ.

The May 2007 issue of *Frontiers* was on ecology in an era of globalization, with four articles and workshop reports. First, an American and a Mexican ecologist collaborated on "A Strategy for Ecology in an Era of Globalization;" Jeffrey Herrick from the USDA Jornada Experimental Range, Las Cruces, New Mexico and José Sarukhán from the Instituto de Ecología, Universidad Nacional Autónoma de México, argued that economic globalization "can lead to increases in the rate and scale of environmental degradation," often in places having few (if any) professional ecologists who could develop and guide sustainable management systems (Herrick & Sarukhán 2007:172). Their goal was to explore how ecologists' relevance could increase from local to global levels. They discussed three case histories of nonforested ecosystem transformations associated with globalizations: (1) After the U.S. Civil War, veterans established ranches in the Southwest for food markets in the East and in Europe. High profits led to raising more cattle than the land could sustain. Land degradation led to establishment of experiment stations and development of rangeland ecology—a good response, though reactive. The Millennium Ecosystem Assessment (2005) was proactive, but globalization changes the world faster than ecologists can respond. (2) Brazil cultivated over 8 million ha of savannas (1980–1995) with forces of change similar to those in the United States 125 years ago. Brazil can grow soybeans profitably because of low labor costs and an expanded transportation system. This leads to increased emigration to the Amazonian frontier. (3) In Zacatecas, Mexico, in the early 2000s, marginally productive croplands were being abandoned by workers seeking higher wages in Mexican and U.S. cities; both sets of cities flourished after a reduction in trade barriers. This created opportunities for ecosystem recovery, but only if ecologists could influence land use strategies.

Herrick and Sarukhán thought all three cases were examples of ecological opportunities that have the potential to slip away because ecologists are not prepared to respond quickly. They urged for the development of an Ecological Knowledge System (EKS) to "facilitate the local and global dissemination and interpretation of ecological information" (2007:175). Their EKS included three tools: (1) an ecological information acquisition tool; (2) an ecological knowledge tool; and (3) a synthesis, interpretation, and application tool. The difference between the information and knowledge tools is illustrated in a diagram of the EKS (Figure 7.3). Global problems often require collaboration among individuals of diverse professions, drawing upon equally diverse databases that exist online, in libraries, and elsewhere. An EKS can help assemble these resources quickly enough to be of practical use.

Another *Frontiers* article, by Elena Bennett from the McGill School of Environment, McGill University, Montreal and Patricia Balvanera from the Centro de Investigaciones en Ecosistemas, Universidad Nacional Autónoma de México, Morelia, addressed "The Future of Production Systems in a Globalized World." To meet the demands for food and fiber, humans have rapidly changed ecosystems, which "radically alter provision of other ecosystem services, both now and in the future, usually by diminishing them" (2007:191). Deforestation and agriculture have

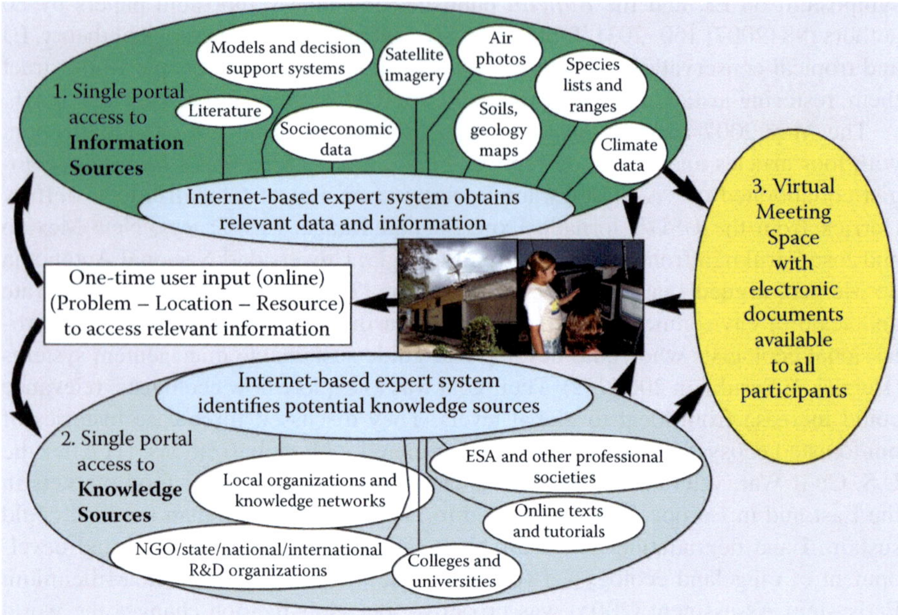

FIGURE 7.3 Three components of an Ecological Knowledge System, with sample elements for each. (From Herrick & Sarukhán 2007:176. With permission.)

increased rates of species extinction 1,000 times higher than in the fossil record. Fertilizer and manure runoff have reduced productivity of aquatic ecosystems and created hypoxic zones in many estuaries. Increasing demands for products from land and aquatic environments leads to both intensification and extensification: efforts to produce more products from land/aquatic areas already in use and bringing more land/aquatic areas into use. However, such increased production can "lead to detrimental and often unexpected impacts on other ecosystem services" (Bennett & Balvanera 2007:193). From recent literature, they compiled a table of unintended results coming from increased productivity and provisioning services. Ecological research was needed to determine how to minimize the side effects that come from increased productivity (Figure 7.4).

Debra Peters from the USDA Jornada Experimental Range, Las Cruces, New Mexico and Osvaldo Sala from the Center for Environmental Studies, Brown University, Providence, Rhode Island, organized a workshop that included three other ecologists to explore the "Cascading Events in Linked Ecological and Socioeconomic Systems," which met at the ESA International Conference, Merida, Mexico, January 8 to 12, 2006, with the publication appearing in *Frontiers* in May 2007. They discussed four cascading events: (1) 1930s Dust Bowl, (2) 1880s fire history, (3) 1910s fire history, and (4) hurricanes. Peters and different coauthors had provided an orientation for this workshop in an article, "Cross-Scale Interactions, Nonlinearities, and Forecasting Catastrophic Events" (Peters et al. 2004). The workshop's unsurprising conclusion was that "Understanding cascading events can improve our ability to forecast future events in the presence of similar conditions or

Some Unintended Results of Efforts to Increase Productivity and Provisioning Services of Production Systems

Primary Scale of Action*	Unintended Effects of Efforts to Cope with Changes in Drivers of Production Systems (i.e., increased demand)	Action	References
G	Increased greenhouse emission	Deforestation and fertilizer use	Robertson et al., 2000
G/R	Decreased water availability for ecosystem processes and changes in water vapor flows	Crop irrigation	Gordon et al., 2005; Vorosmarty et al., 2006
G/R	Biodiversity loss: reductions in the number of species, number of populations, genetic diversity	Land-use change, overexploitation, species introductions, GMO introductions, dam construction	Bellon and Berthaud, 2006; Mace et al., 2006
G/R	Transformation of global and regional nutrient cycles	Fertilizer use	Bennett et al., 2001
G/R	Salinization of irrigated lands	Crop irrigation in dry areas	Rengasamy, 2006
R	Regional climate change due to changes in surface energy and water balances	Changes in albedo due to land-use change and changes in water cycling	Foley et al., 2003
R	Desertification	Feedbacks between land use and water cycles, land-use change	Reynolds and Stafford-Smith, 2002
R/L	Soil erosion	Land-use change	Lavelle et al., 2005
L	Lake eutrophication and consequent effects on trophic linkages	Land-use change and fertilizer use, manure production	Vorosmarty et al., 2006
L	Reduced nutrient availability and fertility	Erosion, soil compaction, consequences of fire and machinery on nutrient retention mechanisms	Lavelle et al., 2005
L	Changes in species' relative abundances and dominant life-history traits with consequences for most ecosystem processes	Changes in disturbance regime, nutrient and water availability	Chapin et al., 2000

* G = global, G/R = global to regional, R = regional, R/L = regional to local, L = local.

FIGURE 7.4 Unintended results from increased productivity. (From Bennett & Balvanera 2007:195. With permission. Note that their bibliography provides full reference information on the literature cited here.)

behavior" (Peters et al. 2007:223), but the workshop also recommended two lines of new research: "(1) development of a general framework for cascading events that explicitly links ecological and socioeconomic-political systems; and (2) new experiments and predictive models to identify the conditions leading to cross-scale interactions and catastrophic events" (Peters et al. 2007:224).

A well-known factor affecting global warming is the extent of forests. Demands for timber and agricultural products are major causes of deforestation. Jonathan Foley from the Center for Sustainability and Global Environment, University of Wisconsin–Madison, and 10 coauthors (from several disciplines) published in *Frontiers* "Amazonia Revealed: Forest Degradation and Loss of Ecosystem Goods and Services in the Amazon Basin." One reason for their study was because "The loss and degradation of tropical rainforests and the services they provide may be greater and more widespread than previously reported" (Foley et al. 2007a:25). Locals defend clearing tropical forests on the grounds of the benefits to people living in the Amazon Basin. To protect forests from these takings, outsiders who might invest in saving forests needed to know of all the ecosystem services the forests provide. The short-term benefits of rain forest clearance needed to be evaluated against the loss of long-term "carbon storage in forests and soils, regulation of water balance and river flow, modulation of atmospheric circulation and regional climate, and the amelioration of infectious diseases" (Foley et al. 2007a:30).

Foley et al.'s article (2007a) provoked two similar multi-authored responses. First, Marielos Peña-Claros, Zulma Villegas, and Rudy Guzman from Santa Cruz, Bolivia and Lourens Poorter from Wageningen University, Netherlands, regretted that Foley et al. "do not make a clear distinction between forest management and unplanned logging, and seem to argue that selective logging is equivalent to deforestation" (Peña-Claros et al. 2007:237). The other letter, by Francis Putz and Jack Ewel, Department of Botany, University of Florida, Gainesville, and Ariel Lugo, USDA Forest Service, Puerto Rico, objected "That some of the logging detected from satellites is being carried out according to government-approved management plans is not even mentioned" (2007:237). Foley et al. responded (2007b:238) that "good forest management practices are the way forward, but this hasn't yet happened at the large, nearly continental scale that we aimed to address in our paper." When two different groups of ecologists lodge the same general complaint, it is fair to say that Foley et al. (2007a) had not made their argument as precise as it should have been, and the exchange in the June *Frontiers* issue emphasized how important all three groups of ecologists viewed the subject.

A guest editorial in the August 2007 *Frontiers* was by Professor Larry Li, Chair, EcoSummit International Scientific Committee, University of California, Riverside, and Professor Rusong Wang, Chair, EcoSummit 2007 Organizing Committee, Chinese Academy of Sciences, Beijing, China. They reported on the Third EcoSummit in Beijing in May. The first two had been organized by editors of three ecological journals published by Elsevier and held in 1996 in Copenhagen and in 2000 in Halifax, Canada. This Third EcoSummit was co-sponsored by the Ecological Society of China, Elsevier, and 29 other ecological organizations and attracted 1,400 participants from over 70 countries. Emphasis was on practical applications of ecological knowledge, and participants adopted the Beijing Ecological Declaration, "A World

Role for Ecology: The Key to Life," which included the statement: "Ecology can help us design, plan, manage, and protect our environment to ensure that we can all enjoy healthier lives and pass on a good environment for our grandchildren to appreciate."

The EcoSummit was a giant step in the right direction, but Katherine Ellison's "Finishing Lines" commentary ending the October 2007 *Frontiers* issue was on "Crisis and Opportunity in China." The crisis was that 300 million Chinese lack safe drinking water and China was building a new coal power plant every seven to ten days without pollution controls. The opportunity, of course, was to use ecological understanding to solve both problems.

Robert Treat Paine (1933–)

(From Root 1979:156. With permission.)

From Cambridge, Massachusetts, Paine earned a B.A. degree in zoology at Harvard University (1954) and M.S. and Ph.D. degrees from the University of Michigan (1959, 1961). At Michigan he studied under Nelson Hairston, Sr., and Frederick Smith, who published a notable theoretical paper (with Lawrence Slobodkin), "Community Structure, Population Control, and Competition" (1960). Paine's dissertation, chaired by Fred Smith, was on an obscure brachiopod, but simultaneously he explored a coastal waters mollusk community (Stolzenburg 2008:16–26). In 1961, he went to the Scripps Institute of Oceanography for postdoctoral research and thought about removing starfish from a coastal waters community to discover starfish impact upon natural communities. In 1962, he became an assistant professor at the University of Washington, where he remained and advanced to professor. Leading a field trip for students in 1963, he discovered that Mukkaw Bay fit his requirements for experiments removing starfish from coastal waters communities. He discovered that starfish play a major role in maintaining species diversity in coastal waters communities (Paine 1966). In a subsequent note, he named such top predators as the starfish "keystone species" (Paine 1969). Paine and his student Robert Vadas then decided to remove sea urchins from a tidal pool and see what happened (1969). They found that a pool practically devoid of kelp with urchins present became a kelp jungle after urchin removal. In the summer of 1971, Paine visited Amchitka, an Aleutian Island, and suggested that a visiting graduate student, James A. Estes, who was studying sea otters, compare coastal waters where sea otters were present with coastal waters where they were absent to discover the difference (Stolzenburg 2008:53–63). Estes did so and discovered that sea urchins eliminated kelp forests when sea otters were absent and kelp forests flourished where sea otters were present and ate sea urchins (Estes et al. 1978). Sea otters are also a keystone species. Paine was elected ESA president from 1979 to 1980 (Root 1979), and he received the MacArthur Award in 1983. In 1989, the Ecology Institute in Germany awarded him its Excellence in Ecology Prize. In 1996, the American Society of Naturalists awarded him its Sewall Wright Award (Wilbur 1998) and in 2009 conferred upon him its Honorary Lifetime Membership Award (Anon. 2009). In 2013, he was awarded the Cosmos Prize. In 1999, Marian Kohn drew Paine's academic family tree, which accompanies Ed Yong's "Dynasty: Bob Paine Fathered an Idea—and an Academic Family—that Changed Ecology" (2013).

The October 2008 *Bulletin* has a report on a symposium from ESA's August meeting on "Tropical Deforestation and Climate Change." None of the 11 coauthors of "Amazonia Revealed" (2007) were among the five participants in the 2008 symposium. Douglas Boucher of the Union of Concerned Scientists, organized the symposium but was not one of the four speakers. David Schwartzman from Howard University, discussed evidence for the interaction of the Earth's biota, atmosphere, and temperature during four billion years, as the Earth gradually cooled, until the recent global warming, partly due to tropical deforestation. Tracy Johns from Woods Hole, discussed tropical deforestation as an element in a global climate treaty. Tropical deforestation is the second-largest source of global warming, after the use of fossil fuels. The Kyoto Protocol did not address the consequences of deforestation in developing countries, but it would be included in negotiations at the coming meeting at Copenhagen in December 2009. Dana Roth from the USDA Forest Service International Program, worked in Latin America to develop sustainable forestry and biodiversity conservation. She described helping the Honduran Forest Service develop a plan concerning mahogany trees and assisting Mexico to develop strategies to cope with hurricanes and forest fires. In Central America, conservation of biological diversity can succeed if it is accompanied by economic development of small farms. Elizabeth Shapiro from the University of California, Berkeley, described assisting Mexico to design a compensation program for Reduced Emissions from Deforestation and Degradation (REDD). It paid landowners not to cut their forests, and it required monitoring, outreach, and education.

A different approach to human alteration of the world was "Putting People in the Map: Anthropogenic Biomes of the World," by Erle Ellis and Navin Ramankutty. They discovered that "More than 75% of Earth's ice-free land showed evidence of alteration as a result of human residence and land use, with less than a quarter remaining as wild lands, supporting just 11% of terrestrial net primary production" (2008:439). Ecologists needed to know the 18 anthropogenic biomes that they identified from human usage (Figure 7.5). They provided a color world map, with a

Anthropogenic Biome Descriptions

Group	Biome	Description
Dense settlements		*Dense settlements with substantial urban area*
11	Urban	Dense built environments with very high populations
12	Dense settlements	Dense mix of rural and urban populations, including both suburbs and villages
Villages		*Dense agricultural settlements*
21	Rice villages	Villages dominated by paddy rice
22	Irrigated villages	Villages dominated by irrigated crops
23	Cropped and pastoral villages	Villages with a mix of crops and pasture
24	Pastoral villages	Villages dominated by rangeland
25	Rainfed villages	Villages dominated by rainfed agriculture
26	Rainfed mosaic villages	Villages with a mix of trees and crops
Croplands		*Annual crops mixed with other land uses and land covers*
31	Residential irrigated cropland	Irrigated cropland with substantial human populations
32	Residential rainfed mosaic	Mix of trees and rainfed cropland with substantial human populations
33	Populated irrigated cropland	Irrigated cropland with minor human populations
34	Populated rainfed cropland	Rainfed cropland with minor human populations
35	Remote croplands	Cropland with inconsequential human populations
Rangeland		*Livestock grazing; minimal crops and forests*
41	Residential rangelands	Rangelands with substantial human populations
42	Populated rangelands	Rangelands with minor human populations
43	Remote rangelands	Rangelands with inconsequential human populations
Forested		*Forests with human populations and agriculture*
51	Populated forests	Forests with minor human populations
52	Remote forests	Forests with inconsequential human populations
Wildlands		*Land without human populations or agriculture*
61	Wild forests	High tree cover; mostly boreal and tropical forests
62	Sparse trees	Low tree cover, mostly cold and arid lands
63	Barren	No tree cover; mostly deserts and frozen land

FIGURE 7.5 Anthropogenic biome descriptions. (From Ellis & Ramankutty 2008:442. With permission.)

color-keyed legend and also a table of verbal descriptions of their biomes, along with two other color diagrams concerning characteristics of the biomes. They did not advocate abandoning other biome systems based on climate, geology, and terrain, but their system added an additional dimension of knowledge.

If ecologists research global ecological problems, what about priorities? They could be influenced by previous research experiences when choosing a new project. However, general priorities are important for allocating research funds and organizing large-scale projects. Derek Cabrera, James Mandel, Jason Andreas, and Marie Nydam, all at Cornell University, decided to define and prioritize the world's most pressing problems (2008). They cited Jane Lubchenco's presidential address to AAAS members in 1997 (published 1998) urging them to tackle the world's most pressing problems. She, and others cited, suggested what those problems are. However, to obtain a broader judgment, these faculty members polled the entire Cornell faculty of 4,169 (on an "all-academics" e-mail list) on this issue and tabulated the results. Answers from 350 responders fell into seven categories, with only the first two being specifically ecological: environment and resources, health and disease, education and technology, influence, social institutions, human nature perspective, economics and poverty. But even the issues under the first two headings were not all ecological. Five environment and resources issues were shortage of potable and clean water, lack of sustainable energy, loss of biodiversity and species extinction, climate change and effects on ecosystems, and overpopulation. Three health and disease issues were emerging, infectious diseases, preventable illnesses in the Third World, and the global HIV pandemic. There were few, if any, surprises in their findings.

Invasive species is one of six major challenges that NEON addresses. The topic became a branch of ecology three decades after Englishman Charles Elton (1900–1991) published *The Ecology of Invasions by Animals and Plants* (1958). ESA member Mark Davis published a successful textbook, *Invasion Biology* (2009). South African ecologist David Richardson, specializing in invasion biology, organized in 2008 an international symposium commemorating publication of Elton's book. That remarkable symposium, at Stellenbosch University, had 51 worldwide participants (including 6 Canadians and 18 Americans) who presented 28 papers that became chapters in a landmark *Fifty Years of Invasion Ecology: The Legacy of Charles Elton* (Richardson 2011; plus two concluding chapters).

Daniel Simberloff had published "Community Effects of Introduced Species" in 1981, and when he wrote the second chapter in Richardson's volume, "Charles Elton: Neither Founder nor Siren, But Prophet," he showed (2011a:14–20) that the field of invasion biology did not attract much interest until the late 1980s. He also argued (2011a:12) that Davis et al., "Charles S. Elton and the Dissociation of Invasion Ecology from the Rest of Ecology" (2011), were mistaken in interpreting the development of Elton's ideas on species invasions. Davis' talk at the symposium, "Researching Invasive Species 50 Years after Elton: A Cautionary Tale" (2011) was not a suitable context for a response to Simberloff's criticism.

Daniel S. Simberloff (1942–)

(From *ESAB* 87, 2006, 264. With permission.)

Pennsylvanian Simberloff began collecting insects, which he stored in cigar boxes, at the age of four. At age 11, his family moved to an industrial suburb of New York City, where he excelled academically. He earned a B.A. degree from Harvard University in 1964 and a Ph.D. in 1969, advised by E. O. Wilson. They coauthored a paper from his doctoral dissertation on "Experimental Zoogeography of Islands" in *Ecology* (Simberloff & Wilson 1969), and he won the Mercer Award for 1971. He began his career at Florida State University (1969–1997), then became the first Nancy Gore Hunger Professor in the Department of Ecology and Evolution, University of Tennessee. He expressed his interest in theoretical ecology in "A Succession of Paradigms in Ecology" (1980a) and challenged MacArthur's paradigm of competitively structured communities, championing the use of null models in community ecology (1981). He became ESA's Eminent Ecologist in 2006, being named "a world expert on the threats imposed by invasive species." In 2012, he won the Ramond Margalef Prize in Ecology, bestowed by the Catalonian Government in Spain, with an award of $100,000. Also in 2012, he was elected to the National Academy of Sciences.

The year 2011 was a "hot" year for invasion biology, for another collection of papers also appeared: *Invasive and Introduced Plants and Animals*, edited by Ian D. Rotherham and Robert A. Lambert. The editors wrote introduction and conclusion chapters jointly, and each separately wrote one of the 22 other chapters. Simberloff (2011b) contributed a more generalized version of his argument, not focused on Elton.

Davis did not contribute a chapter in this collection, but his ally, Matt Chew, did. Chew also coauthored a paper in Richardson's collection (Chew & Hamilton 2011).

Publication of these two volumes prompted Davis, Chew, and 17 others to publish "Don't Judge Species on Their Origins" (2011); they did not refer back to Simberloff's criticism. Nevertheless, Simberloff and 140 others responded with "Non-Natives: 141 Scientists Object" (2011). This controversy among American ecologists (within an international context) is not narrowly about science but rather about the practical evaluation of scientific findings.

The June 2008 *Frontiers* issue was devoted to continental-scale ecology, beginning with two guest editorials on ecological networks. Debra Peters, USDA Jornada Experimental Range, Las Cruces, New Mexico, noted in her editorial that continental-scale ecology could only exist if ecologists were connected with networks that facilitated collaboration in multi-agency research. The USDA had been collecting data at many sites for a century, and NSF's LTER sites had been collecting data at its sites for almost 30 years. Other data sets could be found at ecotrends.info. The Consortium for Regional Ecological Observatories (COREO) funded this special issue of *Frontiers*. Steve Carpenter, Center for Limnology, University of Wisconsin–Madison, devoted his editorial to introducing the reports in this issue of *Frontiers* on the understanding of North American ecology developed by using ecological networks.

Peters was also the senior author of a report about using the existing North American network to understand continent-scale events, such as the Dust Bowl. Actually, their data were limited to the 48 contiguous United States, though their map of over 250 ecological research sites also includes Alaska, Hawaii, and Puerto Rico (Peters et al. 2008:235). However, for the sake of illustrating their point about broad-scale data usage, 48 states sufficed. They argued that integrated ecological networks are essential for understanding and forecasting "causes and consequences of anthropogenic global environmental change" (Peters et al. 2008:236). A team of five ecologists led by Todd Crowl from Utah State University, Logan, used ecological networks to explore "The spread of invasive species and infectious disease as drivers of ecosystem change." They noted that "Invasive species are the second leading cause of extinction in the US and cost approximately $120 billion annually" and that "Disease vectors and pathogens are spreading across continents due to human transport, land-use change, and climate change" (Crowl et al. 2008:238). Continent-wide ecological networks are essential for monitoring and combatting invasives (Figure 7.6).

Other reports in this June issue of *Frontiers* explored lakes and streams impacted by environmental changes (Williamson et al. 2008), effects of sea-level rise and windstorms on coastal and inland ecosystems (Hopkinson et al. 2008), impacts of urbanization and pollution (Grimm et al. 2008), and predicting continental ecosystem responses to climate change (Marshall et al. 2008). A final report explained how NEON could contribute to continental ecological understanding (Keller et al. 2008).

In September 2008, Donald Strong from the Department of Evolution and Ecology at the University of California, Davis, wrote a guest editorial in *Frontiers* on "Ecologists and Environmentalism," arguing that we must defend what we study (just as art historians and archeologists defend what they study) (2008a). In response, Carlos de la Rosa from Catalina Island Conservancy, Avalon, California, agreed

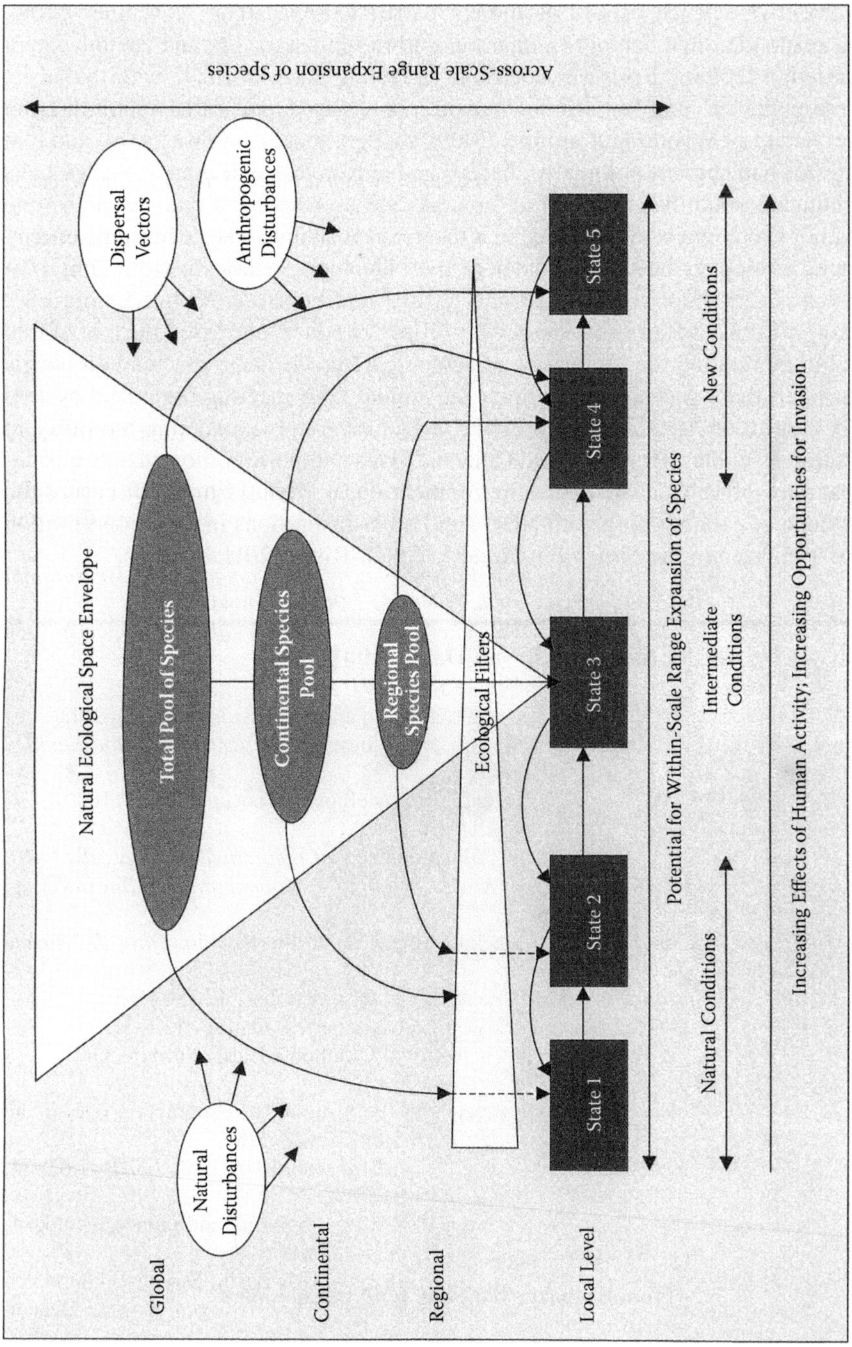

FIGURE 7.6 Factors influencing the spread of invasive species and diseases. (From Crowl et al. 2008:244. With permission.)

with the idea of ecologists protecting environments they study, but he also worried that if ecologists accepted the label of "environmentalist" they risked having their objective science viewed as merely partisan propaganda (November 2008). Strong responded that ecologists function within human affairs and should accept that situation (2008b). Rosa was not alone in feeling uncomfortable with the labels "environmentalist" and "environmentalism." Indy Burke and Bill Lauenroth from the University of Wyoming, Laramie (2009), thought that Strong was naïve and that these terms had become politically charged and only polarize discourse. Such labels, they thought, would be catastrophic for ESA. Strong responded that although "one should not shout 'environmentalist' in a theater crowded with suits from the energy industry," ecologists need not pander to the "ideological claptrap from conservative groups" (2009:240). In the February 2010 *Frontiers* issue, Nadine Lymn, ESA Director of Public Affairs and Sue Silver, *Frontiers* editor, discussed the role of that journal in facilitating the application of ecological knowledge to practical problems. A year later, there was a symposium at the annual ESA meeting organized by Jahi Chappell et al. on "Revolutionary Ecology: Defining and Conducting Stewardship and Action as Ecologists and Global Citizens." This symposium showed how to sidestep the unresolvable debate about environmentalism: "[Paul] Ehrlich discussed the importance of collaborating with NGOs and other institutions in their conservation efforts" (reported in the *Bulletin* by Ricardo Colón-Rivera 2011:417).

Margaret Bryan Davis (1931–2014)

(From Brubaker 1987:490. With permission.)

Bostonian Davis earned a B.A. degree from Radcliffe College (1953). A Fulbright Fellowship enabled her to study paleobotany at Copenhagen (1953–1954), and then she earned a Ph.D. at Harvard University (1957). She was on the faculty of Yale University (1960–1961) and the University of Michigan (1961–1983), at both, she made decisive advances in the use of pollen sediments to understand changes in plant communities over long periods (Davis 1969, 1981; Langenheim 1996:17–19). In 1983, she joined the faculty of the University of Minnesota, St. Paul, as Regents Professor in the Department of Ecology, Evolution, and Behavior. She became a "leader in the use of palynological data to study past vegetation dynamics, ecosystem processes, and response to climate change" and ESA president in 1987 (Brubaker 1987:490). She is a member of numerous professional societies, including the National Academy of Sciences, and has served as presidents of ESA and of the American Quaternary Association. In 1993, she received ESA's Eminent Ecologist Award: "She has served on numerous national and international committees, panels, and advisory groups. She has consistently been praised as an exemplary teacher and has guided the research of many graduate students and post-doctoral associates. Her publication record is outstanding" (Burgess 1993b, 1996:35). She was the senior author of a paper, "Evolutionary Responses to Climate Change" (Davis, Shaw, & Etterson 2005) that won ESA's William S. Cooper Award in 2011.

Coincidentally, two multi-authored discussions of ecosystem services appeared in 2009. However, they were quite different in scope, format, and purpose. One was a very large and opulent "coffee-table book" for a broad audience (McNeely et al. 2009). Its goal was a comprehensive description of ecosystem services in a less academic format than a textbook provides, thus encouraging readers to take both intellectual and emotional interests in the subject. (If you could afford to buy the book, you could afford to give generously to The Nature Conservancy, World Wildlife Fund, and others.)

The other discussion was in a special issue of *Frontiers* (February 2009), which focused upon evaluating ecosystem service projects. Erik Nelson and 13 coauthors, from both academic and preservation-conservation institutions, used the Willamette Basin, Oregon, to model ecosystem services at a landscape scale using a spatially explicit modeling tool, Integrated Valuation of Ecosystem Services and Tradeoffs (InVEST). They found that "scenarios that received high scores for a variety of ecosystem services also had high scores for biodiversity, suggesting there is little tradeoff between biodiversity conservation and ecosystem services" (Nelson et al. 2009:4).

Four California ecologists—Heather Tallis, Rebecca Goldman, Melissa Uhl, and Berry Brosi—studied 103 ecosystem service projects from 37 countries conducted by either The Nature Conservancy or the World Wildlife Fund to evaluate the integration of conservation and development. They classified the projects into nine categories according to who pays for ecosystem services, and within each category they indicated initial land use, final land use, and targeted ecosystem service (Tallis et al. 2009:16). Evaluation of the success of each project depended upon monitoring performance, but only 37% of projects had arranged for monitoring (Tallis et al. 2009:20). Hopefully, Tallis et al.'s study would inspire greater efforts at monitoring new projects.

Another report on ecosystem services, by Patrick Bohlen and five coauthors, was on "Paying for Environmental Services from Agricultural Lands" (2009) and focused upon the northern Everglades, near Lake Okeechobee in Florida. Three coauthors were from Florida and two from conservation organizations in Washington, DC. In the Florida Ranchlands Environmental Services Project (FRESP), state agencies paid eight ranchers to store water and reduce phosphorus runoff. FRESP was a pilot program begun in 2005. One of the maps provided showed that the greatest concentration of beef cattle in Florida was in that region (Figure 7.7). "A study conducted by WWF and partners concluded that ranchers—with appropriate incentive—could

(a) (b) (c) (d)

FIGURE 7.7 Florida ranches have important habitat for wildlife, including threatened species: (a) the burrowing owl (*Athene cunicularia*) and (b) the crested caracara (*Polyborus plancus*); and the not threatened (c) green tree frog (*Hyla cineria*) and (d) white-tailed deer (*Odocoileus virginianus*). (From Bohlen et al. 2008:51. With permission.)

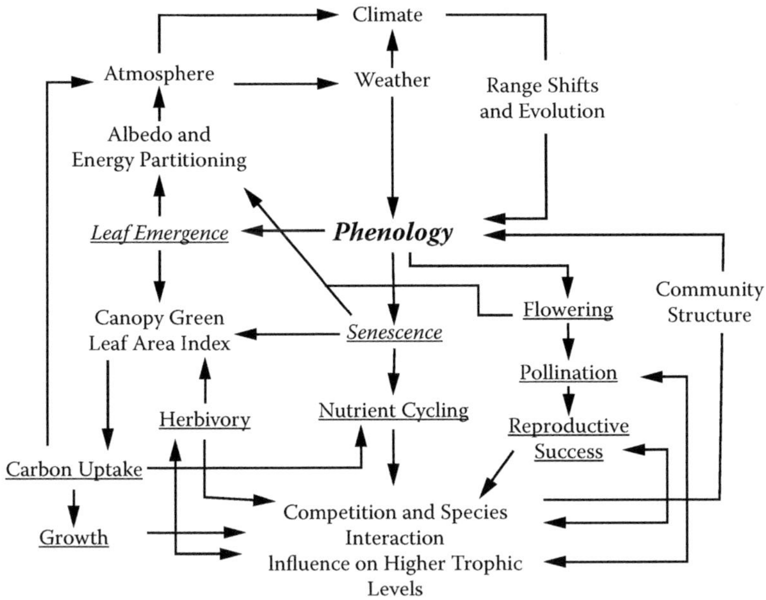

FIGURE 7.8 A model that shows the ways in which plant phenology in temperate climates are linked to environmental factors. (From Morisette et al. 2009:254. With permission.)

harness the extensive canals and ditches, berms, and water control structures, which were originally designed for drainage and irrigation, to retain more water, rehydrate drained wetlands, and reduce P loads" (Bohlen et al. 2009:49–50).

That global warming affects phenology was well understood before 2009, but how should ecologists research the situation? Jeffrey Morisette and 10 coauthors provided a complex answer: with instruments, at different scales (plot, landscape, regional, continental, global), and modeling. "Advanced field monitoring devices that can provide phenological information are becoming less expensive, easier to use, and smaller in size, even as they have expanded in capacity" (Morisette et al. 2009:255). A U.S. Phenology Network (USA-NPN) had been developing since 2007 to coordinate data collected across a range of spatial and temporal scales. However, the Network did not yet know "how the multivariate influence of meteorological conditions (e.g., temperature, precipitation, solar radiation) drives phenology" (Morisette et al. 2009:257) (Figure 7.8). This is an example of a volunteer organization of focused ecologists with specialized equipment and modeling know-how achieving progress in understanding complex situations.

In the December 2009 issue of *Frontiers*, Fred Meyerson from the Department of Natural Resources Science, University of Rhode Island, Kingston, published a guest editorial on human population versus biodiversity. He thought that the United Nations projections for human population leveling off during the 21st century were overly optimistic, and therefore he lamented the absence of population growth on the agenda for the 2009 Copenhagen conference on climate: "To give biodiversity a fighting chance, ecologists and everyone else must focus attention on managing the population of our own species" (Meyerson 2009). Four responses appeared in the March 2010 issue of *Frontiers* (Butler 2010; Feeney 2010; Pimentel 2010; Rosa, Dietz & York 2010), along with his response to them. There is perhaps a tendency for readers who disagree with an author to respond more readily than readers who agree, but all four of these respondents expressed agreement, wishing only to expand upon aspects of his discussion. Meyerson's response (2010) provided an opportunity for him to cite three of his own articles relevant to the discussion.

Four environmental scientists near the U.S. and Mexican border—Laura López-Hoffman, Robert Varady, Karl Flessa, Patricia Balvanera—two at the University of Arizona, Tucson, one at the Centro de Investigaciones en Ecosistemas, Universidad Nacional Autónoma de México, Morelia, and one at both universities, explored the "Ecosystem Services across Borders" (López-Hoffman et al. 2010) and illustrated the need for a framework for trans-boundary conservation policy. Three case histories involved: (1) Groundwater seeping from a U.S. canal into Mexico, where it is a valuable resource that San Diego wants to capture by lining the canal; (2) Endangered long-nosed bats as pollinators of blue agave grown at Mexican plantations, but with bats living on both sides of the border; and (3) Migrating monarch butterflies from Canada and the United States wintering on Mexican mountains and being a tourist attraction. In these (and other) cases, diplomatic understanding and monitoring provisions were needed.

What impact will global warming have on the prairie–forest border in North America? Lee Frelich and Peter Reich from the Department of Forest Resources, University of Minnesota, St. Paul, explored this question (Figure 7.9). They concluded that global warming would increase drought in the center of North America, which means that trees would grow more slowly and forests would experience more frequent fires. There could be some migration of trees adapted to warmer climates northward, but generally the conditions would favor prairie vegetation migrating northeastward and forests retreating in the same direction. Furthermore, environmental conditions would favor tree parasites: "Survival of insects and disease organisms is generally greater when winters are less severe, and their populations can increase faster with higher survival and longer breeding seasons, leading to greater rates of spread" (Frelich & Reich 2010:374). The predicted environment would also favor invasive European earthworms, which "have been linked to dieback and reproductive failure in mature northern hardwood trees, and loss of native plant species richness through a cascade of ecological effects" (Frelich & Reich 2010:375). Three Canadian ecologists—Norman Henderson, Elaine Barrow, and Mark Johnston—endorsed the general goals and conclusions of Frelich and Reich, but they raised "considerations concerning confidence levels in this prediction, the possibility of important regional variations, and the critical role of human choice and management in counteracting probable forest retreat" (2011:207).

A 2006 symposium on environmental justice (EJ) is discussed above. No great improvements in conditions on the ground had occurred by September 2010 when Douglass Rohrman used his commentary in *Frontiers* to review the history of EJ since 1970. Around 1970, the U.S. Public Health Service acknowledged that lead-based paint was peeling from walls in African American and Hispanic housing, and slowly a concern emerged about the public health of minorities. In 1979, North

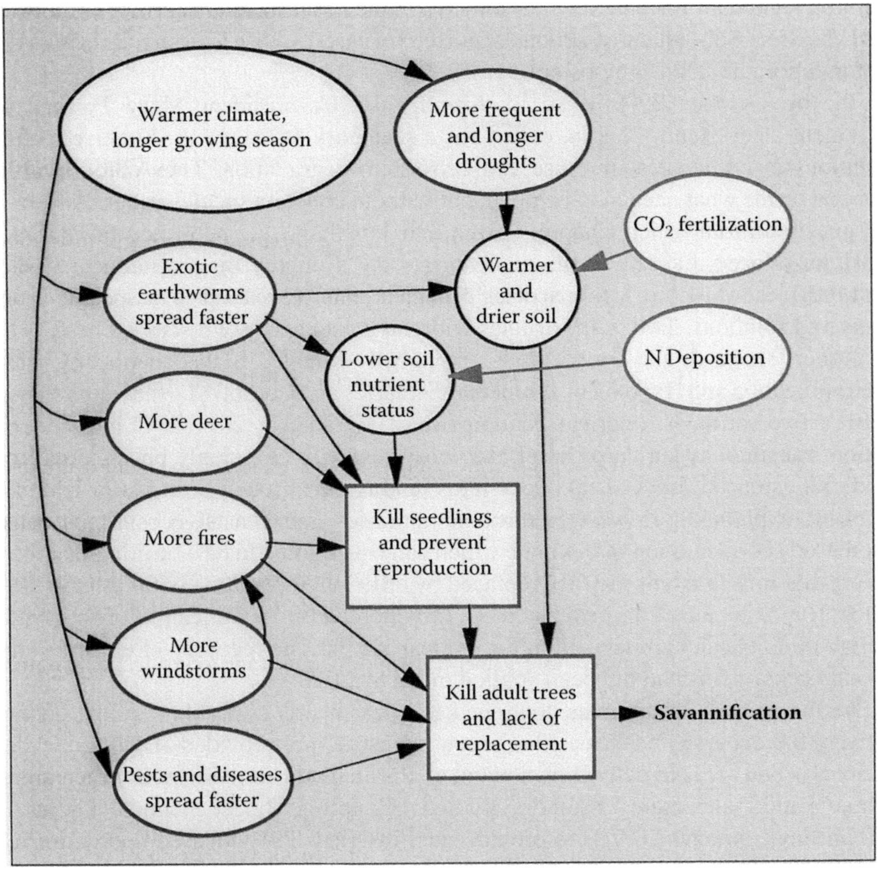

FIGURE 7.9 Interactions between global warming and other drivers of change affecting the prairie–forest border of central North America, and their impact on trees. (From Frelich & Reich 2010:376. With permission.)

Carolina applied 60,000 tons of soil contaminated by polychlorinated biphenyls to unpaved roads for dust control in rural Shocco Township, which had a 75% African American population and ranked 97 out of 100 counties in income. The United Church of Christ objected, demonstrations followed, and 500 people were arrested for civil disobedience. The Church and environmental groups pressured state and federal agencies, and the site was decontaminated—after only 21 years. In 1986, Congress passed the Emergency Planning and Community Right to Know Act, requiring dissemination of information to persons previously uninformed about environmental risks. Sociologist Robert D. Bullard published the first text on EJ, *Dumping in Dixie: Race, Class, and Environmental Quality* (1990). In 1991, the First National People of Color Environmental Leadership Summit wrote *Environmental Justice in the 21st* Century, and in 2002, its Summit II wrote *Environmental Justice Timeline—Milestones.* In 1992, EPA had established its Office of Environmental Justice, and Congress had passed an Environmental Justice Act. In 1994, President

Clinton signed an Executive Order on environmental justice, concerning minorities and children. Subsequently, actions occurred on case-by-case lawsuits. Rohrman did not mention the 2006 symposium.

In the October 2009 issue of *Frontiers*, ESA president Mary Power and president-elect Stuart Chapin, called for a planetary stewardship initiative to find solutions to human resource use and ecosystem degradation. They made clear in general terms what needed to be done, but without concrete specifications. A follow-up guest editorial which Chapin, Power, and Jonathan Cole published in February 2011 introduced a special issue of *Frontiers* on "Coupled Biogeochemical Cycles and Earth Stewardship." Nine articles provided concrete examples of specific problems and solutions. Two of the articles with the most pages are discussed here.

Robert Howarth and six coauthors explained "Coupled Biogeochemical Cycles: Eutrophication and Hypoxia in Temperate Estuaries and Coastal Marine Ecosystems" (2011). Two-thirds of American estuaries were degraded by 2011, from nutrient pollution, causing oxygen depletion. Lake ecosystems are commonly phosphorus limited, but estuaries and coastal ecosystems tend to be nitrogen limited (for complex reasons, explained). However, eutrophic estuaries and coastal ecosystems cause increased sedimentation of diatoms, which can cause silica to be a limiting factor for living diatoms. Diatoms are then replaced by other phytoplankton with faster growth rates. Hypoxia causes an increase of CO_2, which causes acidification of the water, which inhibits calcium deposition for mollusk shells. The recovery of estuaries and coastal ecosystems had not been studied much by 2011.

As the world's urban areas continued to grow, it was desirable to know quantitatively the ecosystem services that green infrastructure provided. That knowledge enables urban areas to determine how much they can afford to invest in green areas. Diane Pataki and eight coauthors studied "Coupling Biogeochemical Cycles in Urban Environments" (2011) to provide such insights. They focused upon climate-, water-, and atmosphere-regulating services provided by urban green spaces (forests, parks, gardens, etc.). However, they also studied potential disservices (increased allergens, promoted invasive species, host pathogens, etc.) to obtain net effectiveness (Figure 7.10). Unfortunately, available data were insufficient for precise conclusions.

Jane Lubchenco was one of the planners for the Sustainable Biosphere Initiative in the late 1980s, a report that ESA accepted in 1990 and published in 1991. That led to a meeting to establish an International Sustainable Biosphere Initiative in Mexico in 1991. ESA sponsored meetings, workshops, reports, and initiatives to advance SBI, and Lubchenco published "Reflections on the Sustainable Biosphere Initiative: SBI + 20 Celebration" (2012) to explain its successes and future challenges. This successful effort to disseminate ecological understanding to policy makers, media, the private sector, nongovernmental organizations (NGOs), and communities shifted federal policy toward more holistic and integrated approaches toward the environment. However, while some problems are better managed than before, new and serious problems remain as future challenges for SBI.

Levity is generally inappropriate in scientific journals. Science is a serious enterprise. However, the *Bulletin* is a quasi-scientific journal—partly science, partly ESA concerns. ESA concerns are sometimes less serious than science. Stephen S. Hale's "Field Guide to Scientific Conferences: An Ecological View" (*Bulletin* 2012) is

Commonly Discussed Urban Ecosystem Services/Disservices Associated with Biogeochemical Cycles, with Their Potential Magnitudes (Relative to the Scope of the Associated Environmental Problem) and Uncertainty Levels

Ecosystem Service	Potential Magnitude	Current Level of Uncertainty
C sequestration	Low	Low
Net GHG emissions	Moderate	High
Local cooling	High	Moderate
Stormwater mitigation	High	Moderate
Water-quality mitigation	High	High
Air-quality mitigation	Low	High
General human health	Moderate	Moderate
Ecosystem Disservice	**Potential Magnitude**	**Current Level of Uncertainty**
Water use	High	Moderate
Net GHG emissions	Moderate	High
Source of allergens	High	Low
VOC emissions	Moderate	Moderate

Notes: GHG emissions are listed as both a service and disservice because the impacts of plants or soils may be either positive (net cooling) or negative (net warming) in hot climates. VOC = volatile organic compounds, which are precursors to the formation of ozone pollution.

FIGURE 7.10 Urban ecosystem services and disservices associated with biogeochemical cycles. GHG = greenhouse gas. (From Pataki et al. 2011:31. With permission.)

for graduate students, but other ecologists could read it as a critique of conference behavior. When, as a grad student, you pop down a rabbit hole and find yourself at an ecological conference, what do you do? Scientists at conferences behave like a troop of baboons: status is important. Attendees are in three classes: Eminents, Lowlies, and Everyone Else. (Grad students, of course, are Lowlies.) People with a scientific bent are often introverts with modest social skills, but even so, graduate students should attend such meetings, because there are many ecosystem goods and services: stimulating ideas, new methods and techniques, making yourself known, getting burning questions answered, seeing old friends, meeting new people, finding a job, field trips, interesting restaurants, and such swag as insulated beer-can holders. "Confusion comes from the subtexts: pursuit of power, status, food, comfort, shelter, and reproduction (of ideas, of course)" (Hale 2012:223). Hale's literature cited consists of two works by Victorian mathematician Lewis Carroll, *Alice in Wonderland* (1865) and *Through the Looking-Glass* (1871).

The September 2013 issue of *Frontiers* was a special issue on the "Interdisciplinary Approaches to Earth Stewardship." Since previous issues were on practically the same subject, what new was there to say? An NSF-funded Workshop on the Ecological Dimensions of Earth Stewardship provided the answers. Seven researchers introduced the September issue, arguing that the practice of science must change. They offered seven recommendations (slightly abridged: Sayre et al. 2013):

1. Expand the concept of stewardship to encompass all lands and waters.
2. Work alongside various stakeholders.
3. Include people not only as variables affecting ecosystems but also as participants in ecosystems and in doing science.
4. Acknowledge that social justice and environmental health are not separate concerns.
5. Understand processes and practices at multiple scales, interacting in complex ways.
6. Value diverse ways of knowing, including indigenous knowledge.
7. Embrace stewardship as an ethical guide to practicing science.

Some decades earlier, ecologists might not have recognized this list as pertaining to science, or to their science. They might have said it reads more like social work or an Audubon Society agenda.

These recommendations are followed by seven multi-authored articles that illustrate the new practice. The professional affiliations of these authors indicate that a majority of authors were not ecologists. For example, the authors of the first article, "Global Assemblages, Resilience, and Earth Stewardship in the Anthropocene" (Ogden et al. 2013) were affiliated with five departments at five universities: the Department of Global and Sociocultural Studies, the Department of Geography, the Department of Anthropology, the Department of Natural Resources, and the Nelson Institute for Environmental Studies. The last two affiliations may indicate ecology faculty members, but the other three apparently do not. But this was appropriate for the multidisciplinary approach that the September issue's authors envisioned for Earth Stewardship.

Ogden et al. (2013:341) noted that Tansley, in his ecosystem paper advised that "We cannot confine ourselves to the so-called 'natural' entities and ignore the processes and expressions of vegetation now so abundantly provided us by the activities of man" (1935:304). The boundaries of professional disciplines are matters of convenience that can be ignored when appropriate and when multidisciplinary teams can work together to define and resolve particular problems involving humans, human activities, and nature. The time had come to go beyond programmatic statements to the nitty-gritty of real cases. Ogden et al. (2013) used three case histories to illustrate indigenous peoples caught between the environment and economics. Coffee growing by the Gimi peoples in Papua New Guinea paid US$0.15 an hour to produce coffee that sold for $12 a pound at Starbucks. The Iñupiat on the North Slope of Alaska depended upon marine mammals for subsistence during the winter, which they hunted on the ice. Global warming was undermining their economy by diminishing the ice. In the Pacific lowland rain forests of Columbia, Afro-Columbian peasant farmers and fishermen protected the biodiversity in 5 million ha, which they collectively owned, but armed terrorists massacred natives in order to force their kinsmen to flee their lands so that palm oil plantations could be established. These cases were offered "to improve dialogue between ecologists and those of us who are committed to understanding the human dimensions of environmental change" (Ogden et al. 2013:346).

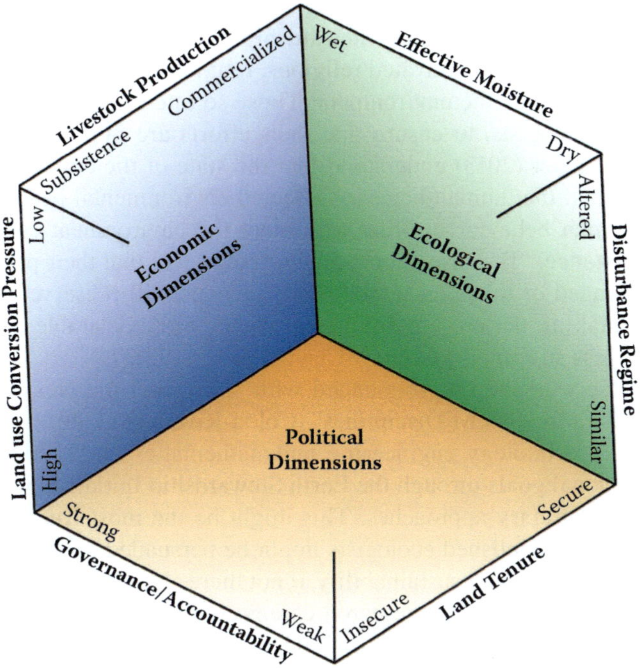

FIGURE 7.11 Ecological, economic, and political gradients that structure the prospects for effective Earth Stewardship of rangelands worldwide. (From Sayre et al. 2013:349. With permission.)

Nathan Sayre and four coauthors conceptualized the "Earth Stewardship of Rangelands" in terms of "coping with ecological, economic, and political marginality" (2013:348). "Rangelands constitute 30–40% of Earth's land surface and contribute to the livelihoods of 1–2 billion people." These authors explored two dimensions, each of their three aspects of rangeland dilemmas, and then discussed their interactions (Figure 7.11). They provided a neat diagram of these factors, which supported their advice that "Rangelands should be understood and valued positively for the full suite of economic and ecological services they provide, rather than as a residual category that invites their continued loss to more intensive uses" (Sayre et al. 2013:353).

Two articles followed on ecology and environmental design for cities. One by Frederick Steiner et al. (2013) focused upon the impacts of natural disasters, such as Hurricane Katrina upon New Orleans and Superstorm Sandy upon New York City. Structures should be designed for resilience. The article by Felson et al. (2013) focused upon urban design experiments. A new 500-ha suburb in Tuxedo, New York, was designed with assistance from Rutgers University to avoid important amphibian migration routes and to use rain gardens instead of lawns to minimize stormwater flow and recharge groundwater. An ecologist helped plan The New York City Afforestation Project, a public–private endeavor to improve the environment by planting a million native trees. It began in 2010 with 10,000 trees planted on 56 plots, testing urban tree health, soil health, and species recruitment.

Next came two attitudinal articles. Gregory Hitzhusen and Mary Tucker explored "The Potential of Religion for Earth Stewardship" (2013). Global warming and other environmental crises have motivated religions, religious groups, and religion scholars to take an interest in the environment. Those so interested are mostly receptive to dialog with ecologists, to ensure that their efforts are most effective. Clayton, Litchfield, and Geller (2013) understood that the state of the environment affected mental health, and that human behavior affected environmental health. Yet, efforts to improve human behavior in order to protect the environment could encounter emotional resistance. "Because people do not like to feel that their personal control is being challenged or taken away, the best strategies will preserve the perception of choice by making the pro-environmental choice more desirable, rather than by punishing the less sustainable option" (Clayton et al. 2013:379).

Finally, these workshop reports ended with "Fostering the Next Generation of Earth Stewards in the STEM Disciplines" (Colón-Rivera et al. 2013). Graduate students in science, technology, engineering, and mathematics could add civic concerns to their professional goals through the Earth Stewardship Initiative, which also promoted cross-disciplinary approaches. This might be the most fruitful approach of recruitment; some established ecologists might be persuaded to shift their research focus toward environmental sustainability, if not there already, but graduate students who have not yet defined their research career could lay the foundation for their research with sustainability as a priority if the opportunity were presented to them at the start. If they become involved with interdisciplinary projects, they gain networking skills that enhance their appeal when seeking employment.

Our last example of innovative biosphere studies is from the February 2014 issue of *Frontiers*, which was devoted to macrosystems ecology. The guest editorial, "Macrosystems Ecology: Big Data, Big Ecology," was by Patricia Soranno from the Department of Fisheries and Wildlife at Michigan State University and David Schimel from the Jet Propulsion Laboratory at California Institute of Technology. They explained that the NSF established a MacroSystems Biology program, and the nine multi-authored papers in this issue were funded by that NSF program (Soranno & Schimel 2014). Soranno was the coauthor of four papers; Schimel was the coauthor of none. The motivation for studying macrosystems is the large-scale practical environmental problems. Six of the papers focus on methodology and three on particular environments. Two of these papers illustrate this level of analysis, one on methodology and one on a particular environment.

J. B. Heffernan and 18 coauthors (2014) reported on "Understanding Ecological Patterns and Processes at Continental Scales." Their goal was to integrate such biological factors as migrations, range expansions, metapopulations, pathogen outbreaks, invasions, land clearings, and extinctions with such geophysical factors as atmospheric deposition, carbon sequestration, droughts, groundwater recharge, land–ocean transport, major river discharges, and fire regimes, and such sociocultural factors as natural resources extraction, land use change, environmental attitudes, and national environmental policy (Figure 7.12). They argued that macroscale patterns must be linked to local and global scales across space and time. This kind of study is only practicable with authors who have different training and experience.

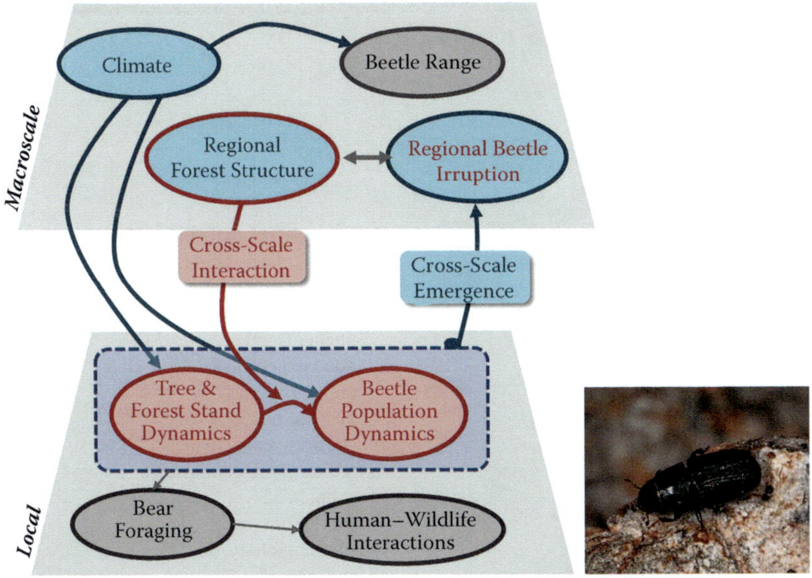

FIGURE 7.12 Some key interactions within and across scales that influence the frequency and extent of bark beetle outbreaks in forests. (Photo: A mountain pine beetle *Dendroctonus ponderosae*.) (From Heffernan et al. 2014:12. With permission.)

Kevin McCluney and seven coauthors focused upon "Riverine Macrosystems Ecology: Sensitivity, Resistance, and Resilience of Whole River Basins with Human Alterations" (2014). Because dams, pollution, deforestation, agriculture—with or without erosion or irrigation—and urbanization can have major impacts upon rivers, the magnitude of anthropogenic factors must be evaluated when trying to understand the riverine environment and life. A comparison of such factors in different watersheds led to a series of complex conclusions (McCluney et al. 2014:57) that would be impractical to summarize here.

No one expects all ecologists to jump on the sustainability bandwagon. ESA is large enough to have ecologists who have attained expertise in a great many traditional research areas to continue their excellent research while others take up new challenges that address sustainability issues. Science in general and ecology in particular expand their scope in new directions while maintaining fruitful research elsewhere.

8 ESA History
Conclusions

Anyone writing an organization's history will find reasons to praise that organization. A search for reasons to praise the Ecological Society of America is especially easy, because it has been very successful from the start in 1915. This history is a reminder of the many achievements, large and small, that have contributed to that success.

Ecology in America arose in the Midwest, where institutions of higher education were growing faster than in the East; thus, it was easier to add ecologists to faculties in the Midwest. The initiative to establish ESA also came from Midwestern academic ecologists. ESA included Canada from the start, but a focus on also including Mexico and other Latin American countries only occurred in 2006. Annual summer and winter meetings were held from the start, usually with Association for the Advancement of Science (AAAS) and/or other biological organizations.

The early volumes of the *Bulletin*, beginning in 1917, document ESA's humble origins and slow initial growth. However, its third issue (March 1917) was a *Handbook* that included a directory, and directories were issued periodically throughout the 1900s, being replaced by an online directory in the 2000s. *Ecology* began in 1920 as a fine scientific journal in its first issue and has maintained a high standard ever since then. The Depression and World War II stopped ESA's early momentum, but without stopping ESA activities and publications. *Ecological Monographs* was being planned in 1928 and first appeared in 1931. *Ecological Applications* began in 1991, *Frontiers in Ecology and the Environment* in 2000, and now ESA publishes seven journals.

By 1949, ESA membership began a rapid increase from about 800 to about 2,000 in 1959, to over 10,000 currently. Postwar prosperity in the United States and Canada was good for science, because many veterans were able to go to college who might not have gone otherwise, and there were research funds for faculty and their graduate student assistants. Gene Odum and Art Hasler also received National Science Foundation (NSF) funding for building research institutes at the universities of Georgia and Wisconsin.

The first annual award was the George Mercer Award, first presented in 1948. Others were gradually added until today there are 11. There are 25 topical sections, the first being Aquatic Ecology, established in 1961. There are seven geographic chapters, the first being Western, established in 1936, and the second being Minnesota, established in 1972. The Institute of Ecology was established in 1971 and lasted until 1984.

Frank McCormick suggested professional certification in 1978, which was debated until a poll of members was taken in 1980; to which they agreed. The first list of certified ecologists appeared in 1982, when the cost was $40. By 1999, the cost had risen to $125 and only 5% of ESA members were certified.

Spinoff organizations began with the International Biological Program (IBP) in Europe in 1963 and in Canada and the United States in 1967. The IBP was very successful and produced many publications. That momentum encouraged the establishment of Long Term Ecological Research (LTER) sites beginning in 1982, and *LTER Network News* in 1987. The Sustainable Biosphere Initiative began in 1992, and in 2012 Jane Lubchenco pronounced it a success. In 2004, the National Ecological Observatory Network (NEON) began to establish a nationwide network of ecological sensors to "document and forecast changes in ecology, at both regional and continental scales, caused by climate and land use change" (Melillo et al. 2005:351). Still other such organizations proliferated over time.

In a presidential address, Art Cooper suggested opening a Washington office to give ESA a public voice, educate members, and influence government. In January 1983, marine biologist Elliott Norse became Director of Public Affairs and published reports in the *Bulletin*. In a letter to members by Jane Lubchenco in 1993, she suggested uniting the Office of Public Affairs and the Business Office (which at the time was in Arizona) into one ESA headquarters in Washington, with an executive director. The matter was discussed at the annual meeting in August and passed.

By 1987, the role of women in ESA became a topic of discussion, at a time when women were becoming more numerous as ESA members, and soon also as officers. In the early 1980s, an environmental justice (EJ) movement arose, inspired by civil rights and feminist movements, to combat pollution dumping in poor minority communities.

After reading the early manuscript version of this history, Ed Johnson sent me some additional thoughts worth considering. In my opinion, his important points are about the history of American ecology rather than the history of ESA, but either way, his thoughts (August 31, 2013) are worth pondering.

In the 1950s and 1960s ecology saw two major conceptual changes:

1. The introduction of quantitative methodology replacing the old species list and descriptions. It shows up in a large number of papers starting as early as 1949. Although some individuals play bigger roles it seems to have developed over a wide front.
2. Field experiments and lab experiments began playing a much more important role. Paine and Connell both play an important role. This resulted in a big controversy about whether there could be such things as natural experiments.

> Also in the 1960s there was an increase in importance of both mathematics in general but specifically statistics as a necessary part of an ecologist's training. In the 1970s and 1980s, David Tilman combined the development of models started by Robert MacArthur with the field experiments of Paine and statistical design. In the 1990s we see a revolution which is still going on of the introduction of geoscience into ecology, particularly in biogeochemistry, but also in ecosystem under the new field biogeoscience. This period also saw the explicit shift from empirical descriptions to understanding processes. This worked well [with] the increasing use of mathematics, particularly the ability of computers to solve nonlinear partial differential equations. Should metabolic scaling theory be mentioned in final chapter?

During its first century, the Ecological Society of America was an outstanding organization, and what its members achieved collectively is quite remarkable—a wonderful example for the second century.

Appendix A: Ecological Society of America (ESA) Officers

President	Vice President
1916 Victor E. Shelford	William M. Wheeler
1917 Ellsworth Huntington	John W. Harshberger
1918 Henry C. Cowles	Robert E. Coker
1919 Barrington Moore	T. L. Hankinson
1920 Barrington Moore	George E. Nichols
1921 Stephen A. Forbes	Edgar N. Transeau
1922 Forest Shreve	Henry E. Crampton
1923 Charles C. Adams	Gustaf A. Pearson
1924 Edgar N. Transeau	W. Clyde Allee
1925 Arthur S. Pearse	John E. Weaver
1926 John W. Harshberger	Raymond C. Osburn
1927 Chancey Juday	William S. Cooper
1928 Homer L. Shantz	Royal N. Chapman
1929 W. Clyde Allee	Walter P. Taylor
1930 John E. Weaver	George P. Burns
1931 A. O. Feese	Francis Ramaley
1932 George E. Nichols	Joseph Grinnell
1933 Edwin B. Powers	Herbert C. Hanson
1934 George D. Fuller	Paul S. Welch
1935 Walter P. Taylor	E. Lucy Braun
1936 William S. Cooper	James G. Needham
1937 Robert E. Coker	Howard De Forest
1938 Herbert C. Hanson	Lee R. Dice
1939 Charles T. Vorhies	Clarence F. Korstian
1940 Francis Ramaley	Orlando Park
1941 Alfred E. Emerson	Benjamin C. Tharp
1942 Clarence F. Korstian	Claude E. ZoBell
1943 Orlando Park	Paul B. Sears
1944 Robert F. Griggs	Alfred C. Redfield
1945 Alfred C. Redfield	John M. Aikman
1946 John M. Aikman	Aldo Leopold
1947 Aldo Leopold	Paul B. Sears
1948 Paul B. Sears	William A. Dreyer
1949 Zeno P. Metcalf	Charles E. Olmsted
1950 E. Lucy Braun	R. V. Truitt
1951 S. Charles Kendeigh	Fred W. Albertson
1952 Frank C. Gates	David E. Davis
1953 Lee R. Dice	Stanley A. Cain

Secretary–Treasurer

1916–1919 Forest Shreve
1920–1930 A. O. Weese
1931 Alfred E. Emerson
1932–1933 Ray Kienholz
1934–1935 Arthur G. Vestal
1936–1937 Orlando Park

Secretary

1938 Orlando Park
1939–1941 W. J. Hamilton, Jr.
1942–1947 William A. Dreyer
1948–1950 William A. Castle
1951–1953 Murray F. Buell
1954–1957 John F. Reed
1958–1961 John E. Cantlon
1962–1964 Paul G. Pearson
1965–1969 Stanley I. Auerbach
1970–1971 William A. Niering
1971–1976 J. Frank McCormick
1976–1979 Edward J. Kormondy
1979–1981 Paul G. Risser
1981–1985 Robert P. McIntosh
1986–1991 Hazel R. Delcourt
1992–1995 Robert K. Peet
1995–1998 Nancy J. Huntley
1998–2001 Lisa J. Grumlich
2001–2004 Jill S. Baron
2004–2010 David W. Inouye
2010–2016 Charles D. Canham

Treasurer

1938–1940 Stanley A. Cain
1941–1943 Royal E. Shanks
1944–1949 Henry J. Oosting
1950 William T. Penfound
1951–1954 Frederick H. Test
1955–1957 Alexander C. Hodson
1958–1959 J. S. Dendy
1959–1962 Kirby L. Hays

President (Cont'd)	**Vice President (Cont'd)**	**Secretary–Treasurer (Cont'd)**
1954 John E. Potzger	Samuel Eddy	1963–1966 Ralph W. Kelting
1955 William J. Hamilton, Jr.	Murray F. Buell	1967–1969 William C. Ashby
1956 Henry J. Oosting	W. Frank Blair	1969–1972 Shelby D. Gerking
1957 William A. Dreyer	William T. Penfound	1972–1975 Forest Stearns
1958 Stanley A. Cain	Frank Preston	1975–1977 Paul Guy Pearson
1959 Thomas Park	Aaron J. Sharp	1977–1981 Lawrence C. Bliss
1960 Charles E. Olmsted	W. Dwight Billings	1981–1984 William A. Reiners
1961 Arthur D. Hasler	Edward S. Deevey	1984–1987 Paul A. Colinvaux
1962 Murray F. Buell	Bostwick H. Ketchum	1987–1990 Rebecca R. Sharitz
1963 F. Frank Blair	Lora M. Shields	1990–1995 Louis F. Pitelka
1964 John F. Reed	LaMont C. Cole	
1965 Eugene P. Odum	John E. Cantlon	
1966 Bostwick H. Ketchum	Robert B. Platt	
1967 Rexford Daubenmire	George M. Woodwell	
1968 Lamont C. Cole	George Sprugel, Jr.	
1969 John E. Cantlon	Pierre Dansereau	
1970 Edward S. Deevey, Jr.	Paul G. Pearson	
1971 F. Herbert Bormann	Robert H. Whittaker	
1972 Stanley J. Auerbach	Forest W. Sterns	
1973 Robert B. Platt	Frank B. Golley	
1974 Frederick E. Smith	Charles R. Goldman	
1975 Richard S. Miller	Arthur W. Cooper	
1976 Forest W. Sterns	Gordon H. Orians	
1977 Frank B. Golley	Lawrence C. Bliss	
1978 George M. Woodwell	Robert T. Paine	
1979 W. Dwight Billings	Gene Likens	
1979 Robert T. Paine	Richard B. Root	
1980 Arthur W. Cooper	Jean H. Langenheim	
1981 Gene E. Likens	Ralph E. Good	
1982 Lawrence C. Bliss	Richard T. T. Forman	
1983 Francis C. Evans	Dennis H. Knight	
1984 Paul G. Risser	Harold A. Mooney	
1985 Richard B. Root	Earl E. Werner	
1986 Jean H. Langenheim	H. Ronald Pulliam	
1987 Margaret B. Davis	James H. Brown	
1988 Harold A. Mooney	Jane Lubchenco	
1989 Dennis H. Knight	Barbara Lee Bentley	
1990 Simon A. Levin	Rebecca R. Shantz	
1991 H. Ronald Pulliam	Judy Lynn Meyer	
1992 Jane Lubchenco		
1993 Jerry F. Franklin	Patrice A. Morrow	
1994 Judy Lynn Meyer	Robert K. Colwell	
1995 Gordon H. Orians	VP Ed HR: 1995–2000 Alan R. Berkowitz	VP Fin: 1995–1996 Louis F. Pitelka
1996 James H. Brown	VP Ed HR: 2000–2006 Carol A. Brewer	VP Fin: 1996–2002 William H. Schlesinger

President (Cont'd)

1997 James A. MacMahon
1998 Katherine L. Gross
1999 Diana H. Wall
2000 Stephen R. Carpenter
2001 Pamela A. Matson
2002 Ann M. Bartuska
2003 William H. Schlesinger
2004 Jerry M. Melillo
2005 Nancy B. Brimm
2006 Alan P. Covich
2007 Norman L. Christensen Jr.
2008 Alison G. Power
2009 Mary E. Power
2010 Francis S. Chapin III
2011 Steward T. A. Pickett
2012 Scott L. Collins
2013 Jill S. Baron
2014 David W. Inouye
2015 Monica Turner

Vice President (Cont'd)

VP Ed HR: 2000
Alan R. Berkowitz
VP Ed HR: 2001–2006
Carol A. Brewer
VP Ed HR: 2006–2012
Margaret D. Lowman
VP Ed HR: 2012–2015
Julie Reynolds
VP Pub Aff: 1995–1999
Ann Marie Bartuska
VP Pub Aff: 1996
Anthony C. Janetos
VP Pub Aff: 1996–1999
Marie Bartuska
VP Pub Aff: 1999–2002
Timothy D. Schowalter
VP Pub Aff: 2002–2005
Alison Gail Power
VP Pub Aff: 2005–2008
Richard V. Pouyat
VP Pub Aff: 2008–2011
Laura F. Huenneke
VP Pub Aff: 2011–2014
Sharon K. Collinge
VP Pub Aff: 2014–2015
Scott L. Collins
VP Sci: 1995–1998
Steward T. A. Pickett
VP Sci: 1998–2004
James S. Clark
VP Sci: 2004–2007
Gaius R. Shaver
VP Sci: 2007–2011
Robert B. Jackson
VP Sci: 2011–2013
Deborah Goldberg
VP Sci: 2013–2016
Valerie Eviner
VP Fin: 2002–2005
Norman L. Christensen Jr.
VP Fin: 2005–2011
William J. Parton
VP Fin: 2011–2014
Leslie A. Real
VP Fin: 2014–2017
Evan H. DeLucia

Business Manager

1920–1943 Charles Stuart Gager
1944 George M. Reed
1945–1946 Henry K. Svenson
1947 Brooklyn Botanical Gardens
1948–1954 Eric Duncan
1954–1968 Henry J. Oosting
1969 W. Dwight Billings
1969–1973 William C. Ashby
1973–1979 Ralph E. Good
1979–1995 Duncan T. Patten

Appendix B: Seven of the ESA Awards

MERCER AWARD

1948 Herbert C. Hansen
1949 Edsko J. Dyksterhuis
1950 Henry S. Fitch
1951 Helmut K. Buechner
1952 Robert B. Platt
1953 Frank Alois
1954 F. Herbert Bormann
1955 Shelby D. Gerking
1956 Eugene P. Odum, Howard T. Odum
1957 John J. Christian
1958 Jerry S. Olsen
1959 Robert H. MacArthur
1960 Calvin McMillan
1961 Robert A. Norris
1962 W. Dwight Billings, Harold A. Mooney
1963 Joseph H. Connell
1964 Orie L. Loucks
1965 Kenneth S. Norris
1966 Crawford S. Holling
1967 William A. Niering, Robert H. Whittaker
1968 Edward Broadhead, Anthony Wapshere
1969 Lynn White
1971 Daniel S. Simberloff, Edward O. Wilson
1972 Joel E. Cohen
1973 Carl F. Jordan
1974 Paul K. Dayton
1975 Peter L. Marks
1976 William E. Neill
1977 Douglas G. Sprugel
1978 Earl E. Werner
1979 Jane Lubchenco, Bruce A. Menge
1980 Thomas G. Whitham
1981 Wayne P. Sousa
1982 Svata M. Louda
1983 Kenneth P. Sebens
1984 Phyllis D. Coley

1985 Peter J. Morin
1986 Douglas W. Schemske
1987 Curtis M. Lively
1988 May R. Berenbaum
1989 Russell J. Schmitt
1990 Richard Karban
1991 James S. Clark
1992 Kirk O. Winemiller
1993 Naomi Cappuccino
1994 J. Timothy Wootton
1995 Shahid Naeem
1996 Lars Hedin
1997 Charles D. Canham, Richaard K. Kobe, Stephen W. Pacala, Eric Ribbens, John Saponara, John Silander
1998 John P. Buonaccorsi, Greg Dwyer, Joseph S. Elkinton
1999 Mark A. McPeek
2000 David Hooper
2001 Brian Enquist
2002 Jonathan Levine
2003 Jean M. L. Richardson
2004 Heather Fried, Richard Osman, Robert Whitlatch
2005 Daniel Bolnick, John H. David, James Fordyce, Matthew Forister, Darrin Hulsey
2006 Anurag Agrawal
2007 Jordi Bascomple
2008 Jonathan M. Chase
2009 Richard A. Lankau
2010 Megan A. Duffy, Spencer R. Hall
2011 Tracy Langkilde
2012 Sally Archibald, Carla Staver
2013 James T. Hoverman, Pieter T. J. Johnson
2014 Douglas Rasher, Andrew S. Hoey, Mark E. Hay

EMINENT ECOLOGIST AWARD

1954 Henry S. Conard
1955 Albert H. Wright
1956 George B. Rigg
1957 Karl P. Schmidt
1958 Arthur W. Sampson
1959 Henry A. Gleason
1960 Walter P. Cottam
1961 Charles S. Elton
1962 G. Evelyn Hutchinson
1963 William S. Cooper
1964 Lee R. Dice

1965 Paul B. Sears
1966 Alfred C. Redfield
1967 Alfred E. Emerson
1968 Victor E. Shelford
1969 Stanley A. Cain
1970 Murray F. Buell
1971 Thomas Park
1972 Ruth M. Patrick
1973 Robert H. MacArthur
1974 Eugene P. Odum
1975 Cornelius H. Muller
1976 Alton A. Lindsay
1977 Walter B. McDougal
1978 S. Charles Kendeigh
1979 Rexford F. Daubenmire
1980 Donald W. Tinkle
1981 Robert H. Whittaker
1982 Edward S. Deevey, Jr.
1983 W. Thomas Edmondson
1984 John L. Harper
1985 Joseph H. Connell
1986 Evelyn C. Pielou
1987 Archie F. Carr, Jr.
1988 Herbert G. Andrewartha, Louis C. Birch
1989 George C. Williams
1990 William E. Ricker
1991 Nelson G. Hairston, W. Dwight Billings
1992 Frank A. Pielka
1993 Margaret B. Davis
1994 Edward O. Wilson
1995 F. Herbert Bowmann, Gene E. Likens
1996 Harold A. Mooney
1997 Frances C. James
1998 Gordon H. Orians
1999 Crawford S. Holling
2000 Robert T. Paine
2001 Paul R. Ehrlich
2002 Charles K. Krebs
2003 Richard B. Root
2004 Samuel J. McNaughton
2005 Lawrence B. Slobotkin
2006 Daniel Simberloff
2007 Otto Langae
2008 Michael Leo Rosenzweig
2009 Stephen P. Hubbell
2010 Simon A. Levin

2011 Thomas G. Whitham
2012 Robert J. Naiman
2013 William A. Reiners
2014 Jane Lubchenco

DISTINGUISHED SERVICE CITATION

1975 Jack Major
1976 George Sprugel, Jr.
1978 Frank E. Egler
1979 Forest W. Stearns, Paul Guy Pearson
1981 W. Dwight Billings
1982 John F. Reed, Paul Guy Pearson
1983 David G. Frey
1984 Arthur W. Cooper
1985 Josephine K. Doherty, Stanley I. Auerbach
1986 Frank W. Preston
1987 Francis C. Evans
1988 Robert L. Burgess
1989 Ralph E. Good
1990 Donald B. Lawrence
1993 Paul G. Risser
1994 Duncan T. Patten
1995 Robert K. Peet
1996 Lee N. Miller
1997 Jane Lubchenco
1998 Simon A. Levin
1999 Robert K. Colwell
2000 William Robertson IV
2001 Louis F. Pitelka
2002 H. Ronald Pullian
2003 Allen M. Solomon
2004 O. James Reichman
2005 James A. MacMahon
2006 Margaret Ann Palmer
2007 Frank N. Egerton
2008 Katherine Lynn Gross
2010 Rosina M. Bierbaum
2011 Donald R. Strong
2012 Janet Lanza
2013 Wes Jackson

ROBERT H. MacARTHUR AWARD

1983 Robert T. Paine
1984 Robert M. May

1986 Thomas W. Schoener
1988 Simon A. Levin
1990 William M. Murdoch
1992 Peter M. Vitousek
1994 Henry M. Wilbur
1996 David Tilman
1998 Robert V. O'Neill
2000 Stephen R. Carpenter
2002 James H. Brown
2004 May R. Berenbaum
2006 Alan Hastings
2008 Monica Turner
2010 Stephen W. Pacala
2012 Anthony R. Ives
2014 Mercedes Pascual

WILLIAM S. COOPER AWARD

1985 William H. Romme
1986 Edward A. Johnson
1987 Leroy H. Wullstein, Ronald P. Nelson
1988 James S. Clark
1989 David Tilman
1990 James C. Ritchie
1991 F. Ian Woodard
1992 Peter M. Vitousek
1993 Cliff R. Hupp
1994 Alwyn H. Gentry
1995 Joseph R. McAuliffe
1996 W. Carter Johnson
1997 Christopher C. Fastie
1998 Andrea H. Lloyd, Lisa J. Granmlich
1999 Serge Payette, Dominique Arseneault
2000 Paul K. Dayton, Peter Edwards, Kristin Riser, Mia Tegner
2001 Julio Betancourt, Thomas Swetnam
2002 J. W. Terborgh, M. R. Silman, Nigel C. A. Pitman, W. E. Palacios, P. V. Nunez, D. A. Neill, C. E. Cern, M. Aulestia
2003 Benjamin Slater, Glen Motzkin, David R. Foster
2004 John W. Williams, Thompson Webb III, Bryan N. Shuman
2005 Kenneth Lertzman, David Gavin, Linda Brubaker
2006 Stephen P. Hubbel
2007 Italo Mesones, Paul Fine, Phyllis D. Coley
2008 Campbell Webb, Mark A. McPeek, Michael Donaghue, David Ackerly
2009 Kirsten Ward, Glenn Motzkin, Jerry Jenkins
2010 Guy S. Robinson, Katherine B. Lininger, Stephen T. Jackson, Jacquelyn L. Gill, John W. Williams

2011 Ruth G. Shaw, Julie R. Etterson, Margaret B. David
2012 Maeiej A. Zwieniecki, Jung-Eun Lee, Taylor S. Field, Tim J. Brodribb, C. Kevin Boyce
2013 Philippe Vernet, John Thompson, Stephanie Roset, Pierre-Heari Gouyon, Anne Carpenter, Faustine Charmasson, Bruno Buatois, Guillaume Bouquet
2014 Scott L. Wing, Caroline A. E. Strömberg, Leo J. Hickey, Fleur Tiver, Brian Willis, Robyn J. Burnham, and Anna K. Behrensmeyer

E. LUCY BRAUN AWARD

1988 Nancy L. Munn
1989 Silvia I. Strauss-Debenedetti
1990 Barbara A. Roy
1991 Amy D. Rosemond
1992 Jennifer H. Mattei
1993 Rebecca A. Reed
1994 Karen L. Kandl
1995 Stephan M. Ogle
1996 Andrew P. Beckerman
1997 Caren C. Dymond
1998 Kara L. Webster
1999 Dylan Perry
2000 David E. Lytie
2001 Antonio Golubski
2002 Timothy L. Dickson
2003 Sean T. Michaletz
2004 Pedro Flombaum
2005 Phoebe L. Zametske
2006 Daniel Laughlin
2008 Jacquelyn L. Gill
2009 Emily Farrer
2010 Janise M. Bauman
2011 Joseph E. Fader
2012 Emane H. M. Vieira Neto
2013 Tony Kovach

EUGENE P. ODUM AWARD

2000 Peter Feinsinger
2001 James H. Brown
2002 Margaret D. Lowman
2003 Alan R. Berkowitz
2004 Richard B. Root
2005 James W. Porter
2006 Claudia Lewis

2007 Carol Ann Brewer
2008 Stuart Fisher
2009 Christine Flanagan
2010 Kenneth M. Klemow
2011 John C. Moore
2012 Charlene D'Avanzo
2013 Martin B. Main
2014 Manuel C. Molles, Jr.

ESA Bibliography

SPECIAL ABBREVIATIONS

ANB: American National Biography
BBGM: Brooklyn Botanic Garden Memoirs
BDAC: *Biographical Dictionary of American and Canadian Naturalists and Environmentalists* (Sterling et al. 1997)
DAB: *Dictionary of American Biography*
DSB: *Dictionary of Scientific Biography*
ESA: Ecological Society of America
ESAB: *Ecological Society of America Bulletin*
Frontiers: *Frontiers in Ecology and the Environment*
JHB: *Journal of the History of Biology*
NASBM: National Academy of Sciences Biographical Memoirs
QRB: *Quarterly Review of Biology*

LITERATURE

Adams, C. C. 1913. *Guide to the study of animal ecology.* Macmillan, New York. 1977. Arno Press, New York.
Adams, C. C., G. P. Burns, T. L. Hankinson, B. Moore & N. Taylor. 1920. Plants and animals of Mount Marcy, New York. *Ecology* 1:71–94 + plt., 204–233, 274–288.
Aikman, J. M. & F. C. Gates, eds. 1952. *Ecology: Thirty year index (Volumes 1–30, 1920–1949).* Ecological Society of America, Washington, DC.
Aldous, A. E. 1930. The yield and vigor of prairie grass vegetation. *Ecology* 11:752–759.
Aldrich, M. 1974. Nelson Horatio Darton (1865–1948), geologist. *DAB* Suppl. 4:218–219.
Allee, W. C. 1931. *Animal aggregations: A study in general sociology.* University of Chicago, IL.
Allee, W. C. 1932. *Animal life and social growth.* Williams & Wilkins, Baltimore, MD.
Allee, W. C. 1938. *The social life of animals.* Norton, New York.
Allee, W. C. 1939. Review: Just 1939. *Ecology* 20:418–421.
Allee, W. C., A. E. Emerson, O. Park, T. Park & K. Schmidt. 1949. *Principles of animal ecology.* Saunders, Philadelphia.
Allee, W. C. & T. Park. 1939. Concerning ecological principles. *Science* 89:166–169.
Allen, T. F. H. 1976. Review: Pielou 1974. *Ecology* 57:402–403.
Allen, T. F. H. 1977. Review: May 1976. *Ecology* 58:942–943.
Allen, T. F. H. & T. W. Hoekstra. 1992. *Toward a unified ecology.* Columbia University Press, New York.
Allen, T. F., H., S. Will-Wolf & P. H. Zedler. 2009. Memorial Resolution of the Faculty of the University of Wisconsin–Madison on the Death of Professor Emeritus Grant Cottam. University of Wisconsin–Madison Faculty Document 2134, October 5, 2009. 2 pages.
Allison, S. K., W. J. Ehmann, A. M. Ellison & J. Mull. 1996. A synthetic review of several major ecology textbooks. *ESAB* 77:93–99.
Allred, B. W. 1954. Review: Weaver 1954. *Ecology* 35:591.

Andrewartha, H. G. & L. C. Birch. 1954 *The distribution and abundance of animals.* University of Chicago, Chicago.

Anonymous. 1955. Dr. Henry J. Oosting—President. *ESAB* 36:105–106.

Alvo, R. & S. Ponomarenko. 2001. Biodiversity and ecosystem function: Alternate hypotheses or a single theory? *ESAB* 82:88–89.

Anonymous. 1994a. Frederic Edward Clements. Pp. 105–118 in Burgess 1995.

Anonymous. 1994b. Rexford F. Daubenmire. Pp. 173–177 in Burgess 1994.

Anonymous. 2003. ESA's SEEDS program. *ESAB* 84:53.

Anonymous. 2009. Honorary Lifetime Membership Award: Richard T. Paine. *Amer. Nat.* 174(3):iii.

Appel, T. A. 2000. *Shaping biology: The National Science Foundation and American biological research, 1945–1975.* Johns Hopkins University Press, MD.

Armitage, K. C. 2013. The soil doctor: Hugh Bennett, soil conservation, and the search for a democratic science. Pp. 87–102 in Jørgensen et al. 2013.

Aronova, E., K. S. Baker & N. Oreskes. 2010. Big science and big data in biology: From the International Geophysical Year through the International Biological Program to the Long Term Ecological Research (LTER) Network, 1957–present. *Hist. Stud. Nat. Sci.* 40:183–224.

Arrhenius, S. A. 1896. On the influence of carbonic acid in the air upon the temperature of the ground. *Philos. Mag.* 41:237–276.

Ashworth, W. 2006. *Ogallala blue: Water and life on the High Plains.* Norton, New York.

Auerbach, S. L. 1995. George M. Van Dyne—A reminiscence. Pp. xxvii–xxx in Patten & Jørgensen 1995.

Auerbach, R. L. Burgess & R. V. O'Neill. 1977. The biome program: Evaluating an experiment. *Science* 193:902–904.

Ayres, P. 2012. *Shaping ecology: The life of Arthur Tansley.* Wiley-Blackwell, Oxford, UK.

Bailes, K. E. 1990. *Science and Russian culture in an age of revolutions: V. I. Vernadsky and his scientific school, 1863–1945.* Indiana University Press, Bloomington, IN.

Baker, D. Z. 1990. Eighty years of research at Iowa Lakeside Laboratory: A bibliography. *Iowa Acad. Sci. J.* 97:190–199.

Baker, R. G. 1991. Vegetational patterns. *ESAB* 72:47–49.

Baker, T. L. 1980. Turbine-type windmills of the Great Plains and Midwest. *Ag. Hist.* 54:38–51.

Bandler, M. J. 1970. Dr. Ruth Patrick: A diagnostician of water pollution. *Smithsonian* 1(5):12–15.

Banks, E. M. 1985. Warder Clyde Allee and the Chicago school of animal behavior. *J.H. Behavioral Sciences* 21:345–353.

Barash, D. P. 1976. Review: Wilson, 1975. *Ecology* 57:399–400.

Barash, D. P. 2008. *Natural selections: Selfish altruism, honest liars, and other realities of evolution.* Belleview Literary Press, New York.

Barbour, M. G. 1996. American ecology and American culture in the 1950s: Who led whom? *ESAB* 77:44–51.

Barrett, G. W. 2003. Resolution of respect: Eugene P. Odum: Pioneer of ecosystem science, 1913–2002. *ESAB* 84:11–12.

Barrett, G. W. & T. L. Barrett, eds. 2001. *Holistic science: The evolution of the Georgia Institute of Ecology.* Taylor & Francis, New York.

Bartlett, R. A. 1962. *Great surveys of the American West.* University of Oklahoma Press, Norman, OK.

Baskin, Y. 1997. Center seeks synthesis to make ecology more useful. *Science* 275:310–311.

Bates, C. G. & R. Zon. 1922. Research methods in the study of forest environment. *USDA Bull.* 1059, Washington, DC.

Bates, M. 1960. *The forest and the sea: A look at the economy of nature and the ecology of man*. Random House, New York.
Bates, M. & Editors of *Life*. 1964. *The land and wildlife of South America*. Time, New York.
Bateson, P. 2005. The ethics of hunting. *Frontiers* 3:393–394.
Baumann, P. C., J. F. Kitchell, J. J. Magnuson & T. B. Keyes. 1974. Lake Wingra, 1837–1973: A case history of human impact. *Wis. Acad. Sciences, Arts & Lett. Trans.* 62:57–94.
Baxter, B. 2007. *A Darwinian worldview: Sociobiology, environmental ethics, and the work of Edward O. Wilson*. Ashgate, Aldershot, UK.
Bazzaz, F. A. 1982. Lawrence C. Bliss, president 1982–83. *ESAB* 63:221–222.
Bazzaz, F. A. 2001. Cooling the diversity debate. *ESAB* 82:136–137.
Beard, J. S. 1944. Climax vegetation in tropical America. *Ecology* 25:127–158.
Beard, J. S. 1955. The classification of tropical American vegetation types. *Ecology* 36:89–101.
Bebbington, W. P. 1990. *History of Du Pont at the Savannah River Plant*. E. I. Du Pont de Nemours, Wilmington, DE.
Beckel, A. L., ed. 1987. *Breaking new waters: A century of limnology at the University of Wisconsin. Wis. Acad. Sciences, Arts & Letters Trans.,* Special Issue.
Beebe, C. W. 1907. Geographical variation in birds with especial reference to the effects of humidity. *Zoologica* 1:3–41.
Beebe, C. W. 1916. Fauna of four square feet of jungle debris. *Zoologica* 2:107–119.
Beebe, C. W. 1924. *Galapagos: World's end*. G.P. Putnam's Sons, New York.
Beebe, C. W. 1928. *Beneath tropical seas: A record of diving among the coral reefs of Haiti*. G.P. Putnam's Sons, New York.
Beeton, A. M. & D. C. Chandler. 1963. The St. Lawrence Great Lakes. Pp. 535–558 in Frey 1963a.
Beidleman, R. G. 2009. To California with Jepson's "Phyto–jogs" in 1913. *Madroño* 56:49–56.
Beltrán, E. 1955. Paul B. Sears: Botanico, ecologo, conservacioniste. *Rivista de la Sociedad Mexicana de Historia Natural* 16:35–39.
Beltrán, E. 1956. Veinte Años de vida de la Sociedad Mexicana de Histoira Natural. *Revista de la Sociedad Mexicana de Historia Natural* 17:17–35.
Bendell, J. F. 1954. Review: Woodbury 1954. *Ecology* 35:585–587.
Bennett, E. M. & P. Balvanera. 2007. The future of production systems in a globalized world. *Frontiers* 5:191–198.
Bennett, H. H. 1939. *Soil conservation*. McGraw-Hill, New York. 1970. Arno Press, New York.
Benson, N. G. 1970a. *A century of fisheries in North America*. The American Fisheries Society, Washington, DC.
Benson, N. G. 1970b. The American Fisheries Society, 1920–1970. Pp. 13–24 in Benson 1970a.
Bergman, C. & I. T. Adams. 1993. Examining differences between ecology and other sciences: crisis? What crisis? *ESAB* 74:365–370.
Berra, T. M. 1977. *William Beebe: An annotated bibliography*. Archon Books, Hamden, CT.
Beyer, N. 1979. Letter on need for ecologists to receive more training in applied ecology for employment opportunities. *ESAB* 60:3.
Billings, W. D. 1957. Physiological ecology. *Ann. Rev. Plant Physiology* 8:375–392.
Billings, W. D. 1964. *Plants and the ecosystem*. Wadsworth, Belmont CA. Edition 3, 1978.
Billings, W. D. 1970, December. Resolution of respect: Henry J. Oosting, 1903–1968. *ESAB* 51(4):16–17.
Billings, W. D. 1994. Letter of 9 Nov. 56: Autobiography and bibliography. Pp. 47–53 in Burgess 1994.
Billings, W. D., E. S. Hathaway & C. L. Newcombe. 1951. Resolution of respect: Forrest Shreve, 1878–1950. *ESAB* 32:5–6.
Birch, L. C. 1955. Review: Clarke 1954. *Ecology* 36:369.

Birchard, B. 2005. *Nature's keepers: The remarkable story of how the Nature Conservancy became the largest environmental organization.* Jossey-Bass/Wiley, San Francisco, CA.

Birge, E. A. 1898. Plankton studies on Lake Mendota. II. The Crustacea of the plankton from July 1894 to December 1896. *Wis. Acad. Sciences, Arts & Letters Trans.* 11:274–448. Reprint in Birge & Juday 1977.

Birge, E. A. & C. Juday. 1922. The inland lakes of Wisconsin: The plankton. I. Its quantity and chemical composition. *Wis. Geol. & Nat. Hist. Surv.* Bull. 64. Reprint in Birge & Juday 1977.

Birge, E. A. & C. Juday. 1977. *Limnology in Wisconsin.* F. N. Egerton, compiler. Arno Press, New York.

Birkhead, T. R., J. Wimpenny & Bob Montgomerie. 2014. *Ten thousand birds: Ornithology since Darwin.* Princeton University Press, Princeton, NJ.

Blair, J. M. 1998. Global letter to world leaders from scientists about population growth and resource use. *ESAB* 79:127.

Blair, W. F. 1977. *Big biology: The US/IBP.* Dowden, Hutchinson & Ross, Stroudsburg, PA.

Blanquet, J. 1932. *Plant sociology: The study of plant communities.* G. D. Fuller & H. S. Conard, transl. & revised. 1965. Hafner, New York.

Blew, R. D. 1996. On the definition of ecosystems. *ESAB* 77:171–173.

Bliss, L. C., ed. 1977. *Truelove Lowland, Devon Island, Canada: A high Arctic ecosystem.* University of Alberta Press, Canada. Paperback ed. 1987.

Bliss, L. C., ed. 1981. *Tundra ecosystems: A comparative analysis.* Cambridge University Press, New York.

Bliss, L. C. 1994. Autobiographical letter, November 14, 1961. Pp. 57–60 in Burgess 1994.

Blumer, K. 1984. Current trends of women in ecology: View from the 1983 ESA Annual Meeting. (Abstract) 65:204.

Bocking, S. 1990. Stephen Forbes, Jacob Reighard, and the emergence of aquatic ecology in the Great Lakes region. *JHB* 23:461–498.

Bocking, S. 1995. Ecosystems, ecologists, and the atom: Environmental research at Oak Ridge National Laboratory. *JHB* 28:1–47.

Bocking, S. 1997a. *Ecologists and environmental politics: A history of contemporary ecology.* Yale University Press, New Haven, CT.

Bocking, S. 1997b. Donald Strathearn Rawson (1905–1961), limnologist. *BDAC* 666–668.

Bocking, S. 2007. Science and spaces in the northern environment. *Environ. Hist.* 12:867–894.

Bocking, S. 2013. Situated yet mobile: Examining the environmental history of Arctic ecological science. Pp. 164–178 in Jørgensen et al. 2013.

Boffey, P. M. 1976. International Biological Program: Was it worth the cost and effort? *Science* 193:866–868.

Bohlen, P. J. et al. 2009. Paying for environmental services from agricultural lands: An example from the northern Everglades. *Frontiers* 7:46–55.

Bohn, D. 1967. *Glacier Bay: The land and the silence.* Ballantine Books/Sierra Club, New York.

Bolsenga, S. J. & C. H. Herdendorf, eds. 1993. *Lake Erie and Lake St. Clair: Handbook.* Wayne State University Press, Detroit, MI.

Bonnifield, P. 1979. *The Dust Bowl: Men, dirt and depression.* University of New Mexico Press, Albuquerque, NM.

Bonta, M. M. 1991. *Women in the field: America's pioneering women naturalists.* Texas A & M University Press, College Station, TX.

Bormann, F. H. 1953. Factors determining the role of loblolly pine and sweetgum in early old-field succession in the Piedmont of North Carolina. *Ecol. Monogr.* 23:339–358.

Bormann, F. H. 1970. Subtraction by multiplication. *ESAB* 51:3–10.

Bormann, F. H. 1971, September. President's report. *EASB* 52(3):2–4.

Bormann, F. H. 1996. Ecology: A personal history. *Ann. Rev. Energy Environ.* 21:1–29.

Bormann, F. H. & G. E. Likens. 1967. Nutrient cycling. *Science* 155:424–429.

Bormann, F. H. & G. E. Likens. 1979. *Pattern and process in a forested ecosystem.* Springer-Verlag, New York.

Bormann, F. H. & P. G. Pearson. 1975, December. Murray Fife Buell, 1906–1975, *ESAB* 56(4):26.

Bormann, F. H., P. R. Shafer & D. Mulcahy. 1958. Fallout on the vegetation of New England during the 1957 atom bomb test series. *Ecology* 39:376–378.

Borradaile, L. A. 1923. *The animal and its environment: A text-book of the natural history of animals.* Frowde, Hodder & Stoughton, London.

Boucher, D. H. 1985. *The biology of mutualism: Ecology and evolution.* Oxford University Press, New York.

Boucher, D. H. et al. 2008. Tropical deforestation and climate change. *ESAB* 89:475–480.

Bowen, E. S. 1931. The role of the sense organs in aggregations of *Ameiurus melas. Ecol. Mono.* 1:1–35.

Bowen, M. 2008. *Censoring science: Inside the political attack on Dr. James Hansen and the truth of global warming.* Dutton, New York.

Bowers, J. E. 1988. *A sense of place: The life and work of Forrest Shreve.* University of Arizona Press, Tucson, AZ.

Braun, E. L. 1921. Composition and source of the flora of the Cincinnati (Ohio) region. *Ecology* 2:161–180.

Braun, E. L. 1926. National Parks and National Monuments. Pp. 20–27 in Shelford & Shreve 1926.

Braun, E. L. 1950. *Deciduous forests of eastern North America.* Blakiston, Philadelphia, PA. 1964. Hafner, New York.

Braun, E. L. 1955. The phytogeography of unglaciated Eastern United States. *Bot. Rev.* 21:297–375.

Braun, E. L. 1958. The development of association and climax concepts, their use in interpretation of the deciduous forest. Pp. 329–339 in Steere 1958.

Braun, E. L. 1961. *The woody plants of Ohio.* Ohio State University Press, Columbus, OH.

Braun, E. L. 1969. *An ecological survey of the vegetation of Fort Hill State Memorial, Highland County, Ohio and annotated list of vascular plants.* Ohio Biological Survey, Columbus, OH.

Braun, E. L. 1994. Autobiographical letter, 1953. Pp. 61–67 in Burgess 1994.

Brett, M. T., L. Brouwer & L. A. Brett. 1999. Research productivity and reputational ratings at United States ecology, evolution, and behavior programs. *ESAB* 80:250–256.

Brewer, C. A., K. M. Doyle, J. J. Honaker et al. 1992. The Sustainable Biosphere Initiative: A student critique and call to action. *ESAB* 73:23–25.

Breymeyer, A. I. & G. M. Van Dyne, eds. 1980. *Grasslands, systems analysis and man.* Cambridge University Press, New York.

Bridges, W. 1974. *A gathering of animals: An unconventional history of the New York Zoological Society.* Harper & Row, New York.

Brock, T. 1985. *A eutrophic lake: Lake Mendota, Wisconsin.* Springer-Verlag, New York.

Bronstein, J. L. 1991. Mutualism studies and the study of mutualism. *ESAB* 72:6–8.

Brown, J. H. 1999. The legacy of Robert MacArthur: From geographical ecology to mammalogy. *J. Mammalogy* 80:333–344.

Brown, M. T. 2004. A picture is worth a thousand words: Energy systems language and simulation. *Ecological Modelling* 178:83–100.

Brown, M. T. & C. A. S. Hall, eds. 2004a. *Through the macroscope: The legacy of H. T. Odum. Ecological Modelling* 178(1–2):1–294.

Brown, M. T. & C. A. S. Hall. 2004b. III. General systems, ecological modeling, and symbols. *Ecological Modelling* 178:81–82.

Brown, M. T., C. A. S. Hall & S. E. Jørgensen. 2004. Eulogy. Pp. 1–10 in Brown & Hall 2004a.

Brubaker, L. B. 1987. Margaret B. Davis, president 1987–1988. *ESAB* 68:490–491.

Brynildson, O. M. 1958. *Lime treatment of brown-stained lakes and their adaptability for trout and largemouth bass.* Ph.D. dissertation, University of Wisconsin–Madison, WI.

Buchholtz, C. W. 1983. *Rocky Mountain National Park.* University Press of Colorado, Niwot, CO.

Buell, M. F. 1963. Eminent Ecologist for 1963: William Skinner Cooper. *ESAB* 44:108–110.

Buell, M. F., P. Pearson & E. S. Deevey. 1971. Eminent Ecologist, 1971, Thomas Park. *ESAB* 52(4):15.

Bullen, K. E. 1976. Alfred Lothar Wegener (1880–1930), meteorology, geophysics. *DSB* 14:214–217.

Burgess, R. L. 1977. The Ecological Society of America: Historical data and some preliminary analyses. Pp. 1–24 in Egerton 1977.

Burgess, R. L. 1981a. United States. Pp. 67–101 in Kormondy & McCormick 1981.

Burgess, R. L. 1981b. The ecology photograph, IV International Congress of Plant Sciences. *ESAB* 62:203–207.

Burgess, R. L. 1981c. Sources of biographical information on American ecologist. *ESAB* 62:236–255.

Burgess, R. L. 1983. 1982 Banquet: Reunion of past presidents and Eminent Ecologists. *ESAB* 64:17–21.

Burgess, R. L. 1984. Historical Records Committee report. *ESAB* 65:486–487.

Burgess, R. L. 1990. Ninety years of ecological instruction, research, and application: The role of the Ecological Society of America. Unpublished talk, 75th ESA Annual Meeting, ESA. 19 pages.

Burgess, R. L. 1993a. John Thomas Curtis: Botanist, ecologist, conservationist. Pp. 1–43 in Fralish et al. 1993.

Burgess, R. L. 1993b. Eminent Ecologist Award: Margaret Bryan Davis. *ESAB* 74:278.

Burgess, R. L., ed. 1994. *The Murray Fife Buell Seminar on World Ecologists.* SUNY College of Environmental and Forest Biology, Syracuse, New York.

Burgess, R. L. 1996. American ecologists: A biographical bibliography. *Huntia* 10:5–116.

Burgess, R. L. 1997. Resolution of respect: Frank Edwin Egler, 1911–1996. *ESAB* 78:193–194.

Burgess, R. L. 2009. Paul B. Sears and the Ecological Society of America. *Ohio J. Sci.* 109:104–108.

Burke, I. & B. Lauenroth. 2009. "Environmentalist" label not in our best interest. *Frontiers* 7:240.

Burns, T. P. 1992. Ecosystem: A powerful concept and paradigm for ecology. *ESAB* 73:36–43.

Burrows, W. 1940. Review: Moulton 1939. *Ecology* 21:404–405.

Butler, C. 2010. Population and climate change—A response to Meyerson. *Frontiers* 8:65.

Cabrera, D., J. T. Mandel, J. P. Andreas & M. L. Nydam. 2008. What is the crisis? Defining and prioritizing the world's most pressing problems. *Frontiers* 6:469–475.

Cain, S. A. 1944. *Foundations of plant geography.* Harper & Brothers, New York.

Cain, S. A. 1959. Henry Allan Gleason, Eminent Ecologist, 1959. *ESA Bulletin* 40:105–110.

Cairns, J., Jr. 2014. Resolution of respect: Ruth Patrick, 1907–2013. *ESAB* 95:11–13.

Callahan, J. T. 1987. Perspectives on LTER. *LTER Network News* 1:2–4.

Callendar, G. S. 1938. The artificial production of carbon dioxide and its influence on climate. *Quart. J. Royal Meteorological Soc.* 64:223–240.

Callicott, J. B. 1998, January. A critical examination of "Another look at Leopold's land ethic." *J. Forestry* 96(1):20–26.

Campbell, B. 1995. *Human ecology: The story of our place in nature from prehistory to the present.* Aldine & Gruyter, New York. Edition 1, 1983.

Campbell, D. H. 1926. *An outline of plant geography.* Macmillan, New York.

Cantlon, J. E. 1970, December. Ecological bridges. *ESAB* 51(4):5–10.

Cantlon, J. E. 2001. Ecology and economics. *ESAB* 82:155–156.

Cappuccino, N. 1993. Review: Peters 1991. *Ecology* 74:1907–1908.

Carpenter, J. R. 1939. The biome. Pp. 75–91 in Just 1939.

Carpenter, S. 2008. Emergence of ecological networks. *Frontiers* 6:228.

Carpenter, S. & J. Kitchell. 2001. Resolution of respect: Arthur Davis Hasler, 1908–2001. *ESAB* 82:172–173.

Carr, A. F., Jr. 1994. *A naturalist in Florida: Celebration of Eden.* M. H. Carr, ed. Yale University Press, New Haven, CT.

Carson, R. L. 1941. *Under the sea wind: A naturalist's picture of ocean life.* Simon & Schuster, New York.

Carson, R. L. 1951. *The sea around us.* Oxford University Press, New York.

Carson, R. L. 1955. *The edge of the sea.* Houghton Mifflin, Boston, MA.

Carson, R. L. 1962. *Silent spring.* Houghton Mifflin, Boston, MA.

Cassidy, V. M. 2007. *Henry Chandler Cowles: Pioneer ecologist.* Kedzie Sigel Press, Chicago, IL.

Cattell, J. M. & D. R. Brimhall, eds. 1920. *American men of science: A biographical directory. Science P.,* Garrison, New York.

Chambers, C. D. 1997. Lee Raymond Dice, 1887–1977. *BDAC* 206–208.

Chapin, F. S. III, J. T. Randerson et al. 2008. Changing feedbacks in the climate-biosphere system. *Frontiers* 6:313–320.

Chapman, R. N. 1928. The quantitative analysis of environmental factors. *Ecology* 9:111–122.

Chapman, R. N. 1931. *Animal ecology with especial reference to insects.* McGraw-Hill, New York.

Chazdon, R. L. 2001. Diversity, stability, and evolution of professional ecologists: A report on the ESA membership 2000. *ESAB* 82:181–186.

Cherif, A. H. 1990. Mutualism—The forgotten concept in teaching science. *American Biology Teacher* 52:206–208. Cited from Bronstein 1990.

Cherrett, J. M., ed. 1989. *Ecological concepts: The contribution of ecology to an understanding of the natural world.* Blackwell Scientific Publications, Oxford, UK.

Chew, M. K. 2011. Anekeitaxonomy: Botany, place and belonging. Pp. 137–151 in Rotherham & Lambert 2011.

Chew, M. K. & A. L. Hamilton. 2011. The rise and fall of biotic nativeness: A historical perspective. Pp. 35–47 in Richardson 2011.

Child, C. M. 1947. Biographical memoir of Francis Bertody Sumner, 1874–1945. *NASBM* 25:147–173 + portrait.

Cittadino, E. 1993a. A "Marvelous cosmopolitan preserve": The dunes, Chicago, and the dynamic ecology of Henry Chandler Cowles. *Perspectives on Sci.* 1:520–559.

Cittadino, E. 1993b. The failed promise of human ecology. Pp. 251–283 in Shortland 1993.

Cittadino, E. 1997. Frederic Edward Clements, 1874–1945, plant ecologist, botanist. *BDAC* 156–158.

Clark, R. B., compiler. 1994. George Elwood Nichols. Pp. 423–427 in Burgess 1994.

Clarke, G. L. 1954. *Elements of ecology.* J. Wiley & Sons, New York.

Clayton, S., C. Litchfield & E. S. Geller. 2013. Psychological science, conservation, and environmental sustainability. *Frontiers* 11:377–362.

Clements, E. S. 1960. *Adventures in ecology: Half a million miles...from mud to macadam.* Pagent Press, New York.

Clements, F. E. 1905. *Research methods in ecology.* University Publishing, Lincoln, NE. 1977. Arno Press, New York.

Clements, F. E. 1916. *Plant succession: An analysis of the development of vegetation.* Carnegie Institution, Washington, DC.

Clements, F. E. 1917. The development and structure of biotic communities. *J. Ecology* 5:120–121. Abstract of a talk in 1916.
Clements, F. E. 1936. Nature and structure of the climax. *J. Ecology* 24:252–284. Reprint in Real & Brown 1991:59–97.
Clements, F. E. 1949. *Dynamics of vegetation.* B. W. Allred & E. S. Clements, eds. H. W. Wilson, New York.
Clements, F. E. & R. W. Chaney. 1924. Methods and principles of paleo-ecology. P. 319 in *Carnegie Institution of Washington Yearbook* for 1923.
Clements, F. E. & R. W. Chaney. 1936. *Environment and life in the Great Plains.* Carnegie Institution, Washington, DC.
Clements, F. E. & G. W. Goldsmith. 1924. *The phytometer method in ecology: The plant and community as instruments.* Carnegie Institution, Washington, DC.
Clements, F. E. & V. E. Shelford. 1939. *Bio-ecology.* John Wiley & Sons, New York.
Clements, F. E. & J. E. Weaver. 1924. *Experimental vegetation: The relation of climaxes to climate.* Carnegie Institution, Washington, DC.
Clements, F. E., J. E. Weaver & H. C. Hanson. 1929. *Plant competition: An analysis of community functions.* Carnegie Institution, Washington, DC. 1977. Arno Press, New York.
Cody, M. L. & J. M. Diamond, eds. 1975. *Ecology and evolution of communities.* Harvard University Press, Cambridge, MA.
Cohn, J., compiler. 1994. Henry Allen Gleason. Pp. 255–265 in Burgess 1994.
Coker, R. E. 1947. *This great and wide sea.* University of North Carolina Press, Chapel Hill, NC.
Cole, G. A. 1975. *Textbook of limnology.* C. V. Mosby, St. Louis, MO.
Cole, J. R. & H. Zuckerman. 1987, February. Marriage, motherhood, and research performance in science. *Sci. Amer.* 256(2):119–125.
Cole, L. C. 1950. Review: Allee et al. 1949. *Ecology* 31:153–155.
Cole, L. C. 1955. Review: Andrewartha & Birch 1954. *Ecology* 36:538.
Cole, L. C. 1971. The ecosphere. Pp. 11–16 in Ehrlich et al. 1971.
Coleman, D. C. 2010. *Big ecology: The emergence of ecosystem science.* University of California Press, Berkeley, CA.
Coleman, D. C., M. I. Dyer, J. E. Ellis, N. R. French, J. H. Gibson, G. S. Innis, J. K. Marshall, F. M. Smith & G. M. Van Dyne. 1973, June. Grassland ecosystem evaluation. *ESAB* 54(4):8–10.
Coleman, D. C., J. M. Oades & G. Uehara, eds. 1989. *Dynamics of soil organic matter in tropical ecosystems.* University of Hawaii Press, Honolulu, HI.
Colinvaux, P. A. 2007. *Amazon expeditions: My quest for the Ice Age equator.* Yale University Press, New Haven, CT.
Collias, N. E. 1953. Review: Tinbergen 1953a. *Ecology* 34:810–811.
Collias, N. E. 1955. Review: Tinbergen 1953b. *Ecology* 36:371–372.
Collias, N. E. 1991. The role of American zoologists and behavioral ecologists in the development of animal sociology, 1934–1964. *Animal Behavior* 41:613–631.
Collins, S., compiler. 1994a. Charles Christopher Adams. Pp. 1–10 in Burgess 1994.
Collins, S., compiler. 1994b. Lee Raymond Dice. Pp. 187–201 in Burgess 1994.
Colón-Rivera, R. J. 2011. Revolutionary ecology: Defining and conducting stewardship and action as ecologists and global citizens. *ESAB* 92:415–419.
Colón-Rivera, R. J., K. Marshall, F. J. Soto-Santiago, D. Ortiz-Torres & C. E. Flower. 2013. Moving forward: Fostering the next generation of Earth Stewards in the STEM disciplines. *Frontiers* 11:383–391.
Colwell, R. K. & D. J. Futuyma. 2011. Resolution of respect: Lawrence B. Slobodkin, 1928–2009. *ESA B* 92:9–32.
Connell, J. H. 1971. Review: Odum & Pigeon 1970. *Science* 172:831–832.

Cook, R. E. 1977. Raymond Lindeman and the trophic-dynamic concept in ecology. *Science* 198:22–26.

Cook, S. G. 1980. *Cowles Bog, Indiana and Henry Chandler Cowles (1869–1939): A study in historical geography and the history of ecology.* 70 pp. (mimeographed). Indiana Dunes National Lakeshore, Chesterton, IN.

Cook, W. C. 1930. Review: Shelford 1929. *Ecology* 11:611–614.

Cooper, A. W. 1982. Why doesn't anyone listen to ecologists and what can ESA do about it? *ESAB* 63:346–356.

Cooper, A. W. & M. K. Wali. 1989. Robert Lewis Burgess: Distinguished Service Citation. *ESA Bulletin* 70:34–35.

Cooper, W. S. 1908. Alpine vegetation in the vicinity of Long's Peak, Colorado. *Bot. Gaz.* 45:319–337. Reprint in Egerton 1977b.

Cooper, W. S. 1912. The ecological succession of mosses as illustrated upon Isle Royale, Lake Superior. *Plant World* 15:197–213. Reprint in Egerton 1977b.

Cooper, W. S. 1913. The climax forest of Isle Royale, Lake Superior, and its development. *Bot. Gaz.* 55:1–44 + map, 115–140, 189–235. Reprint in Egerton 1977b.

Cooper, W. S. 1923. The recent ecological history of Glacial Bay, Alaska. *Ecology* 4:93–128 + map, 223–246, 355–365.

Cooper, W. S. 1926. The fundamentals of vegetation change. *Ecology* 7:391–413.

Cooper, W. S. 1927. "Ecology" or "plant ecology." *Ecology* 8:490–491.

Cooper, W. S. 1928. Seventeen years of successional change upon Isle Royale, Lake Superior. *Ecology* 9:1–5 + 2 plts.

Cooper, W. S. 1931. A third expedition to Glacier Bay, Alaska. *Ecology* 12:61–95.

Cooper, W. S. 1935. Henry Chandler Cowles. *Ecology* 16:281–283.

Cooper, W. S. 1994. Autobiographical letter, February 28, 1953. Pp. 127–138 in Burgess 1994.

Cottam, G. 1971. Foreword to the second printing. Pp. v–vi in Curtis 1971.

Coupland, R. T., R. Y. Zacharuk & E. A. Paul. 1969. Procedures for study of grassland ecosystems. Pp. 25–47 in Van Dyne 1969.

Court, F. E. 2012. *Pioneers of ecological restoration: The people and legacy of the University of Wisconsin Arboretum.* University of Wisconsin Press, Madison, WI.

Cowles, H. C. 1899. The ecological relations of vegetation on the sand dunes of Lake Michigan. *Bot. Gaz.* 27:95–117, 167–202, 281–308, 361–391. Reprints in Cassidy 2007:100–170, Egerton 1997b, Real & Brown 1991:28–58.

Cowles, H. C. 1901a. The influence of underlying rocks on the character of the vegetation. *Bull. Amer. Bureau of Geogr.* 2:163–176, 376–388. Reprint in Cassidy 2007:204–220.

Cowles, H. C. 1901b. The physiographic ecology of Chicago and vicinity. *Bot. Gaz.* 31:73–108, 145–182. Reprint in Egerton 1977b.

Cowles, H. C. 1901c. The plant societies of Chicago and vicinity. *Bull. Geogr. Soc. Chicago* 2:1–76. Reprint in Cassidy 2007:171–203.

Cowles, H. C. 1911. The causes of vegetative cycles. *Bot. Gaz.* 51:161–183. Reprints in Cassidy 2007:221–240, Egerton 1977b.

Cox, D. L. 1979. *Charles Elton and the emergence of modern ecology.* Ph.D. dissertation, Washington University, Saint Louis, MO.

Cox, D. L. 1980. A note on the queer history of "niche." *ESAB* 61:201–202.

Cox, D. W. 1971. *Pioneers of ecology.* Hammond, Maplewood, NJ.

Craig, P. 2005. *Centennial history of the Carnegie Institution of Washington: Vol. 4: The Department of Plant Biology.* Cambridge University Press, New York.

Craige, B. J. 2001. *Eugene Odum: Ecosystem ecologist and environmentalist.* University of Georgia Press, Athens, GA.

Crawford, E., T. Shinn & S. Sörlin, eds. 1993. *Denationalizing science: The contexts of international scientific practice.* Kluwer Academic, Boston, MA.

Croker, R. A. 1991. *Pioneer ecologist: The life and work of Victor Ernest Shelford, 1877–1968.* Smithsonian Institution Press, Washington, DC.
Croker, R. A. 2001. *Stephen Forbes and the rise of American ecology.* Smithsonian Institution Press, Washington, DC.
Crossley, D. A., Jr. 1963. Movement and accumulation of radiostrontium and radiocesium in insects. Pp. 103–105 in Schultz & Klement 1963.
Crowl, T. A. et al. 2008. The spread of invasive species and infectious diseases as drivers of ecosystem change. *Frontiers* 6:238–246.
Cuddington, K. & B. E. Beisner, eds. 2005a. *Ecological paradigms lost: Routes of theory change.* Elsevier Academic, Boston, MA.
Cuddington, K. & B. E. Beisner. 2005b. *Kuhnian paradigms lost: Embracing the pluralism of ecological theory.* Pp. 419–427 in Cuddington & Beisner 2005a.
Cunfer, G. 2005. *On the Great Plains: Agriculture and environment.* Texas A & M Press, College Station, TX.
Cunha, A. A. & M. V. Vierira. 2004. Two bodies cannot occupy the same place at the same time, or the importance of space in the ecological niche. *ESAB* 85:25–26.
Curtis, J. T. 1959. *The vegetation of Wisconsin.* University of Wisconsin Press, Madison, WI.
Curtis, J. T. 1971. *The vegetation of Wisconsin.* University of Wisconsin Press, Madison, WI. Reprint of 1959.
Curtis, J. T. & R. P. McIntosh. 1950. The interrelations of certain analytic and synthetic phytosociological characters. *Ecology* 31:434–455.
Daily, G. C., ed. 1997. *Nature's services: Societal dependence on natural ecosystems.* Island Press, Washington, DC.
Dale, E. E., Jr. 1967, November. Resolution of respect: John Ernst Weaver, 1884–1966. *ESAB* 48(3):107–109.
Dale, V., G. W. Barrett et al. 2000. ESA's professional certification program: Let's make it work. *ESAB* 81:255–287.
Dana, S. T. 1967. Barrington Moore (1883–1966). *J. Forestry* 65:69–72.
Dansereau, P. 1938. Etudes sur les hybrides de cistes. III. Statistique des caracteres du *Cistus monspeliensis. Bull. Soc. Bot. France* 85:668–678.
Dansereau, P. 1939. Monographie du genre *Cistus* L. *Boissiera* 4:1–90.
Dansereau, P. 1940. Etudes sur les hybrides de cistes. II. Releves caryologiques. *Ann. Epiphyt. Phytogen.* 6:13–26.
Dansereau, P. 1941a. Notes sur les cistes. I. La collection de Bailey Hortorium. *Contrib. Inst. Bot. U. Montreal* 37:1–14.
Dansereau, P. 1941b. Etudes sur les hybrides de cistes. IV. Correlation des caracters du *C. salviafolium* L. *Contrib. Inst. Bot. U. Montreal* 38:1–13.
Dansereau, P. 1941c. Etudes sur les hybrides de cistes. VI. Introgression dans la section Ladanium. *Can. J. Research* C. 19:59–67.
Dansereau, P. 1943a. Etudes sur les hybrides de cistes. V. Le comportement du X. *Cistus florentinus* Lam. *Candollea* 10:9–22.
Dansereau, P. 1943b. Gaspé maple groves may indicate cooling climate. *Pollen Analysis Circ.* 3:2–3.
Dansereau, P. 1946. Os planos de biogeografia. *Revists Brasileira de Geografia* 8:189–210.
Dansereau, P. 1947a. Zonation et succession sur la resting de Rio de Janerio: I. Halosere. *Revue Can. De Biologie* 6:448–477.
Dansereau, P. 1947b. The vegetation of Brazilian sand dunes. *Trans. Roy. Soc. Canada* 41:200.
Dansereau, P. 1947c. Notes sobre a biogeografia de una parte de serra do Mar. *Rev. Bras. Geogr.* 9:497–520.
Dansereau, P. 1957. *Biogeography: An ecological perspective.* Ronald Press, New York.

Dansereau, P. 2005. *Autobiographie: Vol. 1: La Lancée, 1911–1936.* Éditions Multimones, Sainte–Foy, Quebec.

Dansereau, P. & F. Segadas-Vianna. 1947. The high mountains vegetation of southern Brazil. *ESAB* 28(4):48.

Darton, N. H. 1899. Geology and water resources of Nebraska west of the one hundred and third meridian. *U.S. Geological Survey Annual Report* 19:719–785.

Darwin, C. R. 1859. *The origin of species by means of natural selection, or the preservation of favoured races in the struggle for life.* John Murray, London. 1964. Harvard University Press, Cambridge, MA.

Daubenmire, R. F. 1938. Merriam's life zones of North America. *QRB* 13:327–332.

Daubenmire, R. F. 1947. *Plants and environment: A textbook of autecology.* John Wiley & Sons, New York.

Davidson, D. W., compiler. 1994. Grant Cottam. Pp. 139–146 in Burgess 1994.

Davidson, J. 1918. Suggests members of committee on preservation write descriptions of areas proposed for preservation. *ESA Bull.* 2(6–10):3.

Davis, F. R. 2007. *The man who saved sea turtles: Archie Carr and the origins of conservation.* Oxford University Press, New York.

Davis, Margaret B. 1969. Climatic changes in southern Connecticut recorded in pollen deposition at Roger's Lake. *Ecology* 50:409–422. Reprinted in Real & Brown 1991:650–663.

Davis, Margaret B. 1981. Quarternary history and stability of forest communities. Pp. 132–153 in West et al. 1981.

Davis, Margaret B., R. G. Shaw & J. R. Etterson. 2005. Evolutionary responses to climate change. *Ecology* 86:1704–1714.

Davis, Mark A. 2009. *Invasion biology.* Oxford University Press, New York.

Davis, Mark A. 2011. Researching invasive species 50 years after Elton: A cautionary tale. Pp. 269–276 in Richardson 2011.

Davis, Mark A., M. Chew & 17 others. 2011. Don't judge species on their origins. *Nature* 474:153–154.

Davis, Mark A., K. Thompson & J. P. Grime. 2011. Charles S. Elton and the dissociation of invasion ecology from the rest of ecology. *Diversity & Distributions* 7:97–102.

Davis, T. T. 2004. Biography of David Tilman. *Nat. Acad. Sci. Proc.* 101:10851–10853.

Deevey, E. S., Jr. 1971. Inter-American Institute of Ecology: Twenty questions. *ESAB* 52:5–10.

Deevey, E. S., Jr. 1972. What an ecologist isn't. *ESAB* 53:5–6.

De la Rosa, C. 2008. Ecologists, environmentalists, and the problem with labels. *Frontiers* 6:467–468.

Dennis, D. 1992. Idealism to realism: Applying ecological science to society's environmental problems—The Sustainable Biosphere Initiative Project office creates implementation plan. *ESAB* 73:219–222.

Dethier, V. G. 1954. Evolution of feeding preferences in phytophagous insects. *Evolution* 8:33–54.

Dexter, R. W. 1953. Review: Dice 1952. *Ecology* 34:438–440.

Dexter, R. W. 1978. History of the Ecologists' Union—Spin-off from the ESA and prototype of the Nature Conservancy. *ESAB* 59:146–147.

Dice, L. R. 1943. *The biotic provinces of North America.* University of Michigan Press, Ann Arbor, MI.

Dice, L. R. 1952. *Natural communities.* University of Michigan Press, Ann Arbor, MI.

Dice, L. R. & P. M. Blossom. 1937. *Studies of mammalian ecology in southwestern North America, with special attention to the colors of desert mammals.* Carnegie Institution, Washington, DC.

Dietz, T., E. A. Rosa & R. York. 2007. Driving the human ecological footprint. *Frontiers* 3:13–18.

Diffendal, R. E., Jr. 2004. Ogallala Aquifer. Pp. 857–858 in *Encyclopedia of the Great Plains*. D. J. Wishart, ed. University of Nebraska Press, Lincoln, NE.

Di Mento, J. F. C. & P. Doughman, eds. 2007. Climate change: What it means for us, Pp. 181–196 in *Climate Change: What it means for us, our children, and our grandchildren*. MIT Press, Cambridge, MA.

Doherty, J. & A. W. Cooper. 1990. The short life and early death of the Institute of Ecology: A case study in institution building. *ESAB* 71:6–17.

Douglass, A. E. 1919–1936. *Climatic cycles and tree-growth*. Carnegie Institution, Washington, DC.

Douglass, A. E. 1920. Evidence of climate effects in the annual rings of trees. *Ecology* 1:24–32.

Dreyer, W. A. 1954. Review: Storer 1953. *Ecology* 35:588.

Dreyer, W. A. 1955. Review: Frisch 1955. *Ecology* 36:772.

Drouin, J. -M. 1993. *L écologie et son histoire: réinventer la nature*. Flammarion, Paris, France. Edition 2.

Drude, O. 1896. *Deutschlands Pflanzengeographie: Ein geographes Charakterbild der Flora von Deutschland*. J. Engelhorn, Stuttgart.

Duncan, D. & K. Burns. 2012. *The Dust Bowl: An illustrated history*. Chronicle Books, San Francisco, CA. Companion volume to a TV series.

Dunlap, T. R. 1988. *Saving America's wildlife*. Princeton University Press, Princeton, NJ.

DuShane, G. P. 1946. Review: F. Sumner 1945. *Ecology* 27:94–95.

Duvigneaud, P. & S. Denaeyer-De Smet. 1970. Biological cycling of minerals in temperate deciduous forests. Pp. 199–225 in Reichle 1970.

Edmondson, W. T. 1934. Investigations of some Hispaniolan Lakes. I. Rotatoria, *Arch. Hydrobiol.* 26:465–471.

Edmondson, W. T. 1960. Reproductive rate of rotifers in natural populations. *Mem. 1st Ital. Idrobiol.* 12:21–77.

Edmondson, Y. H. 1971. Some components of the Hutchinson legend. *Limnol. & Oceanogr.* 16:157–161.

Edmondson, W. T. 1991. G. Evelyn Hutchinson, 1903–1991. *ESAB* 72:212–216.

Edwards, C. A., D. E. Reichle & D. A. Crossley, Jr. 1970. The role of soil invertebrates in turnover of organic matter and nutrients. Pp. 147–172 in Reichle 1970.

Egan, T. 2006. *The worst hard times*. Mariner Books/Houghton Mifflin, New York.

Egerton, F. N. 1973. Changing concepts of the balance of nature. *QRB* 48:322–350.

Egerton, F. N., ed. 1977a. *History of American ecology*. Arno Press, New York.

Egerton, F. N., compiler. 1977b. *American plant ecology: 1897–1917*. Arno Press, New York.

Egerton, F. N. 1985. *Overfishing or pollution? Case history of a controversy on the Great Lakes*. Great Lakes Fishery Commission, Ann Arbor, MI.

Egerton, F. N. 1987. The Wisconsin limnology community. Pp. 85–122 in Beckel 1987.

Egerton, F. N. 1993. The history and present entanglements of some general ecological perspectives. Pp. 9–23 in McDonald & Pickett 1993.

Egerton, F. N. 1995. Review: Allen & Hoekstra 1992. *Hist. Philos. Life Sci.* 17:116–117.

Egerton, F. N. 1999a. Edward Asahel Birge (1851–1950), lilmnologist and president of the University of Wisconsin. *ANB* 2:809–810.

Egerton, F. N. 1999b. Chancey Juday (1871–1944), limnologist. *ANB* 12:294–295.

Egerton, F. N. 1999c. Victor Ernest Shelford (1887–1968), ecologist. *ANB* 19:785–786.

Egerton, F. N. 2007. Understanding food chains and webs, 1700–1970. *ESAB* 88:50–69.

Egerton, F. N. 2008a. Arthur Davis Hasler (1908–2001), fish physiology, limnology. *New Dictionary of Scientific Biography* 3:260–262.

Egerton, F. N. 2008b. History of ecological sciences, part 27: Naturalists explore Russia and the North Pacific during the 1700s. *ESAB* 89:39–60.

Egerton, F. N. 2009. Homage to Frederic E. Clements, historian of plant succession studies. *ESAB* 90:43–79.

Egerton, F. N. 2010. History of ecological sciences, part 37: Charles Darwin's voyage on the *Beagle*. *ESAB* 91:398–431.

Egerton, F. N. 2012a. *Roots of ecology: Antiquity to Haeckel*. University of California Press, Berkeley, CA.

Egerton, F. N. 2012b. Bibliographic essay: Relationships in nature, a history of ecology. *Choice* 50:607–619.

Egerton, F. N. 2013. History of ecological sciences, part 48: Formalizing plant ecology, about 1870 to mid-1920s. *ESAB* 94:341–378.

Egerton, F. N. 2014a. History of ecological sciences, part 49: Formalizing animal ecology. *ESAB* 95:59–81.

Egerton, F. N. 2014b. History of ecological sciences, part 50: Formalizing limnology. *ESAB* 95:131–153.

Egerton, F. N. 2014c. History of ecological sciences, part 51: Formalizing marine ecology. *ESAB* 95:347–430.

Egerton, F. N. 2016. History of ecological sciences, part 57: Limnology, 1925–90: Hutchinson and Hasler. *ESAB* 97.

Egler, F. E. 1951. A commentary on American plant ecology based on the textbooks of 1947–49. *Ecology* 32:673–694.

Egler, F. E. 1960. Review: Bates 1960. *Ecology* 41:605–606.

Egler, F. E. 1972. ESA needs a code of ethics and a certification program. *ESAB* 53:2–4.

Egler, F. E. Review: Whittaker 1973 & Knapp 1974. *Ecology* 56:249–251.

Ehrlich, P. R. 1968. *The population bomb*. Ballantine, New York.

Ehrlich, P. R. 2000. Life support systems. Pp. 16–27 in Newbold 2000.

Ehrlich, P. R. & A. H. Ehrlich. 1996. *Betrayal of science and reason: How anti-environmental rhetoric threatens our future*. Island Press, Washington, DC.

Ehrlich, P. R., J. P. Holdren & R. W. Holm, comp. & commentaries. 1971. *Man and the ecosphere: Readings from* Scientific American. W.H. Freeman, San Francisco, CA.

Ehrlich, P. R. & P. H. Raven. 1964. Butterflies and plants: A study in coevolution. *Evolution* 18:586–608.

Ellenberg, H., ed. 1973. *Oekosystemforschung: Ergebnisse von Symposien der Deutschen Botanischen Gesellschaft und der Gesellschaft für Angewandt Botanik in Innsbruck, July 1971*. Springer–Verlag, Berlin.

Ellis, E. C. & N. Ramankutty. 2008. Putting people in the map: Anthropogenic biomes of the world. *Frontiers* 6:439–447.

Ellison, K. 2005. Light at the end of the tunnel. *Frontiers* 3:232.

Ellison, K. 2007. Crisis and opportunity in China. *Frontiers* 5:452.

Elton, C. S. 1927. *Animal ecology*. Sidgwick & Jackson, London.

Elton, C. S. 1940. Scholasticism in ecology. *J. Animal Ecology* 9:151–152.

Elton, C. S. 1942. *Voles, mice and lemmings: problems in population dynamics*. ClarendonPress, Oxford, UK.

Elton, C. S. 1958. The ecology of invasions by animals and plants. Methuen, London. Partly reprinted in Lomolimo et al. 2004:575–593.

Emerson, A. E. & T. Park. 1955. Warder Clyde Allee: Ecologist and ethologist. *Science* 121:686–687.

Engel, J. R. 1983. *Sacred sands: The struggle for community in the Indiana Dunes*, Wesleyan University Press, Middletown, CT.

Engelmann, M. 1970. Resolution of respect: Orlando Park, 1901–1969. *ESAB* 51:16–20.

England, J. M. 1982. *A patron for pure science: The National Science Foundation's formative years*. National Science Foundation, Washington, DC.

Erdtman, G. 1927. Literature on pollen statistics published before 1927. *Forhandlingar; Geologiska Föreningen i Stockholm* 49:196–211.

Erdtman, G. 1929. Literature on pollen statistics during the years 1927–1928. *Förh. Geol. Fören. Stockholm* 52:191–213.
Erdtman, G. 1931. Pollen statistics: A new research method in paleoecology. *Science* 73:399–401.
Erdtman, G. 1943. *An introduction to pollen analysis.* Chronica Botanica, Waltham, MA.
Erickson, R. F. 1999a. Henry Chandler Cowles (1869–1939), botanist. *ANB* 5:617–618.
Erickson, R. F. 1999b. John William Harshberger (1869–1929), botanist. *ANB* 10:230–232.
Errington, P. L. 1973. *The red gods call.* C. Errington, ed. Iowa State University Press, Iowa City, IA.
ESA (Ecological Society of America). 1917. The Ecological Society of America Meeting at New York City, December 27–29, 1916. *J. Ecology* 5:119–128.
ESA (Ecological Society of America). March 1917. Handbook of the Ecological Society of America. *ESAB* 1:1–56.
ESA (Ecological Society of America). 1978. TIE moves. *ESAB* 59:141.
ESA (Ecological Society of America). 1979. Ecological Society of America certification program. *ESAB* 60:72–74.
ESA (Ecological Society of America). 1980. Ecological Society of America certification program. *ESAB* 61:40–43.
ESA (Ecological Society of America). 1982. Ecological Society of America registry of certified ecologists, 1982. *ESAB* 63:48–50.
ESA (Ecological Society of America). 1983a. The Institute of Ecology moves to the Washington area [&] TIE report. *ESAB* 64:24–26.
ESA (Ecological Society of America). 1983b. Ecological Society of America opens Washington office. *ESAB* 64:29.
ESA (Ecological Society of America). 1983c. Policy and related decisions regarding the ESA Washington office. *ESAB* 64:198–201.
ESA (Ecological Society of America). 1988. Nominations for awards. *ESAB* 69:229–230.
Estes, J. A., N. S. Smith & J. F. Palmisano. 1978. Sea otter predation and community organization in the Western Aleutian Islands, Alaska. *Ecology* 59:822–833.
Evans, E. E. & C. M. Wolf. 2005. The ethics of hunting. *Frontiers* 3:394–395.
Ewan, J. A., ed. 1969a. *A short history of botany in the United States.* Hafner, New York.
Ewan, J. A. 1969b. Plant geography. Pp. 115–123 in Ewan 1969a.
Ewan, J. A. 1971. Frederic Edward Clements (1874–1945), botany. *DSB* 3:317–318.
Ewel, J. J. 1988. Eminent Ecologist [Archie Fairly Carr, Jr.]. *ESAB* 69:24–25.
Farnworth, E. G. & F. B. Golley, eds. 1974. *Fragile ecosystems: Evaluation of research and applications in the neotropics.* Springer-Verlag, New York.
Fauth, J. E. 1997. Working toward operational definitions in ecology: Putting the system back into ecosystems. *ESAB* 78:295–297.
Fedoseyev, I. A. 1976. Vladimir Ivanovich Vernadsky (1863–1945), minerology, geochemistry, biogeochemistry. *DSB* 13:616–620.
Feeney, J. 2010. Population and species extinction—A response to Meyerson. *Frontiers* 8:66–67.
Felson, A. J., M. A. Bradford & T. M. Terway. 2013. Promoting Earth Stewardship through urban design experiments. *Frontiers* 11:362–367.
Fernald, M. L. 1911. Review: Harshberger 1911. *Rhodora* 13:213–224.
Fernald, M. L. 1926. Campbell's *Outlines of plant geography. Ecology* 7:510–516.
Fine, P. & L. DuPont. 2012, September 6. Scientist focused on the collaboration between plants and humans. *Globe and Mail.* [obituary: P. Dansereau.]
Fiscus, D. 2001. The ecosystemic life hypothesis, I: introduction and definitions. *ESAB* 82:248–250.

Fiscus, D. 2002a. The ecosystemic life hypothesis, II: four connected concepts. *ESAB* 83:94–96.
Fiscus, D. 2002b. The ecosystemic life hypothesis, III: the hypothesis and its implications. *ESAB* 83:146–149.
Fleming, J. R. 2007. *The Callendar effect: The life and work of Guy Stewart Callendar (1898–1964), the scientist who established the carbon dioxide theory of climate change.* American Meteorological Society, Boston, MA.
Fletcher, M. & D. Allen. 1997. The Baruch Marine Field Laboratory. *ESAB* 78:274–279.
Foley, J. A. et al. 2007a. Amazonia revealed: Forest degradation and loss of ecosystem goods and services in the Amazon Basin. *Frontiers* 5:25–32.
Foley, J. A. et al. 2007b. The authors reply. *Frontiers* 5:238.
Forbes, S. A. 1887. The lake as a microcosm. *Peoria Scientific Assoc. Bull.* 77–87. 1925 reprint. *Illinois State Nat. Hist. Survey Bull.* 15:537–550. 1977 reprint in *Forbes, Ecological Investigations.* 1991. Pp. 14–27 in Real & Brown 1991.
Forbes, S. A. *Ecological investigations*, F. N. Egerton, compiler. Arno Press, New York.
Forman, R. T. T., P. Dayton & F. James. 1982. Distinguished Service Citation [Billings]. *ESAB* 63:12–13.
Fortescue, J. A. C. & G. G. Marten. 1970. Micronutrients: Forest ecology and systems analysis. Pp. 173–198 in Reichle 1970.
Fosberg, F. R. 1963. Review: Carson 1962. *Ecology* 44:624.
Fournier, R. O. 1982. Gordon Riley—The elegant scientist. Pp. 331–332 in Riley 1982.
Fraenkel, G. S. 1959. The raison d'etre of secondary plant substances. *Science* 129:1466–1470.
Fralish, J. S., R. P. McIntosh & O. L. Loucks, eds. 1993. *John T. Curtis: Fifty years of Wisconsin plant ecology.* Wisconsin Academy of Sciences, Arts & Letters, Madison, WI.
Frank, P. W. 1954. Review: Odum 1953. *Ecology* 35:297–298.
Freckman, D. 1989. Letter on women and minorities. *ESAB* 70:174–175.
Fred, E. B. 1928. Review: Waksman 1927. *Ecology* 9:94–96.
Freeman, B. 2013. *A history of the Nature Conservancy of Canada.* Oxford University Press, New York.
Frelich, L. E. & P. B. Reich. 2010. Will environmental changes reinforce the impact of global warming on the prairie–forest border of central North America? *Frontiers* 8:371–378.
Fretwell, S. D. 1975. The impact of Robert MacArthur on ecology. *Ann. Rev. Ecol. Syst.* 6:1–13.
Frey, D. G., ed. 1963a. *Limnology in North America.* University of Wisconsin Press, Madison, WI.
Frey, D. G. 1963b. Wisconsin: The Birge–Juday era. Pp. 3–54 in Frey 1963a.
Frisch, K. von. 1950. *Bees—Their vision, chemical senses, and language.* Cornell University Press, Ithaca, NY.
Frisch, K. von. 1955. *The dancing bees.* Harcourt Brace, New York.
Fuller, G. D. 1928. Review: McDougall 1927. *Ecology* 9:99.
Galishoff, S. 1999. Selman Abraham Waksman (1888–1973), microbiologist. *ANB* 22:465–467.
Geddes, P. 2005a. The dangers of scientific consensus. *Frontiers* 3:242.
Geddes, P. 2005b. Moral approach to climate change. *Frontiers* 3:469.
Gilbert, J. J. 1981. Gene E. Likens, president 1981–1982. *ESAB* 62:197–198.
Gleason, H. A. 1917. The structure and development of the plant association. *Torrey Bot. Club Bull.* 44:463–481. Reprint in Egerton 1977b.
Gleason, H. A. 1920. Some applications of the quadrat method. *Torrey Bot. Club Bull.* 47:21–37.
Gleason, H. A. 1922. The vegetational history of the Middle West. *Assoc. Amer. Geogr. Ann.* 12:39–85.

Gleason, H. A. 1926. The individualistic concept of the plant association. *Torrey Bot. Club Bull.* 53:7–26. Reprint in Real & Brown 1991:98–117.

Gleason, H. A. 1929. Plant associations and their classification: A reply to Dr. Nichols. *International Congress of Plant Sciences Proceedings* 1:643–646. Gleason, H. A. 1953. Autobiographical letter. *ESAB* 34:40–42 + cover photo.

Gleason, H. A. 1953. Autobiographical letter. *ESAB* 34:40–42 + cover photo.

Gleason, H. A. 1975, December. [Letter on history of American ecology, February 18, 1952] *ESAB* 56(4):7–10.

Goetzmann, W. H. 1967. *Exploration and empire: The explorer and the scientist in the winning of the American West.* Knopf, New York.

Goldburg, R. & R. Naylor. 2005. Future seascapes, fishing, and fish farming. *Frontiers* 3:21–28.

Golley, F. B. 1993. *A history of the ecosystem concept in ecology: More than the sum of the parts.* Yale University Press, New Haven, CT.

Golley, F. B. 2001. Establishing the network. Pp. 13–27 in Barret & Barrett 2001.

Good, R. 1955. Plant geography. Pp. 747–765 in Kessel 1955.

Goodall, D. W. 1952. Quantitative aspects of plant distribution. *Biol. Rev.* 27:194–245.

Gore, A. 1992. *Earth in the balance: Ecology and the human spirit.* Houghton Mifflin, New York.

Gore, A. 2006. *An inconvenient truth: The planetary emergency of global warming and what we can do about it.* Rodale, Emmaus, PA.

Gorham, E. 2014. Influences on my research over sixty-five years. *ESAB* 95:25–26.

Gould, C. G. 2004. *The remarkable life of William Beebe: Explorer and naturalist.* Island Press/Shearwater Books, Washington, DC.

Gowdy, J. M. & K. E. Limburg. 2001. Comment on guide to diplomatic relations with economists. *ESAB* 82:156–157.

Graham, M. H. & P. K. Dayton. 2002. On the evolution of ecological ideas: Paradigms and scientific progress. *Ecology* 83:1481–1489.

Graham, S. A. 1941. Royal Norton Chapman, 1889–1939. *Entomol. Soc. Amer. Annals* 34:521–524 + portrait.

Green, D. 1973. *Land of the underground rain: Irrigation on the Texas High Plains, 1910–1970.* University of Texas Press, Houston, TX.

Greenland, D. 1986. Standardized meteorological measurements for long-term ecological research sites. *ESAB* 67:275–277.

Greig-Smith, P. 1957. *Quantitative plant ecology.* Butterworth's Science, London. Edition 2, 1964.

Grimm, N. B. et al. 2008. The changing landscape: Ecosystem responses to urbanization and pollution across climatic and societal gradients. *Frontiers* 6:264–272.

Grineski, S. F. 2003. Toward an integrative urban ecology: Voices from the IGERTs. *ESAB* 84:200–205.

Grinevald, J. 1987. On a holistic concept for deep and global ecology: The biosphere. *Fundamenta Scientiae* 8:197–226.

Grinevald, J. 1998. Introduction: The invisibility of the Vernadskian Revolution. Pp. 20–36 in Vernadsky 1998.

Grinnell, J. 1917. The niche-relationships of the California thrasher. *Auk* 34:427–433. Reprint in Real & Brown 1991:118–124.

Grinnell, J. 1935. A revised life-zone map of California. *University of California Pub. in Zool.* 40:327–329 + plt. Reprint in Grinnell 1943:229–232 + plt.

Grinnell, J. 1943. *Joseph Grinnell's philosophy of nature: Selected writings of a western naturalist.* University of California Press, Berkeley, CA. 1968. Books for Libraries Press, New York.

Grinnell, J. & H. S. Swarth. 1913. An account of the birds and mammals of the San Jacinto area of Southern California with remarks upon the behavior of geographic races on the margins of their habitats. *University of California Pub. in Zool.* 10:197–406.

Groffman, P. M. et al. 2004. A new governance structure for the Hubbard Brook Ecosystem Study. *ESAB* 85:5–6.

Grover, H. D. 1984. The world after nuclear war: A review. *ESAB* 65:24–25.

Grubb, P. J. & J. B. Whittaker, eds. 1989. *Toward a more exact ecology.* Blackwell Scientific, Oxford UK.

Gurevitch, J. 1988. Differences in the proportion of women to men invited to give seminars: Is the old boy still kicking? *ESAB* 64:155–161.

Haefner, J. W. 1980. Two metaphors of the niche. *Synthese* 43:123–153.

Hagen, J. B. 1988. Organism and environment: Frederic Clements's vision of a unified physiological ecology. Pp. 257–280 in *The American development of biology.* R. Rainger, K. R. Benson & J. Maienschein, eds. University of Pennsylvania Press, Philadelphia, PA.

Hagen, J. B. 1992. *An entangled bank: The origins of ecosystem ecology.* Rutgers University Press, New Brunswick, NJ.

Hagen, J. B. 1999. Frederic Edward Clements (1874–1945), botanist and ecologist. *ANB* 5:53–54.

Hagen, J. B. 2008. Teaching ecology during the environmental age, 1965–1980. *Environ. Hist.* 13:704–723.

Hairston, N. G., Sr. 1991. The literature glut: Causes and consequences: Reflections of a dinosaur. *ESAB* 72:171–174.

Hairston, N. G., Sr., F. E. Smith & L. B. Slobodkin. 1960. Community structure, population control, and competition. *Amer. Nat.* 94:421–425.

Hairston, N. G., Jr. 2000a. Resolution of respect: W. Thomas Edmondson, limnologist, 1916–2000. *ESAB* 81:180–182.

Hairston, N. G., Jr. 2000b. In memoriam: W. Thomas Edmondson: A distinguished limnologist and an intelligent man of nobility, generosity, kindness and honesty. *Archiv für Hydrobiologie* 148:3–8.

Hale, S. S. 2012. Field guide to scientific conferences: An ecological view. *ESAB* 93:223–228.

Hall, C. A. S., ed. 1995. *Maximum power: The ideas and applications of H. T. Odum.* University Press of Colorado, Niwot, CO.

Hall, C. A. S. & D. L. DeAngelis. 1985. Models in ecology: Paradigms found or paradigms lost? *ESAB* 66:339–346.

Halligan, J. P. 1975. Commentary—Too many graduate students. *Ecology* 56:511.

Hamann, S. 1979. Walter Byron McDougall: Perspective on a remarkable life. *Morton Arboretum Quarterly* 15:25–32.

Hamilton, T. H. 1968. Review: MacArthur & Wilson 1967. *Science* 159:71–72.

Hammer, U. T. 2005. Donald Strathearn Rawson (1905–61). *The Encyclopedia of Saskatchewan.* University of Regina, Regina, Saskatchewan.

Hand, C. 1956. Are corals really herbivores? *Ecology* 37:384–385.

Hannibal, M. E. 2012. The spine of the continent: The race to save America's last, best wilderness. Lyons Press, Guilford, CT.

Hansen, H. P. 1947. Postglacial forest succession, climate, and chronology in the Pacific Northwest, American Philosophical Society. *Transactions* 37:1–130.

Hansen, H. P. 1948. Review: Daubenmire 1947. *Ecology* 29:398.

Hansen, J. E. 2009. *Storms of my grandchildren: The truth about the coming climate catastrophe and our last chance to save humanity.* Bloomsbury, New York.

Hansen, J. E., D. Johnson et al. 1981. Climate impact of increasing atmospheric carbon dioxide. *Science* 213:957–966.

Hanson, C. H. 1976. Commentary—Ethics in the business of science. *Ecology* 57:627–628.

Hanson, H. C. 1926. Field symposium of plant ecologists. *Ecology* 7:517–518.
Harmond, R. 1997a. Charles Christopher Adams, 1873–1955, teacher, museum director, ecologist. *BDAC* 8–9.
Harmond, R. 1997b. Edward Asahel Birge, 1851–1950, limnologist and university administrator. *BDAC* 84–86.
Harmond, R. 1997c. Stanley Adair Cain, 1902–1995, botanist, ecologist. *BDAC* 130–132.
Harmond, R. 1999. Paul Bigelow Sears (1891–1990), botanist, ecologist, and conservationist. *ANB* 19:561–562.
Harshberger, J. W. 1900. An ecological study of the New Jersey strand flora. *Acad. Nat. Sciences of Philadelphia Proc.* 623–671. Reprinted in Egerton 1977b.
Harshberger, J. W. 1902. Additional observations on the strand flora of New Jersey. *Acad. Nat. Sciences of Philadelphia Proc.* 642–669. Reprinted in Egerton 1977b.
Harshberger, J. W. 1903. An ecologic study of the flora of mountainous North Carolina. *Bot. Gaz.* 36:241–258, 368–383. Reprinted in Egerton 1977b.
Harshberger, J. W. 1911. *Phytogeographic survey of North America: A consideration of the phytogeography of the North American continent, including Mexico, Central America, and the West Indies, together with the evolution of North American plant distribution.* Wilhelm Engelmann, Leipzig. G. E. Stechart, New York. 1958. Hafner, New York.
Harshberger, J. W. 1918. American heaths and pine heaths. *BBGM* 1:175–186.
Harshberger, J. W. 1924. The 3rd International Phytogeographical Excursion. *Ecology* 5:287–289.
Harshberger, J. W. 1928. *The life and work of John W. Harshberger, Ph.D.: An autobiography.* T.C. Davis & Sons, Philadelphia, PA.
Haskell, E. F. 1940. Mathematical systematization of "Environment," "Organism" and "Habitat." *Ecology* 21:1–16.
Hasler, A. D. 1963. Wisconsin, 1940–1961. Pp. 55–93 in Frey 1963a.
Hasler, A. D. 1966. *Underwater guideposts: Homing of salmon.* University of Wisconsin Press, Madison, WI.
Hasler, A. D. 1971. The Institute of Ecology: Report from the director. *ESAB* 52:9.
Hasler, A. D. 1972. Foreword. P. v in *Man and the living environment.* University of Wisconsin Press, Madison, WI.
Hasler, A. D. 1974. The Institute of Ecology report of director. *ESAB* 55:7–9.
Hasler, A. D., ed. 1975. *Coupling of land and water systems.* Springer-Verlag, New York.
Hasler, A. D., O. M. Brynildson, & T. W. Helm. 1951. Improving conditions for fish in brownwater bog lakes by alkalization. *J. Wildl. Mgt.* 15:347–352.
Hasler, A. D. & S. E. Jones. 1949. Demonstration of the antagonistic action of large aquatic plants on algae and rotifers. *Ecology* 30:359–364.
Hasler, A. D. & A. T. Scholz. 1983. *Olfactory imprinting and homing in salmon.* Springer-Verlag, New York.
Haydon, D. T. & A. L. Lloyd. 1999. On the origins of the Lotka–Volterra equations. *ESAB* 80:205–206.
Hecht, S. & A. Cockburn. 1990. *The fate of the forest: Developers, destroyers, and defenders of the Amazon.* Harper Perennial, New York.
Hedin, L. O. 2012. F. Herbert Bormann (1922–2012). *Science* 337:534.
Heffernan, J. B. & 18 coauthors. 2014. Macrosystems ecology: Understanding ecological patterns and processes at continental scales. *Frontiers* 12:5–14.
Helm, W. T. 1958. *Some notes on the ecology of panfish in Lake Wingra with special reference to the yellow bass.* Ph.D. dissertation, University of Wisconsin, Madison, WI.
Henderson, N., E. Barrow & M. Johnston. 2011. Uncertainty and management options in the prairie–forest ecotone. *Frontiers* 9:207–208.
Henrici, A. T. 1939. The distribution of bacteria in lakes. Pp. 39–64 in Moulton 1939.

Herdendorf, C. E., S. M. Hartley, & L. J. Charlesworth, compilers. 1974. *Lake Erie bibliography in environmental sciences.* Ohio State University, Columbus, OH.

Herrick, J. E. & J. Sarukhán. 2007. A strategy for ecology in an era of globalization. *Frontiers* 5:172–181.

Hiatt, R. W. 1954. *Directory of hydrobiological laboratories and personnel in North America.* University of Hawaii Press, Honolulu, HI.

Hiatt, R. W. 1963. *World directory of hydrobiological and fisheries institutions.* American Institute of Biological Sciences, Washington, DC.

Hill, R. R. 1920. Charting quadrats with a pantograph. *Ecology* 1:270–273.

Hitzhusen, G. E. & M. E. Tucker. 2013. The potential of religion for Earth Stewardship. *Frontiers* 11:368–376.

Hogner, D. C. 1958. *Conservation in America.* J.B. Lippincott, Philadelphia, PA.

Holden, C. 1975. Ruth Patrick: Hard work brings its own (and Tyler) reward. *Science* 188:997–999.

Holland, M. M., ed. 1992. *Profiles of ecologists: Results of a survey of the membership of the Ecological Society of America.* ESA Public Affairs Office, Washington, DC.

Holland, M. M. 1994. Contributions to science policy from the Ecological Society of America: A 10-year retrospective. *ESAB* 75:103–122.

Hölldobler, B. & E. O. Wilson. 2009. *The superorganism: The beauty, elegance, and strangeness of insect societies.* Norton, New York.

Holling, C. S. & G. Orians. 1971. Toward an urban ecology. *ESAB* 52:2–6.

Hopkinson, C. S. et al. 2008. Forecasting effects of sea-level rise and windstorms on coastal and inland ecosystems. *Frontiers* 6:255–263.

Howard, L. O. 1932. Biographical memoir of Stephen Alfred Forbes, 1844–1930. With bibliography by H. C. Oesterling. *National Academy of Sciences of the United States of America Biographical Memoirs* 15:1–54 + portrait. Reprinted in Forbes 1977.

Howarth, R. et al. 2011. Coupled biogeochemical cycles: Eutrophicaton and hypoxia in temperate estuaries and coastal marine ecosystems. *Frontiers* 9:18–26.

Hubbs, C. L. & K. F. Lagler. 1947. *Fishes of the Great Lakes region.* Edition 2. Cranbrook Institute of Science, Bloomfield Hills, MI. Edition 1, 1941.

Hubbs, C. L., K. F. Lagler & G. R. Smith. 2004. *Fishes of the Great Lakes region.* Revised edition. University of Michigan Press, Ann Arbor, MI.

Humphrey, H. B. 1961. *Makers of North American botany.* Ronald Press, New York.

Hunt, L. M. 1997. John William Haarshberger, 1869–1929, botanist, biographer, university professor, conservationist. *BDAC* 349–351.

Huntington, E. 1920. The control of pneumonia and influenza by the weather. *Ecology* 1:6–23.

Huntley, B. J. et al. 1992. A sustainable biosphere: The Global Imperative, the International Sustainable Biosphere Initiative. *ESAB* 73:7–14.

Hurt, R. D. 1981. *The Dust Bowl: An agricultural and social history.* Nelson-Hall, Chicago, IL.

Hutchinson, G. E. 1940. Review: Clements and Shelford 1939. *Ecology* 21:267–268.

Hutchinson, G. E. 1942. *Nati sunt mures et facta est confusion*: Review of *Voles, mice and lemmings* by Charles Elton. *QRB* 17:354–357.

Hutchinson, G. E. 1943. Biogeochemistry of aluminum and certain related elements. *QRB* 18:1–29, 128–153, 242–262, 331–363.

Hutchinson, G. E. 1947. A note on the theory of competition between two social species. *Ecology* 28:319–321.

Hutchinson, G. E. 1948. Circular causal systems in ecology. *NY Acad. Sci. Ann.* 50:221–246.

Hutchinson, G. E. 1953. The concept of pattern in ecology. *Proc. Acad. Nat. Sci. Philadelphia* 105:1–12.

Hutchinson, G. E. 1957. Concluding remarks. *Cold Spring Harbor Symposia on Quantitative Biology* 22:415–427. Reprinted in Hutchinson 2010:194–206, Real & Brown 1991:225–237.

Hutchinson, G. E. 1957–1993. *A treatise on limnology*. 4 vols. John Wiley & Sons, New York.
Hutchinson, G. E. 1959. Homage to Santa Rosalia, or why are there so many kinds of animals? *Amer. Nat.* 93:145–159. Reprinted in Hutchinson 2010:207–221.
Hutchinson, G. E. 1970, September. The biosphere. *Sci. Amer.* 223(3):45–53.
Hutchinson, G. E. 1978. *An introduction to population ecology*. Yale University Press, New Haven, CT.
Hutchinson, G. E. 1979. *The kindly fruits of the earth: Recollections of an embryo ecologist*. Yale University Press, New Haven, CT.
Hutchinson, G. E. 1982. Reminiscences and notes on some otherwise undiscussed papers. Pp. vii–ix in Riley 1982.
Hutchinson, G. E. 1988. W. Thomas Edmondson. *Limnol. Oceanogr.* 33:1231–1233.
Hutchinson, G. E. 2010. *The art of ecology: Writings*. D. K. Skelly, D. M. Post & M. D. Smith, eds. Yale University Press, New Haven, CT.
Hutchinson, G. E. & F. H. Bormann. 1972. Eminent Ecologist, 1972: Ruth Patrick. *ESAB* 53(3):7.
Ilerbaig, J. 1999. Allied sciences and fundamental problems: C. C. Adams and the search for method in early American ecology. *JHB* 32:439–463.
Inger, R. F. 1970. National Institute of Ecology. *ESAB* 51:2, 15.
Inger, R. F. 1971. Inter-American Institute of Ecology. *ESAB* 52:2–4.
Innis, G. S. 1975. *New directions in the analysis of ecological systems*. Society for Computer Simulation, La Jolla, CA.
Innis, G. S., ed. 1978. *Grassland simulation model*. Springer-Verlag, New York.
Jack, H. A. 1945. Biological field stations of the world. *Chronica Botanica* 9:1–73.
Jamison, A. 1993. National political cultures and the exchange of knowledge: The case of systems ecology. Pp. 187–208 in Crawford et al. 1993.
Janetos, A. C. 1992. The Sustainable Biosphere Initiative and the federal agencies. *ESAB* 73:25–28.
Janetos, A. C. 1996. Six years of public affairs activities by ESA: An evolving agenda. *ESAB* 77:228–230.
Janssen, R. E. 1939. *Leaves and stems from fossil forests: A handbook of the paleobotanical collection in the Illinois State Museum*. Illinois State Museum, Springfield, IL. Reprinted 1965.
Jaques, P. J., R. E. Dunlap & M. Freeman. 2008. The organization of denial: Conservative think tanks and environmental skepticism. *Environ. Politics* 17:349–385.
Jenkins, J. et al. 2007. Acid rain in the Adirondacks: An environmental history. Cornell University Press, Ithaca, NY.
Jessop, N. M. 1970. *Biosphere: A study of life*. Prentice Hall, Englewood Cliffs, NJ.
Johansen, B. E. 2010. What's hot: The explosion of knowledge about global warming. *Choice* 47:1217–1228.
Johnson, D. 1974. *Edwin Broun Fred: Scientist, administrator, gentleman*. University of Wisconsin Press, Madison, WI.
Johnson, E. A. 2005. The crossroads of the Society. *ESAB* 84:3.
Johnson, W. D. 1902. The High Plains and their utilizations. *U.S. Geol. Surv. Annual Report* 22:631–669.
Jones, G. T. 1926. Review: Pearse 1926. *Ecology* 7:503–504.
Jones, S. E. 1947. *An ecological study of large aquatic plants in small ponds*. Ph.D. dissertation, University of Wisconsin, Madison, WI.
Jordan, W. R., III., ed. 1981. *The Arboretum: University of Wisconsin–Madison*. University of Wisconsin Arboretum, Madison, WI.
Jørgensen, D., F. A. Jørgensen & S. B. Pritchard, eds. 2013. *New natures: Joining environmental history with science and technology studies*. University of Pittsburgh Press, Pittsburgh, PA.

Juday, C. 1910. Some European biological stations. *Wis. Acad. Sciences, Arts & Letters Trans.* 16:257–277 + 4 plts.
Just, T., ed. 1939. Plant and animal communities. *Amer. Midland Naturalist* 21:1–255. Reprinted. University of Notre Dame Press, Notre Dame, IN.
Kannowski, P. B. 1971. Review: Odum & Pigeon 1970. *Ecology* 52:1121–1132.
Kaspari, M. E. 2008. Thoughts on the 50th anniversary of MacArthur (1958). *ESAB* 89:448–458.
Keddy, P. 1991. Is mutualism really irrelevant to ecology? *ESAB* 71:101–102.
Keddy, P. 1992. Thoughts on a review of a critique for ecology. *ESAB* 73:234–236.
Keddy, P. 1993. On the distinction between ad hominid and ad hominem and its relevance to ecological research: A reply to Scheiner. *ESAB* 74:383–385.
Keever, C. 1985. *Moving on: A way of life.* Author, Charlotte, NC.
Keller, M., D. S. Schimel et al. 2008. A continental strategy for the National Ecological Observatory Network. *Frontiers* 6:282–284.
Kemp, W. M. & W. R. Boynton. 2004. Productivity, trophic structure, and energy flow in the steady-state ecosystems of Silver Springs, Florida. *Ecological Modelling* 178:43–49.
Kendeigh, S. C. 1932. A study of Merriam's temperature laws. *Wilson Bull.* 44:129–143.
Kendeigh, S. C. 1948. Bird populations and biotic communities in northern Lower Michigan. *Ecology* 29:101–114.
Kendeigh, S. C. 1954. History and evaluation of various concepts of plant and animal communities in North America. *Ecology* 35:152–171.
Kendeigh, S. C. 1961. *Animal ecology.* Prentice Hall, Englewood Cliffs, NJ.
Kerasote, T. 2005. The ethics of hunting. *Frontiers* 3:396–397.
Kessel, E. L., ed. 1955. *A century of progress in the natural sciences, 1853–1953.* California Academy of Sciences, San Francisco, CA.
Kevles, D. J. 1977. The National Science Foundation and the debate over postwar research policy, 1942–1945. *Isis* 68:5–26.
Kiger, J. C. 1962. The American Society of Parasitologists: A short history. *J. Parasitology* 48:641–650.
Kimler, W. C. 1988. Warder Clyde Allee (1885–1955), ecology. *DSB* 17:16–17.
Kimler, W. C. 1990. Victor Ernest Shelford (1877–1968), ecology. *DSB* 18:811–813.
Kinchy, A. J. 2006. On the borders of post-war ecology: Struggles over the Ecological Society of America's preservation committee, 1917–1946. *Science in Culture* 15:23–44.
King, Roger J. H. 2005. The ethics of hunting. *Frontiers* 3:392–397.
Kingsland, S. E. 1985. *Modeling nature: Episodes in the history of population ecology.* University of Chicago Press, Chicago, IL. Edition 2, 1995.
Kingsland, S. E. 1991. The battling botanist: Daniel Tremley MacDougal, mutation theory, and the rise of experimental evolutionary biology in America, 1900–1912. *Isis* 82:479–509.
Kingsland, S. E. 1993. An elusive science: Ecological enterprise in the southwestern United States. Pp. 151–179 in Shortland 1993.
Kingsland, S. E. 2005. *The evolution of American ecology, 1890–2000.* Johns Hopkins University Press, Baltimore, MD.
Kingsolver, J. G. & R. T. Paine. 1991. Theses, antitheses, and syntheses: Conversational biology and ecological debate. Pp. 309–317 in Real & Brown 1991.
Klingle, M. 2007. *Emerald city: An environmental history of Seattle.* Yale University Press, New Haven, CT.
Klugh, A. B. 1923. A common system of classification in plant and animal ecology. *Ecology* 4:366–377.
Knapp, R., ed. 1974. *Handbook of vegetation science*, vol. 8: *Vegetation dynamics.* W. Junk, The Hague, the Netherlands.
Knight, R. L. & S. Riedel, eds. 2002. *Aldo Leopold and ecological conscience.* Oxford University Press, New York.

Kofoid, C. A. 1910. The biological stations of Europe. *U.S. Bureau of Education Bulletin 4.* Washington, DC.

Kofoid, C. A. 1930. Review: Shelford 1929. *Ecology* 11:609–611.

Kohler, R. E. 2011. Paul Errington, Aldo Leopold, and wildlife ecology: Residential science. *Historical Studies in the Natural Sciences* 41:216–259.

Kohn, A. J. 1971. Physiology and biogeography of *Hutchinsonia*: G. E. Hutchinson's influence through his doctoral students. *Limnol. & Oceanogr.* 16:173–176.

Kolasa, J. 1993. Review: Allen & Hoekstra 1992. *Ecology* 74:1605–1606.

Kormondy, E. J. & J. F. McCormick, eds. 1981. *Handbook of contemporary developments in world ecology.* Greenwood Press, Westport, CT.

Kramer, D. L. & M. E. Power. 1997. Robert H. Peters, 1946–1996. *ESAB* 78:113–114.

Kramer, P. J. 1974. Burton Edward Livingston (1875–1948), plant physiologist. *DAB* suppl. 4:500–501.

Kricher, J. 2009. The balance of nature: Ecology's enduring myth. Princeton University Press, Princeton, NJ.

Krikorian, A. D. 1973. Burton Edward Livingston (1875–1948), plant physiology, physiological ecology. *DSB* 8:425.

Kwa, C. 1989. Mimicking nature: The development of systems ecology in the United States, 1950–1975. Ph.D. dissertation, University of Amsterdam, Amsterdam, the Netherlands.

Kwa, C. 1993a. Radiation ecology, systems ecology and the management of the environment. Pages 213–249 in *Science and nature: Essays in the history of the environmental sciences.* M. Shortland, editor. British Society for the History of Science, Stanford in the Vale, UK. (Revision of chapter 2 in Kwa 1989.)

Kwa, C. 1993b. Modeling the grasslands. *Historical Studies in the Physical and Biological Sciences* 24:123–155. (Revision of chapter 3 in Kwa 1989.)

Lagerheim, G. 1902. Metoder för Pollenundersökning. *Botaniska Notiser* 75–78.

Lagler, K. F. 1949. *Studies in freshwater fishery biology.* Edition 3. J.W. Edwards, Ann Arbor, MI.

Lamarck, J. B. 1964. *Hydrogeology.* A. V. Carozzi, transl. University of Illinois Press, Urbana, IL.

Langenheim, J. H. 1988. The path and progress of American women ecologists. *ESAB* 69:184–197.

Langenheim, J. H. 1996. Early history and progress of women ecologists. *Ann. Rev. Ecol. & Systematics* 27:1–53.

Langlois, T. H. 1949. *The Biological Station of the Ohio State University: Franz Theodore Stone Lab.* Ohio State University, Columbus, OH.

Langlois, T. H. 1954. *The Western End of Lake Erie and its ecology.* J. W. Edwards, Ann Arbor, MI.

Lannoo, M. J. 1996. *Okoboji wetlands: A lesson in natural history.* University of Iowa Press, Iowa City, IA.

Lannoo, M. J. 2010. *Leopold's shack and Ricketts's Lab: The emergence of environmentalism.* University of California Press, Berkeley, CA.

Lannoo, M. J. 2012. *The Iowa Lakeside Laboratory: A century of discovering the nature of nature.* University of Iowa Press, Iowa City, IA.

Lauff, G. H. 1963. A history of the American Society of Limnology and Oceanography. Pp. 667–690 in Frey 1963a.

Lawrence, D. M., M. M. Holland & D. J. Morrin. 1993. Profiles of ecologists: Results of the membership of the Ecological Society of America. Part I. A snapshot of survey respondents. Part II. Education and employment patterns. *ESAB* 74:21–35, 153–169.

Lawton, J. 1992. Review: Peters 1991. *Nature* 354:444.

Laycock, G. 1966. *The alien animals.* Natural History Press, Garden City, NY.

Laycock, W. A., compiler. 1994. John Ernst Weaver. Pp. 581–594 in Burgess 1994.

Lear, L. 1997. *Rachel Carson: Witness for nature*. Henry Holt, New York.
Lehman, J. T. 1988. Good Professor Edmondson. *Limnol. Oceanogr.* 33:1234–1244.
Leigh, E. G., Jr. 1971. Review: Odum 1971. *Science* 172:664–666.
Leon, B. 1981. New frontiers (again) in applied ecology. *Frontiers* 62:13–14.
Leopold, A. S. 1950. Vegetation zones of Mexico. *Ecology* 31:507–518.
Leopold, A. S. 1955. The conservation of wildlife. Pp. 795–807 in Kessel 1955.
Leopold, R. A. 1931. *Report on a game survey of the North Central States*. Democrat Press for Sporting Arms and Ammunition Manufacturers Institute, Madison, WI.
Leopold, R. A. 1933. *Game management*. Charles Scribner's Sons, New York.
Leopold, R. A. 1949. *A Sand County almanac and sketches here and there*. Oxford University Press, New York.
Leopold, R. A. 1966. Reprint of Leopold 1949.
Leopold, R. A. & S. E. Jones. 1947. A phonological record for Sauk and Dane Counties, Wisconsin, 1935–1945. *Ecol. Monogr.* 17:81–122.
Lévêque, C. 2003. *Ecology from ecosystem to biosphere*. (Translator unnamed.) Science Publishers, Enfield, NH. French ed., 2001.
Levin, S. A. 1992. The problem of pattern and scale in ecology. *Ecology* 73:1943–1967.
Levin, S. A., ed. 2009. *The Princeton guide to ecology*. Princeton University Press, Princeton, NJ.
Levins, R. & R. Lewontin. 1980. Dialectics and reductionism in ecology. *Synthese* 43:47–78. Reprinted in Saarinen 1982.
Lewin, R. 1983a. Santa Rosalia was a goat. *Science* 221:636–639.
Lewin, R. 1983b. Predators and hurricanes change ecology. *Science* 221:737–740.
Lewis, I. F. & E. C. Cocke. 1929. Pollen analysis of the dismal swamp. *Elisha Mitchell Sci. Soc. J.* 45:37–58 + plts. 3–5.
Li, B. L. & R. Wang. 2007. EcoSummit 2007 and beyond. *Frontiers* 5:287.
Likens, G. E. 1983. Address of the past president: A priority for ecological research. *ESAB* 64:234–243.
Likens, G. E. 2002. Arthur Davis Hasler, 1908–2001. *Nat. Acad. Sciences Biog. Mem.* 82:1–14.
Likens, G. E. 2006. Some reflections on ESA then and now. *ESAB* 87:326–328.
Likens, G. E. 2013a. Resolution of respect: F. Herman Bormann, 1922–2012. *ESAB* 94:304–308.
Likens, G. E. 2013b. The Hubbard Brook Ecosystem Study: Celebrating 50 years. *ESAB* 94:336–337.
Likens, G. E. & F. H. Bormann. 1974. Acid rain: A serious regional environmental problem. *Science* 184:1176–1179.
Likens, G. E. & F. H. Bormann. 1975a. Nutrient-hydrologic interactions (Eastern United States). Pp. 1–5 in Hasler 1975.
Likens, G. E. & F. H. Bormann. 1975b. An experimental approach in New England landscapes. Pp. 7–29 in Hasler 1975.
Likens, G. E., F. H. Bormann, & N. M. Johnson. 1972. Acid rain. *Environment* 14:33–40.
Likens, G. E., F. H. Bormann, N. M. Johnson, D. W. Fisher & R. S. Pierce. 1970. Effects of forest cutting and herbicide treatment on nutrient budgets in the Hubbard Brook Watershed-Ecosystem. *Ecol. Monogr.* 40:23–47.
Likens, G. E., F. H. Bormann, R. S. Pierce, J. S. Eaton & N. M. Johnson. 1977. *Biogeochemistry of a forested ecosystem*. Springer-Verlag, New York.
Likens, G. E. & L. O. Hedin. 2014. A biographical memoir: F. Herbert Bormann, March 24, 1922–June 7, 2012. National Academy of Sciences Biographical Memoirs, 1–13.
Lillie, F. R. 1944. *The Woods Hole Marine Biological Laboratory*. University of Chicago Press, Chicago, IL.

Limburg, K. E. 2004. The biogeochemistry of strontium: A review of H. T. Odum's contributions. *Ecological Modelling* 178:31–33.
Lindeman, R. L. 1941a. The development history of Cedar Creek Bog, Minnesota. *Am. Midland Nat.* 25:101–112.
Lindeman, R. L. 1941b. Seasonal food-cycle dynamics in a senescent lake. *Am. Midland Nat.* 26:636–673.
Lindeman, R. L. 1942a. Experimental simulation of winter anaerobiosis in a senescent lake. *Ecology* 23:1–13.
Lindeman, R. L. 1942b. The trophic dynamic aspect of ecology. *Ecology* 23:399–418. Reprinted in Real & Brown 1991:157–176.
Livingston, B. E. 1924. Review: Bates & Zon 1922. *Ecology* 5:99–101.
Livingston, B. E. 1948. Some conversational autobiographical notes on intellectual experiences and development: An auto-obituary. D. B. Lawrence, Introd. *Ecology* 29:227–241 + portrait.
Livingston, B. E. & F. Shreve. 1921. *The distribution of vegetation in the United States, as related to climatic conditions.* Carnegie Institution, Washington, DC.
Livingstone, D. A. 1983. Eminent Ecologist: Edward Smith Deevey. *ESAB* 64:13–14.
Livingstone, D. A. & M. Brenner. 1989. Resolution of respect: Edward S. Deevey, 1914–1988. *ESA Bulletin* 70:137–140.
Loehle, C. 1987. Why women scientists publish less than men. *ESAB* 68:495–496.
Lomolino, M. V., D. E. Sax & J. H. Brown, eds. 2004. *Foundations of biogeography: Classic papers with commentaries.* University of Chicago Press, Chicago, IL.
Lookingbill, B. D. 2001. *Dust Bowl USA: Depression America and the ecological imagination.* Ohio University Press, Athens, OH.
López–Hoffman, L. et al. 2010. Ecosystem services across borders: A framework for transboundary conservation policy. *Frontiers* 8:84–91.
Lorbiecki, M. 1996. *Aldo Leopold: A fierce green fire.* Oxford University Press, New York.
Lotka, A. 1925. *Elements of physical biology.* Williams & Wilkins, Baltimore, MD. 1956. Retitled: *Elements of mathematical biology.* Dover, New York.
Loucks, O. L. 1975. Models linking land–water interactions around Lake Wingra, Wisconsin. Pp. 53–63 in Hasler 1975.
Loucks, O. L. 1986. The United States' IBP: An ecosystems perspective after fifteen years. Pp. 390–405 in *Ecosystem theory and* application. N. Polunin, ed. John Wiley & Sons, New York.
Loucks, O. L. 2001. The rationale for environmental ordination. *ESAB* 82:187–188.
Loucks, O. L. 2007. Resolution of respect: Robert Lewis Burgess, 1931–2002. *ESA Bull.* 88:123–125.
Louda, S. M. 1982a. Distribution ecology: Variation in plant recruitment in relation to insect seed predation. *Ecol. Monogr.* 52:25–41.
Louda, S. M. 1982b. Limitation of the recruitment of the shrub *Haplopappus squarrosus* (Asteraceae) by flower- and seed-feeding insects. *J. Ecology* 70:43–53.
Louda, S. M. 1983. Seed predation and seedling mortality in the recruitment of a shrub, *Haplopappus venatus* (Asteraceae), along a climate gradient. *Ecology* 64:511–521.
Louda, S. M. 2001. Discovering an effect of insect floral herbivory on plant population density and distribution in a "green world." *ESAB* 82:229–231.
Lovejoy, A. O. 1936. *The great chain of being: A study of the history of an idea.* Harvard University Press, Cambridge, MA.
Lovett, G. M. et al. 2007. Who needs environmental monitoring? *Frontiers* 5:253–260.
Lubchenco, J. 1998. Entering the century of the environment: A new social contract for science. *Science* 279:491–497.
Lubchenco, J. 2012. Reflections on the Sustainable Biosphere Initiative: SBI + 20 Celebration. *ESAB* 93:260–267.

Lubchenco, J., K. Gross et al. 1993. Open letter to ESA members. *ESAB* 74:98–106.
Lubchenco, J. & B. Menge. 1978. Community development and persistence in a Low Rocky Intertidal Zone. *Ecol. Monographs* 48:67–94.
Lubchenco, J. K. & B. A. Menge. 1993. Split positions can provide a "same track"—A personal account. *BioScience* 43:243–248.
Lubchenco, J., A. M. Olson et al. 1991. The Sustainable Biosphere Initiative: An ecological research agenda. *Ecology* 72:371–412.
Lugo, A. E. 1995. Contribution of H. T. Odum to tropical ecology. Pp. 23–24 in Hall 1995.
Lugo, A. E. 2004. H. T. Odum and the Luquillo Experimental Forest. *Ecological Modelling* 178:65–74.
Lugo, A. E. & C. Lowe, eds. 1995. *Tropical Forests: Management and ecology.* Springer-Verlag, New York.
Lugo, A. E., W. L. Silver et al. 2001. Managed ecosystems deserve greater attention. *ESAB* 82:91–93.
Lugo, H. L. 1981. Puerto Rico. Pp. 53–65 in Kormondy & McCormick 1981.
Lussenhop, J. 1974. Victor Hensen and the development of sampling methods in ecology. *JHB* 7:319–337.
Lymn, N. & S. Silver. 2010. A fine dividing line. *Frontiers* 8:3.
Lytle, M. H. 2007. *The gentle subversive: Rachel Carson,* Silent Spring, *and the rise of the environmental movement.* Oxford University Press, New York.
MacArthur, R. H. 1955. Fluctuations of animal populations, and a measure of community stability. *Ecology* 36:533–536.
MacArthur, R. H. 1957. On the relative abundance of bird species. *Nat. Acad. Sci. Proc.* 43:293–295.
MacArthur, R. H. 1958. Population ecology of some warblers of northeastern coniferous forests. *Ecology* 39:599–619. Reprinted in Real & Brown 1991:686–706.
MacArthur, R. H. 1962. Some generalized theorems of natural selection. *Proc. Nat. Acad. Sci. U.S.A.* 48:1893–1897.
MacArthur, R. H. 1972. *Geographical ecology: Patterns in the distribution of species.* Harper & Row, New York.
MacArthur, R. H. & J. H. Connell. 1966. *The biology of populations.* John Wiley & Sons, New York.
MacArthur, R. H. & E. O. Wilson. 1967. *The theory of island biogeography.* Princeton University Press, Princeton, NJ.
Magnuson, J. J., T. K. Kratz & B. J. Benson, eds. 2006. *Long term dynamics of lakes in the landscape.* LTER Network Series. Oxford University Press, New York.
Major, J. 1969. Historical development of the ecosystem concept. Pp. 9–22 in Van Dyne 1969.
Major, J. 1975. Review: Ellenberg 1973. *Ecology* 56:248–249.
Margalef, R. 1968. *Perspectives in ecological theory.* University of Chicago Press, Chicago, IL.
Marin, V. H. 1997. General system theory and the ecosystem concept. *ESAB* 78:102–104.
Marsh, G. P. 1864. *Man and nature: Or, physical geography as modified by human action.* Charles Scribner, New York. Harvard University Press, Cambridge, MA. 1965.
Marshall, J. D. et al. 2008. Predicting and understanding ecosystem responses to climate change at continental scales. *Frontiers* 6:273–280.
Martin, G. J. 1973. *Ellsworth Huntington: His life and thought.* Shoestring Press, Hamden, CT.
Martin, W. E., compiler. 1994. John Thomas Curtis. Pp. 151–159 in Burgess 1994.
Martínez, M. L., R. H. Manson et al. 2006. The evolution of ecology in Mexico: Facing challenges and preparing for the future. *Frontiers* 4:259–267.
Marzolf, G. R. 1976. Review: Cole 1975 & Wetzel 1975. *Ecology* 57:400–401.
Mascaro, J. 2005. State of ignorance: Michael Crichton's call to arms. *Frontiers* 3:131.
Matthews, A. & E. A. Norse. 1986. Notes from Washington. *ESAB* 67:247, 280–281.

May, R. M., ed. 1976. *Theoretical ecology: Principles and applications.* W. B. Saunders, Philadelphia, PA. Edition 2, 1981.

May R. M. 1986. Ideas in ecology. *Amer. Sci.* 14:256–257.

May, R. M. 1989. Levels of organization in ecology. Pp. 339–363 in Cherrett 1989.

May, R. M. & A. R. McLean, eds. 2007. *Theoretical ecology: Principles and applications.* Edition 3. Oxford University Press, Oxford, UK.

May, R. M. & J. Seger. 1986. Ideas in Ecology. *Am. Sci.* 74:256–267.

McCluney, K. E. & 7 co-authors. 2014. Riverine macrosystems ecology: Sensitivity, resistance, and resilience of whole river basins with human alterations. *Frontiers* 12:48–58.

McCormick, J. F. 1978. Letter urging ESA certification of professional ecologists. *ESAB* 59:162–163.

McCormick, J. S., compiler. 1994a. William Skinner Cooper. Pp. 127–138 in Burgess 1994.

McCormick, J. S., compiler. 1994b. Frank Edwin Egler. Pp. 215–238 in Burgess 1994.

McCright, A. M. & R. E. Dunlap. 2010. Anti-reflexivity: The American conservative movement's success in undermining climate science and policy. *Theory, Culture & Society* 27:100–133.

McCullough, D. R. 1970. Secondary production of birds and mammals. Pp. 107–130 in Reichle 1970.

McDonnell, M. J. & S. T. A. Pickett, eds. 1993. *Humans as components of ecosystems: The ecology of subtle human effects and populated areas.* Springer-Verlag, New York.

McDougall, W. B. 1927. *Plant ecology.* Lea & Febiger, Philadelphia, PA.

McGinnies, W. G. 1981. *Discovering the desert: Legacy of the Carnegie Desert Botanical Laboratory.* University of Arizona Press, Tucson, AZ.

McGuire, B. 1975. Henry Allan Gleason, January 2, 1882–April 21, 1975. *Torreya* 102:274–282.

McIntosh, R. P. 1958. Plant communities. *Science* 128:115–120.

McIntosh, R. P. 1974. Plant ecology, 1947–1972. *Missouri Bot. Garden Ann.* 61:132–165.

McIntosh, R. P. 1975. H. A. Gleason—"Individualistic ecologist" 1882–1975: His contributions to ecological theory. *Torrey Bot. Club Bull.* 102:253–273. Reprint in Egerton 1977a.

McIntosh, R. P. 1976. Ecology since 1900. Pp. 353–372 in *Issues and ideas in America.* B. J. Taylor & T. J. White, eds. University of Oklahoma Press, Norman OK. Reprint in Egerton 1977a.

McIntosh, R. P. 1980. The background and some current problems of theoretical ecology. *Synthesis* 43:195–255.

McIntosh, R. P. 1983. Pioneer support for ecology. *BioScience* 33:107–112.

McIntosh, R. P. 1985. *The background of ecology: Concept and theory.* Cambridge University Press, New York.

McIntosh, R. P. 1987. Pluralism in ecology. *Ann. Rev. Ecol. Syst.* 18:321–341.

McIntosh, R. P. 1989. Citation classics of ecology. *QRB* 64:31–49.

McIntosh, R. P. 1993. The continuum continued: John T. Curtis's influence on ecology. Pp. 95–122 in Fralish et al. 1993.

McIntosh, R. P. 1999. The succession of succession: A lexical chronology. *ESAB* 80:256–265.

McKibben, B. 1989. *The end of nature.* Random House, New York.

McKibben, B., ed. 2012. *The global warming reader: A century of writing about climate change.* Penguin Books, New York.

McKibben, B. 2013. *Oil and honey: The education of an unlikely activist.* Time Books/Henry Holt, New York.

McNeely, J. A. et al. 2009. *The wealth of nature: Ecosystem services, biodiversity, and human well-being.* ILCP, Arlington, VA.

Meine, C. 1988. *Aldo Leopold: His life and work.* University of Wisconsin Press, Madison, WI.

Meine, C. 1997. (Rand) Aldo Leopold, 1887–1948, naturalist, forester, wildlife ecologist, educator, writer. *BDAC* 458–461.

Meine, C. 1999. Aldo Leopold (1887–1948), conservationist and author. *ANB* 13:496–497.

Meine, C. 2014. Aldo Leopold: Connecting conservation science, ethics, policy, and practice. In *Linking ecology and ethics in a changing world.* R. Rossi, S. T. A. Pickett, C. Palmer, J. J. Armesto & B. Callicott, eds. Springer Science, New York.

Melillo, J. M., N. B. Grimm & A. Covich. 2005. NEON: Lighting the way forward. *Frontiers* 3:351.

Melillo, J. M., N. B. Grimm, & W. H. Schlesinger. 2005. Ecology and the transition to sustainability. *Frontiers* 3:3.

Menge, B. A. & J. Lubchenco. 2001. On the genesis of "Community development and persistence in a low rocky intertidal zone" by Jane Lubchenco and Bruce A. Menge. *ESAB* 82:124–125.

Merriam, C. H. 1890. Results of a biological survey of the San Francisco Mountain region and desert of the Little Colorado, Arizona. *North American Fauna* 3:1–28 + 3 plts. Reprint in Lomolino, Sax & Brown 2004:202–233, and in Merriam 1974.

Merriam, C. H. 1974. *Selected works of Clinton Hart Merriam.* K. B. Sterling, comp. & introd. Arno Press, New York.

Mertz, D. B. 1992. Resolution of respect: Thomas Park, 1908–1992. *ESAB* 73:248–251.

Meyer, B. S. 1958. Awards of certificates of merit at the Fiftieth Anniversary. Pp. 14–32 in Steere 1958.

Meyer, J. S. 1996. Address of the past president: Beyond gloom and doom: Ecology for the future. *ESAB* 77:199–205.

Meyerson, F. 2009. Population, biodiversity, and human well-being. *Frontiers* 7:511.

Meyerson, F. 2010. Linking ecology to human limits. *Frontiers* 8:67.

Michael, E. L. & W. E. Allen. 1921. Problems of marine ecology. *Ecology* 2:84–88.

Middendorf, G. & C. Nilon. 2007. ESA and environmental justice. *ESAB* 88:160–165.

Miller, R. S. 1975. The Tyler Award: Ruth Patrick. *ESAB* 56:2–3.

Mills, E. L. 1989. *Biological oceanography: An early history, 1879–1960.* Cornell University Press, Ithaca, NY.

Mitchell, R., R. A. Moyer & J. Downower. 1976. An evaluation of three biome programs. *Science* 192:859–865.

Mitman, G. A. 1988. From the population to society: The cooperative metaphors of W. C. Allee and A. E. Emerson. *JHB* 21:213–244.

Mitman, G. A. 1992. *The state of nature: Ecology, community and American social thought.* University of Chicago Press, Chicago, IL.

Mitman, G. A. 1997a. Chancey Juday, 1871–1944, limnologist. *BDAC* 415–417.

Mitman, G. A. 1997b. (Ernest) Victor Shelford, 1877–1968, animal and community ecologist, founder of Ecologist's Union (Nature Conservancy). *BDAC* 728–729.

Mitman, G. A. 1999. Warder Clyde Allee (1885–1955), biologist. *ANB* 1:299.

Mittermeier, C. G., ed. 2009. *The wealth of nature: Ecosystem services, biodiversity, and human well-being.* ILCP, Arlington, VA.

Miyanishi, K. 2014. ESA's third decade: The war years. *ESA Bulletin* 95:103–105.

Mladenoff, D. J. & R. L. Burgess. 1993. The pedagogical legacy of John T. Curtis and Wisconsin plant ecology: 1947–1992. Pp. 145–196 in Fralish et al. 1993.

Monk, C. D., compiler. 1994. Pierre Dansereau. Pp. 161–171 in Burgess 1994.

Moon, D. 2013. The grasslands of North America and Russia. Pp. 247–262 in *A companion to global environmental history.* J. R. McNeill & E. S. Mauldin, eds. Wiley-Blackwell, Malden, MA.

Mooney, H. A. & W. D. Billings. 1961. Comparative physiological ecology of Arctic and Alpine populations of *Oxyria digyna. Ecol. Monogr.* 31:1–29.

Mooney, H. A. & S. A. Levin. 1991. Preface. *Ecology* 72:371–372.

Moore, B. 1920. The scope of ecology. *Ecology* 1:3–5.

Moore, B. 1925. The importance of natural conditions in national parks. Pp. 340–355 in *Hunting and conservation: The book of the Boone and Crockett Club.* G. B. Grinnell & C. Sheldon, eds. Yale University Press, New Haven, CT. 1970. Arno Press, New York.
Moore, B. 1938. The beginnings of *Ecology. Ecology* 19:592.
Moore, B. & N. Taylor. 1927. Vegetation of Mount Desert Island, Maine, and its environment. *BBGM* 3:1–151 + map.
Morisette, J. T. et al. 2009. Tracking the rhythm of the seasons in the face of global change: Phenological research in the 21st century. *Frontiers* 7:253–260.
Morrin, D. J., M. M. Holland & D. M. Lawrence. 1993. Profiles of ecologists. Part III. Environmental science capabilities and funding. *ESAB* 74:237–249.
Morris, C. 2009. Milestones in ecology. Pp. 761–773 in Levin 2009.
Mortimer, C. H. 1956. An explorer of lakes. Pp. 165–211 in Sellery 1956.
Moul, E. T. 1954. Review: Odum 1953. *Ecology* 35:297.
Moulton, F. R., ed. 1939. *Problems of lake biology.* Science P. for AAAS, Lancaster, PA.
Muller, C. H., J. T. Tanner & A. A. Lindsey. 1978a. Eminent Ecologist for 1978: S. Charles Kendeigh. *ESAB* 59:168–169.
Muller, C. H., J. T. Tanner, & A. A. Lindsey. 1978b. Distinguished Service Citation for 1978: Frank Edwin Egler. *ESAB* 59:169–170.
Murdoch, W. 1991. Equilibrium and non-equilibrium paradigms. *ESAB* 72:49–51.
Murphy, P. C. 2005. *What a book can do: The publication and reception of* Silent Spring. University of Massachusetts Press, Amherst, MA.
Myerson, L. A., J. Baron et al. 2005. Aggregate measures of ecosystem services: Can we take the pulse of nature? *Frontiers* 3:56–59.
Naeem, S. 2000. Reply to Wardle et al. *ESAB* 81:241–246.
Naeem, S. 2002. Ecosystem consequences of biodiversity loss: The evolution of a paradigm. *Ecology* 83:1537–1552.
Naeem, S., F. S. Chapin III et al. 1999. Biodiversity and ecosystem functioning: Maintaining natural life support systems. *Issues in Ecology* 4:1–12.
Nash, R. 1982. *Wilderness and the American mind.* Edition 3. Yale University Press, New Haven, CT.
Nash, R. 1990. *American environmentalism: Readings in conservation history.* Edition 3. McGraw-Hill, New York.
Needham, J. G. 1941. Fragments of the history of hydrobiology. Pp. 3–11 in Needham et al. 1941.
Needham, J. G. et al. 1941. *A symposium on hydrobiology.* University of Wisconsin Press, Madison, WI.
Nelson, E. et al. 2009. Modeling multiple ecosystem services, biodiversity, conservation, commodity production, and trade-offs at landscape scales. *Frontiers* 7:4–11.
Nelson, E. W. 1930. Methods of studying shrubby plants in relation to grazing. *Ecology* 11:764–769.
Nelson, E. W. & E. A. Goldman. 1926. Mexico. Pp. 574–596 in Shelford & Shreve 1926.
Nelson, F. 2005. The ethics of hunting. *Frontiers* 3:395–396.
Neuhold, J. M. 1975. The Institute of Ecology—TIE. *ESAB* 56:9–12.
Newbold, H., ed. 2000. *Life stories: World-renowned scientists reflect on their lives and the future of life on earth.* University of California Press, Berkeley, CA.
Newton, J. L. 2006. *Aldo Leopold's odyssey: Rediscovering the author of* A Sand County almanac. Island Press, Washington, DC.
Neyman, J., T. Park & E. L. Scott. 1956. Struggle for existence: The *Tribolium* model: Biological and statistical aspects. *Berkeley Symposium on Mathematical Statistics and Probability Proc.* 3:41–79.
Nichols, G. E. 1914. The International Phytogeographic Excursion in America. *Torreya* 14:55–64.

Nichols, G. E. 1917. The interpretation and application of certain terms and concepts in the ecological classification of plant communities. *Plant World* 20:305–319, 341–353.
Nichols, G. E. 1923. A working basis for the ecological classification of plant communities. *Ecology* 4:11–23, 154–179.
Nichols, G. E. 1926. Reviews: Clements & Goldsmith 1924; Clements & Weaver 1924. *Ecology* 7:109–110.
Nichols, G. E. 1928. Plant ecology. *Ecology* 9:267–270.
Nichols, G. E. 1929. Plant associations and their classification. *International Congress of Plant Sciences Proceedings* I:629–641.
Nichols, G. E. 1930. Obituary notice: John William Harshberger, 1869–1929. *Ecology* 11:443–444.
Nicolson, M. 1990. Henry Allan Gleason and the individualist hypothesis: The structure of a botanist's career. *Bot. Rev.* 56:91–160.
Nicolson, M. 2001. "Towards establishing ecology as a science instead of an art": The work of John T. Curtis on the plant community continuum. *Web Ecology* 2:1–6.
Nicolson, M. & R. P. McIntosh. 2002. H. A. Gleason and the individualistic hypothesis revisited. *ESAB* 83:133–142.
Niering, W. A. 1978. Eminent ecologist for 1977: Walter B. McDougall. *ESAB* 59:10–11.
Niering, W. A. 1998. Postscript on an ecological giant: Frank E. Egler. *ESAB* 79:255–256.
Niering, W. A., P. Marks & P. Weigl. 1980. Mercer Award for 1979: Jane Lubchenco and Bruce A. Menge. *ESAB* 61:36–37.
Norse, E. A. 1983–1986. From the Washington office. 64:205, 223–225, 247–248; 65:22–23, 287–289, 310–311, 445–447, 488–490; 66:16–19, 358, 462–465; 67:36–37, 186–187, 242–246.
Norse, E. A. 1985. Letter on boycott of South African science. *ESAB* 66:452–453.
Norse, E. A. 2000. Marine biodiversity. Pp. 93–108 in Newbold 2000.
Northcote, T. G. & P. A. Larkin. 1963. Western Canada. Pp. 451–485 in Frey 1963a.
Nybakken, J. W., ed. 1971. *Readings in marine ecology*. Harper & Row, New York.
Odum, E. P. 1953. *Fundamentals of ecology*. W.B. Saunders, Philadelphia, PA. Edition 2. With H. T. Odum. 1959. Edition 3. 1971.
Odum, E. P. 1964. The new ecology. *BioScience* 14:14–16.
Odum, E. P. 1987. Resolution of respect: S. Charles Kendeigh, 1904–1986. *ESAB* 68:508–509.
Odum, E. P. 2001. Turning points in the history of the Institute of Ecology. Pp. 15–37 in Bartlett & Bartlett 2001.
Odum, E. P. & J. Kuenzler. 1963. Experimental isolation of food chains in old-field ecosystem with the use of phosphorus-32. Pp. 113–120 in Schultz & Klement 1963.
Odum, H. T. 1951. The stability of the world strontium cycle. *Science* 114:407–411.
Odum, H. T. 1957. Trophic structure and productivity of Silver Springs, Florida. *Ecol. Monogr.* 27:55–112.
Odum, H. T. 1971. *Environment, power, and society*. Wiley-Interscience, New York.
Odum, H. T. 1983. Systems ecology: An introduction. Wiley-Interscience, New York.
Odum, H. T. 1995. Tropical forest systems and the human economy. Pp. 343–393 in Lugo and Lowe 1995.
Odum, H. T. & E. P. Odum. 1955. Trophic structure and productivity of a windward coral reef community in Eniwetok Atoll. *Ecol. Monogr.* 25:291–320. Reprinted in Nybakken 1971:381–410.
Odum, H. T. & E. P. Odum. 1956. Corals as producers, herbivores, carnivores, and possibly decomposers. *Ecology* 37:385.
Odum, H. T. & R. F. Pigeon, eds. 1970. *A tropical rain forest: A study of irradiation and ecology at El Verde, Puerto Rico*. Division of Technical Information, U.S. Atomic Energy Comm., Oak Ridge, TN.

Ogden, L., N. Heynen, U. Oslender, P. West, K. Kassam & P. Robbins. 2013. *Frontiers* 11:341–347.

Olmsted, C. E. 1941. George Elwood Nichols. *Ecology* 22:234–236.

Olson, J. S. 1963. Analog computer models for movement of nuclides through ecosystems. Pp. 121–125 in Schultz & Klement 1963.

Oosting, H. J. 1948. *The study of plant communities: An introduction to plant ecology.* W.H. Freeman, San Francisco, CA.

Oreskes, N. & E. M. Conway. 2010. *Merchants of doubt: How a handful of scientists obscured the truth on issues from tobacco smoke to global warming.* Bloomsbury, NY.

Paine, R. T. 1966. Food web complexity and species diversity. *American Naturalist* 100:65–75.

Paine, R. T. 1969. A note on trophic complexity and community stability. *American Naturalist* 102:91–93.

Paine, R. T. 1992. Jane Lubchenco, president, 1992–1993. *ESAB* 73:163–164.

Paine, R. T. 1995. A conversation on refining the concept of keystone species. *Conservation Biology* 9:962–964.

Paine, R. T. 2002. Advances in ecological understanding by Kuhnian revolution or concept evolution? *Ecology* 83:1553–1559.

Palmer, M. A., E. S. Bernhardt et al. 2005. Ecological science and sustainability for the 21st century. *Frontiers* 3:4–11.

Palmer, R. S. 1956. Resolution of respect: Dr. Charles C. Adams (1873–1955). *ESA Bulletin* 37:103–105.

Park, O. 1942. A study in neotropical Pselaphidae. *Northwestern U. Studies in Biology and Medicine* no. 1, 403 pp.

Park, O. 1945. Observations concerning the future of ecology. *Ecology* 26:1–9.

Park, O. 1953. Review: Dice 1952. *Ecology* 34:440–442.

Park, T. 1939a. Analytical population studies in relation to general ecology. *Amer. Midland Nat.* 21:235–255.

Park, T. 1939b. Ecology looks homeward. *QRB* 14:332–336.

Park, T. 1940. Review: G. Simpson & A. Roe. 1939. *Ecology* 21:91–92.

Park, T. 1946. Some observations on the history and scope of population ecology. *Ecol. Monogr.* 16:313–320.

Park, T. 1948. Experimental studies of interspecies competition: Competition between populations of flour beetles, *Tribolium confusum* Duvall and *Tribolium castaneum* Herbst. *Ecol. Monogr.* 18:267–308. Reprinted in Real & Brown 1991, 754–794.

Park, T. 1949. First four decades of the twentieth century. Pp. 43–72 in Allee, Emerson, Park, Park & Schmidt 1949.

Park, T. 1960. An ecologist's view. *ESAB* 41:4–10.

Pataki, D. E. et al. 2011. Coupling biogeochemical cycles in urban environments: Ecosystem services, green solutions, and misconceptions. *Frontiers* 9:27–36.

Patrick, R. M. 1994–2003. *Rivers of the United States.* 6 vols. in 8. John Wiley, New York.

Patrick R. M. 2000. Water pollution. Pp. 85–92 in Newbold 2000.

Patrick, R. M. & C. W. Reimer. 1966–1975. *The diatoms of the United States exclusive of Alaska and Hawaii.* 2 vols. Academy of Natural Sciences of Philadelphia, Philadelphia, PA.

Patten, B. C., Jr., ed. 1971–1975. *Systems analysis and simulation in ecology.* 4 vols. University of Georgia Press, Athens, GA.

Patten, B., compiler. 1994. Paul Bigelow Sears. Pp. 497–512 in Burgess 1994.

Patten, B. 1995. Dedication and remembrance: Complex person, complex ecologist—Father of systems ecology? Pp. xxi–xxv in Patten & Jørgensen 1995.

Patten, B. & G. T. Auble. 1980. Systems approach to the concept of niche. *Synthese* 43:155–181.

Patten, B. & S. E. Jørgensen, eds. 1995. *Complex ecology: The part-whole relation in ecosystems.* Prentice Hall, Englewood Cliffs, NJ.

Pearse, A. S. 1926. *Animal ecology.* McGraw-Hill, New York.
Pearse, A. S. 1952. *Adventure...trying to be a zoologist.* Author, Durham, NC.
Pearson, P. & F. H. Bormann. 1970, December. Eminent Ecologist, 1970: Murray F. Buell. *ESAB* 51(4):11.
Peet, R. K., R. Del Moral, F. C. James, E. A. Johnson, J. E. Keeley, P. L. Marks & T. Webb III. 1990. William S. Cooper Award: David Tilman. *ESAB* 71:25.
Peña-Claros, M. et al. 2007. Amazonia revealed: A perspective from the Amazon. *Frontiers* 5:237.
Penick, J. L., Jr., C. W. Pursell, Jr., M. B. Sherwood & D. C. Swain, eds. 1965. *The politics of American science, 1939 to the present.* Rand McNally, Chicago, IL.
Peters, D. P. C. 2008. Ecology in a connected world: A vision for a "network of networks." *Frontiers* 6:227.
Peters, D. P. C. et al. 2004. Cross-scale interactions, nonlinearities, and forecasting catastrophic events. *Proc. Nat. Acad. Sci. USA* 101:15130–15135.
Peters, D. P. C. et al. 2007. Cascading events in linked ecological and socioeconomic systems. *Frontiers* 5:221–224.
Peters, D. P. C. et al. 2008. Living in an increasingly connected world: A framework for continental-scale environmental science. *Frontiers* 6:229–237.
Peters, R. H. 1980a. Useful concepts for predictive ecology. *Synthese* 43:257–269.
Peters, R. H. 1980b. From natural history to ecology. *Perspectives in Biol. & Med.* 23:191–203.
Peters, R. H. 1991. *A critique for ecology.* Cambridge University Press, New York.
Peterson, K. M. 1997. William Dwight Billings (1910–1997). *Arctic* 50:275–276.
Pianka, E. R. 1999. *Evolutionary ecology.* Edition 6. Addison-Wesley, Boston, MA.
Pianka, E. R. & H. S. Horn. 2005. Ecology's legacy from Robert MacArthur. Pp. 213–232 in Cuddington & Beisner 2005.
Pickett, S. T. A. 2007. The paper trail: W. S. Cooper's "Fundamentals of vegetation change" and a fluent mode of thought for ecology. *ESAB* 88:98–102.
Pickett, S. T. A. & M. J. McDonnell. 1993. Humans as components of ecosystems: A synthesis. Pp. 310–316 in McDonnell & Pickett 1993.
Pielou, E. C. 1974. *Population and community ecology: Principles and methods.* Gordon & Breach Science, New York.
Pimentel, D. 2010. Population and invasives—A response to Meyerson. *Frontiers* 8:66.
Pimentel, D. & W. Dodds. 1999. Human resource use, population growth, and environmental destruction. *ESAB* 80:88–91.
Pimm, S. L. 1991. *The balance of nature? Ecological issues in the conservation of species and communities.* University of Chicago Press, Chicago, IL.
Plass, G. N. 1956a. The carbon dioxide theory of climate change. *Tellus* 8:140–153.
Plass, G. N. 1956b. The influence of the 15u carbon-dioxide band on the atmospheric infrared cooling rate. *Quart. J. Roy. Meteorological Soc.* 82:310–324.
Plass, G. N. 1959, July. Carbon dioxide and climate. *Sci. Amer.* 201:41–47. Reprinted in Ehrlich et al. 1971:173–179.
Platts, W. J. 1994. On the relation of synecology and natural history to the wonder of life: A reply to Sagoff. *ESAB* 75:121–123.
Polis, G. A. & K. Winemiller, eds. 1996. *Food webs: Integration of patterns and dynamics.* Chapman & Hall, New York.
Pomeroy, L. R., D. C. Coleman & D. A. Crossley, Jr. 2007. Resolution of respect: Frank B. Golley, 1930–2006. *ESAB* 88:15–17.
Pool, R. J. 1954. Frederic Edward Clements. *Ecology* 35:109–112.
Pooley, E. 2010. *The climate war: True believers, power brokers, and the fight to save the earth.* Hyperion, New York.
Pound, R. 1954. Frederic E. Clements as I knew him. *Ecology* 35:112–113.

Pound, R. & F. E. Clements. 1900. *The phytogeography of Nebraska*. Edition 2. Botanical Seminar, Lincoln, NE. 1977. Arno Press, New York.
Power, M. E. & F. S. Chapin III. 2009. Planetary stewardship. *Frontiers* 7:399.
Prantner, E., R. Oleszka, M. Vesley & R. Sweeney, compilers. 1974. *Annotated bibliography of limnological and related studies concerning Lake Erie and influent tributaries*. 5 vols. Great Lakes Laboratory, State University College at Buffalo, NY.
Prasad, A. M. G. 1999. Response to Boris Zeide's article "Biodiversity: A mixed blessing." *ESAB* 80:87.
Press, F. 1991. Letter to Jane Lubchenco. *ESAB* 72:168.
Preston, F. W. 1962. The canonical distribution of commonness and rarity, part I. *Ecology* 43:185–215.
Preston, F. W. 1968. Review: MacArthur & Wilson 1967. *Ecology* 49:592–594.
Price, D. de S. 1963. *Little science, big science*. Columbia University Press, New York.
Price, P. W. 1974. Review: Ricklefs 1973. *Ecology* 55:458–459.
Provine, W. B. 1979. Francis B. Sumner and the evolutionary synthesis. *Stud. Hist. Biol.* 3:211–240.
Putz, F. E. et al. 2007. Forest management versus forest degradation and destruction. *Frontiers* 5:237–238.
Raffa, K. F. 1991. Where next for plant-insect interactions? *ESAB* 72:127–130.
Ramsay, C. R. 1984. Glacier Bay. Pp. 16–33 in *Alaska's magnificent parklands*. D. J. Crump & S. L. Fishbein, eds. National Geographic Society, Washington, DC.
Raup, H. M. 1951. Review: Braun 1950. *Ecology* 33:304–307.
Rawson, D. S. 1939. Some physical and chemical factors in the metabolism of lakes. Pp. 9–26 in Moulton 1939.
Read, D. 1997. The ties that bind. *Nature* 388:517–518.
Real, L. A. & J. H. Brown, eds. 1991. *Foundations of ecology: Classic papers with commentaries*. University of Chicago Press, Chicago, IL.
Real, L. A. & S. A. Levin. 1991. Theoretical advances: The role of theory in the rise of modern ecology. Pp. 177–191 in Real & Brown 1991.
Redfield, A. C. 1945. Review: Lillie 1944. *Ecology* 26:208.
Redfield, A. C. 1947. Review: Coker 1947. *Ecology* 28:472.
Rehbock, P. F. 1983. *The philosophical naturalists: Themes in early nineteenth century British biology*. University of Wisconsin Press, Madison, WI.
Reichle, D. E., ed. 1970. *Analysis of temperate forest ecosystems*. Springer-Verlag, New York.
Revelle, R. R. D. & H. E. Suess. 1957. Carbon dioxide exchange between atmosphere and ocean and the question of an increase of atmospheric CO_2 during the past decades. *Tellus* 9:18–27.
Richardson, D. M., ed. 2011. *Fifty years of invasion biology*. Wiley-Blackwell, New York.
Ricklefs, R. E. 1973. *Ecology*. Chiron Press, Portland, OR.
Ricklefs, R. E. 1976. *The economy of nature: A textbook in basic ecology*. Chiron Press, Portland, OR.
Riley, G. A. 1946. Factors controlling phytoplankton populations on Georges Bank. *J. Mar. Res.* 6:54–73. Reprinted in Riley 1982:333–352.
Riley, G. A. 1982. *Selected works*. J. S. Wroblewski, ed. Dalhousie University, Halifax, Nova Scotia.
Riley, G. A. 1984. Reminiscences of an oceanographer. 161 pp. Unpub. MS. Oceanography Department, Dalhousie University, Nova Scotia. Cited from Mills 1989:357.
Risch, S. & D. Boucher. 1978. What ecologists look for. *ESAB* 57:8–9.
Risser, P. G., E. C. Birney, H. D. Blocker, S. W. May, W. L. Parton & J. A. Wiens. 1981. *The true prairie ecosystem*. Hutchinson Ross, Stroudsburg, PA.
Risser, P. G., R. L. Burgess, J. Ghiselin, W. Laycock & J. F. McCormick. 1979. Voluntary certification of professional ecologists. *ESAB* 60:20–21.

Ritter, W. E. 1921. Ellis L. Michael and his scientific work. *Ecology* 2:70–72.

Robertson, G. P. & S. M. Swinton. 2005. Reconciling agricultural productivity and environmental integrity: A grand challenge for agriculture. *Frontiers* 3:38–46.

Robilliard, G. 1977, December. Carrying capacity of ecologists in the environmental consulting niche. *ESAB* 58(4):3–4.

Rohrman, D. F. 2010. Is there no justice? *Frontiers* 8:387.

Romm, J. & J. E. Hansen. 2013, May 27. Courage to fight climate change. *The Nation* 296(21): 15, 17. (Award and acceptance speeches.)

Ronsay, J. de. 1975. *La macroscope: Vers une vision globale*. Editions du Seuil, Paris.

Rooney, T. P. 1998. The lies that bind: Mutualism and parasitism in forests. *ESAB* 79:218.

Root, R. B. 1979. Robert T. Paine, president 1979–1980. *ESAB* 60:156–157.

Root, R. B. 1987. The challenge of increasing information and specialization. *ESAB* 68:538–543.

Rosa, C. de la. 2008. Ecologists, environmentalists, and the problem of labels. *Frontiers* 6:467–468.

Rosa, E. A., T. Dietz & R. York. 2010. Population and consumption—A response to Meyerson. *Frontiers* 8:65–66.

Rosen, R. 1991. *Life itself.* Columbia University Press, New York.

Rotherham, I. D. & R. A. Lambert, eds. 2011. *Invasive and introduced plants and animals: Human perceptions, attitudes and approaches to management.* Earthscan, Washington, DC.

Roughgarden, J. 2001. Guide to diplomatic relations with economists. *ESAB* 82:85–88.

Rowe, S. J. 1997. Defining the ecosystem. *ESAB* 78:95–97.

Rowe, S. J. & B. V. Barnes. 1994. Geo-ecosystems and bio-ecosystems. *ESAB* 75:40–41.

Rumore, G. 2009. A natural laboratory, a National Monument: Carving out a place for science in Glacier Bay, Alaska, 1879–1959. Dissertation, University of Minnesota, Minneapolis, MN.

Rumore, G. 2012. Preservation for science: The Ecological Society of America and the campaign for Glacier Bay National Monument. *JHB* 45:613–650.

Saarinen, E., ed. 1982. *Conceptual issues in ecology.* Reidel, Dordrecht, the Netherlands.

Sabin, P. 2013. *The bet: Paul Ehrlich, Julian Simon, and our gamble over earth's future.* Yale University Press, New Haven, CT.

Sagers, C. 1992. Letter to the editor. *ESAB* 73:211.

Sagoff, M. 1993. Biodiversity and the culture of ecology. *ESAB* 74:374–381.

Sarukhán, J. 1981. Mexico. Pp. 35–51 in Kormondy & McCormick 1981.

Sayre, N. F., R. Kelty & 5 coauthors. Invitation to Earth Stewardship. *Frontiers* 11:339.

Sayre, N. F., R. R. J. McAllister & 3 coauthors. 2013. Earth Stewardship of rangelands: Coping with ecological, economic, and political marginality. *Frontiers* 11:343–354.

Scanland, T. B. 1978. Commentary: Professional ethics as an individual, rather than an institutional responsibility. *ESAB* 59:164–165.

Schantz, H. L. 1945. Frederic Edward Clements (1874–1945). *Ecology* 26:317–319 + portrait.

Scheifer, I. C. 1996. Integrating human dimension in ecoregion/ecosystem studies—A view from the Ecosystem Management Natural Assessments Effort. *ESAB* 77:177–180.

Scheiner, S. M. 1993. Additional thoughts on *A critique for ecology. ESAB* 74:179–180.

Scheiner, S. M. 1994. Why ecologists should care about philosophy: A reply to Keddy's reply. *ESAB* 75:50–52.

Scheiner, S. M., A. J. Hudson & M. A. VanderMeulen. 1993. An epistemology for ecology. *ESAB* 74:17–21.

Schimel, D., W. Hargrove, F. Hoffman & J. McMahon. 2007. Ecological network. *Frontiers* 5:59.

Schmidt, K. P. 1955. Animal geography. Pp. 767–794 in Kessel 1955.

Schmidt, K. P. 1957. Warder Clyde Allee, 1885–1955. *National Acad. Sciences Biog. Memoirs* 30:2–40.
Schneider, D. W. 2000. Local knowledge, environmental politics, and the founding of ecology in the United States: Stephen Forbes on "The lake as a microcosm" (1887). *Isis* 91:681–705.
Schneider, D. W. et al. 1985. Publication of papers reporting research in the Republic of South Africa. *ESAB* 66:330–332.
Schneirla, T. C. 1950. Review: Frisch 1950. *Ecology* 32:562–565.
Schoener, T. W. 1989. The ecological niche. Pp. 79–113 in Cherrett 1989.
Schubert, S. D., M. J. Suarez, P. J. Pegion, R. D. Koster & J. T. Bacmeister. 2004. On the cause of the 1930s Dust Bowl. *Science* 303:1855–1859.
Schulte-Fischedick, K. & T. Shinn. 1993. International Phytogeographical Excursions, 1911–1923: Intellectual convergence in vegetation science. Pp. 107–131 in Crawford et al. 1993.
Schultz, V. & A. W. Klement, eds. 1963. *Radioecology: Proceedings of the 1st National Symposium on Radioecology...September 10–15, 1961.* Reinhold, NY.
Schwarzbach, M. 1986. *Alfred Wegener: The father of continental drift.* Science Tech Publishers, Madison, WI.
Scott, J. P. 1956a. Allee Memorial Number: Introduction. *Ecology* 37:211–213.
Scott, J. P. 1956b. The analysis of social organization in animals. *Ecology* 37:213–221.
Scudo, F. M. 1971. Vito Volterra and theoretical ecology. *Theoretical Population Ecology* 2:1–23.
Scudo, F. M. 1984. The "Golden Age" of theoretical ecology: A conceptual appraisal. *Revue européanne des Sciences sociales* 22:11–64.
Scudo, F. M. & J. R. Ziegler, eds. 1978. *The golden age of theoretical ecology: 1923–40: A collection of works by Volterra, Kostitzin, Lotka & Kolmogoroff.* Springer-Verlag, Berlin.
Sears, P. B. 1930. A record of post-glacial climate in northern Ohio. *Ohio J. Science* 30:205–217.
Sears, P. B. 1931. Pollen analysis of Mud Lake Bog in Ohio. *Ecology* 12:650–655.
Sears, P. B. 1932. Postglacial climate in Eastern North America. *Ecology* 13:1–11.
Sears, P. B. 1935. *Deserts on the march.* University of Oklahoma Press, Norman, OK.
Sears, P. B. 1949. Review: Oosting 1948. *Ecology* 30:109.
Sears, P. B. 1958. Henry Chandler Cowles (1869–1939), botanist, geographer, one of the founders of the new science of ecology. *DAB* Suppl. 2:127–128.
Sears, P. B. 1960. Resolution of respect: Edgar Nelson Transeau, 1875–1960. *ESAB* 41:62–64.
Sears, P. B. 1969. Plant ecology. Pp. 124–131 in Ewan 1969.
Sears, P. B. 1973. Frederic Edward Clements (1874–1945), botanist and pioneer ecologist. *DAB* Suppl. 3:168–170.
Sears, P. B. & E. Janson. 1933. The rate of peat growth in the Erie Basin. *Ecology* 14:348–355.
Seidl, A. 2009. *Early spring: An ecologist and her children wake to a warming world.* Beacon Press, Boston, MA.
Seidl, A. 2011. *Finding higher ground: Adaptation in the age of warming.* Beacon Press, Boston, MA.
Sellery, G. C. 1956. *E. A. Birge: A memoir.* University of Wisconsin Press, Madison, WI.
Seneca, E. D. 1980. Arthur W. Cooper, president, 1980–1981. *ESAB* 61:172–173.
Shane, L. C. K. 2009. Paul B. Sears' contribution to the development of paleoecology. *Ohio J. Sci.* 109:76–87.
Shantz, H. L. 1951. Forrest Shreve, 1878–1950. *Ecology* 32:364–366.
Shantz, H. L. & R. Zon. 1924. *Atlas of American agriculture: The physical basis of agriculture: Natural vegetation.* Government Printing Office, Washington, DC.
Shapiro, A. 1993. Thoughts on a review of *A critique for ecology. ESAB* 74:177–179.

Shapiro, L. P., B. A. Butler et al. 1997. The Oregon Institute of Marine Biology. *ESAB* 78:149–153.
Sharsmith, C. W. 1945. Review: Cain 1944. *Ecology* 26:101–102.
Shaw, C. F. 1930. Potent factors in soil formation. *Ecology* 11:239–245.
Sheail, J. 1987. *Seventy-five years in ecology: The British Ecological Society.* Blackwell Scientific, Oxford, UK.
Sheifer, I. C. 1996. Integrating the human dimension in ecoregion/ecosystem studies—A view from the Ecosystem Management National Assessments Effort. *ESAB* 77:177–180.
Shelford, V. E. 1913. *Animal communities in temperate America as illustrated in the Chicago region: A study in animal ecology.* Geographic Society of Chicago. Edition 2, 1937 reprinted 1977. Arno Press, New York.
Shelford, V. E. 1917, March. The ideals and aims of the Ecological Society of America. *ESA Bulletin* 1(3):1–3.
Shelford, V. E. ed. 1922. *Preservation of natural conditions.* Schnepp & Barnes (printers), Springfield, IL.
Shelford, V. E. 1929. *Laboratory and field ecology: The responses of animals as indicators of current working methods.* Williams & Wilkins, Baltimore, MD.
Shelford, V. E. 1931. Some concepts of bioecology. *Ecology* 12:455–467.
Shelford, V. E. 1938. The organization of the Ecological Society of America. *Ecology* 19:164–166.
Shelford, V. E. 1943. Twenty-five year effort at saving nature for scientific purposes. *Science* 98:280–281.
Shelford, V. E. 1963. *The ecology of North America.* University of Illinois Press, Chicago, IL.
Shelford, V. E. & F. Shreve, eds. 1926. *Naturalist's guide to the Americas.* Williams & Wilkins, Baltimore, MD.
Shelford, V. E. & E. D. Towler. 1925. Animal communities on San Juan Channel and adjacent areas. *Puget Sound Biological Station Publications* 5:33–73.
Shepard, P. & D. McKinley, eds. 1969. *The subversive science: Essays toward an ecology of man.* Houghton Mifflin, Boston, MA.
Shor, E. N. 1999. Charles Edwin Bessey (1845–1915), botanist. *ANB* 2:696–698.
Shortland, M., ed. 1993. *Science and nature: Essays in the history of the environmental sciences.* British Society for the History of Science, Oxford.
Shreve, F. 1916. The Ecological Society of America. *Science* 44:759–762.
Shreve, F. 1929. Report of Committee on Ecological Monographs. *Ecology* 10:261–262.
Shugart, H. H. 1984. *A theory of forest dynamics: The ecological implications of forest succession models.* Springer-Verlag, New York.
Shugart, H. H. & I. R. Noble. 1981. A computer model of succession and fire response of the high-altitude *Eucalyptus* forest of the Brindabella Range. *Australian J. Ecol.* 6:149–164.
Sih, A. 1991. Reflections on the power of a grand paradigm. *ESAB* 72:174–178.
Sih, A., P. Crowley et al. 1985. Predation, competition, and prey communities: A review of field experiments. *Ann. Rev. Ecology & Systematics* 16:269–311.
Sih, A. & K. Nishikawa. 1988. Do men and women really differ in publication rates and contentiousness? An empirical survey. *ESAB* 69:15–18.
Silver, S. 2003. Editorial focus: Exploring the new *Frontiers. Frontiers* 1:3.
Simard, S. W., D. A. Perry et al. 1997. Net transfer of carbon between ectomycorrhizal tree species in the field. *Nature* 388:579–582.
Simberloff, D. 1980a. A succession of paradigms in ecology: Essentialism to materialism and probabilism. *Synthese* 43:3–39. Reprinted in Saarinen 1982.
Simberloff, D. 1980b. Reply. *Synthese* 43:79–93. Reprinted in Saarinen 1982.
Simberloff, D. 1981. Community effects of introduced species. Pp. 53–81 in *Biotic species in ecological and evolutionary times.* M. Nitecki, ed. Academic Press, New York.

Simberloff, D. 2011a. Charles Elton: Neither founder nor siren, but prophet. Pp. 11–24 in Richardson 2011.

Simberloff, D. 2011b. The rise of modern invasion biology and American attitudes towards introduced species. Pp. 121–135 in Rotherham & Lambert 2011.

Simberloff, D. 2012a. Integrity, stability, and beauty: Aldo Leopold's evolving view of non-native species. *Environ. Hist.* 17:487–511.

Simberloff, D. 2012b. Charles Elton: Pioneer conservation biologist. *Environ. & Hist.* 18:183–202.

Simberloff, D. & E. O. Wilson. 1969. Experimental zoogeography of islands: The colonization of empty islands. *Ecology* 50:278–296.

Simberloff, D. et al. 2011. Non-natives: 141 scientists object. *Nature* 475:36.

Simonian, L. 1995. *Defending the land of the jaguar: A history of conservation in Mexico.* University of Texas Press, Austin, TX.

Skinner, B. J. 1986. Can you really believe the evidence? Two stories from geology. *Amer. Sci.* 74:400–409.

Slack, N. G. 2010. *G. Evelyn Hutchinson and the invention of modern ecology.* Yale University Press, New Haven, CT.

Slobodkin, L. B. 1961. *Growth and regulation of animal populations.* Holt, Reinhart & Winston, New York.

Slobodkin, L. B. 2009. My complete works and more. *Evolutionary Ecology Research* 11:327–354.

Slobodkin, L. B. & N. Slack. 1999. George Evelyn Hutchinson: 20th-century ecologist. *Endeavour* 23:24–30.

Small, J. A. 1975. Murray Fife Buell, October 6, 1905–July 2, 1975. *Bull. Torrey Bot. Club* 102:201–206.

Smith, C. C. 1940. The effect of overgrazing and erosion upon the biota of the mixed-grass prairie of Oklahoma. *Ecology* 21:381–397.

Smith, F. E. 1970. Analysis of ecosystems. Pp. 7–18 in Reichle 1970.

Smith, F. E. 1977. TIE. *ESAB* 59:8.

Smith, V. H. 2005. Reply to Geddes' letter. *Frontiers* 3:469–470.

Smith, V. H., M. Marsh & B. Kinzie. 2005. Opinion versus scientific consensus. *Frontiers* 3:302. (Three letters under one title.)

Smocovitis, V. B. 1999. John Merle Coulton (1851–1928), botanist. *ANB* 5:585–586.

Snelders, H. A. M. 1970. Svante August Arrhenius (1859–1927), chemistry, physics. *DSB* 1:296–302.

Snow, C. P. 1959. *The two cultures and the scientific revolution.* Cambridge University Press, Cambridge, UK.

Soranno, P. A. & D. S. Schimel. 2014. Macrosystems ecology: Big data, big ecology. *Frontiers* 12:3.

Souder, W. 2012. *On a father shore: The life and legacy of Rachel Carson.* Crown, New York.

Sprugel, D. G. 1980. A "pedagogical genealogy" of American plant ecologists. *ESAB* 61:197–200.

Srivastava, D. S. & J. L. Ruesink. 1998. Population growth and environmental destruction: It isn't that simple! *ESAB* 79:219–221.

Stamp, N. E. 1992. Theory of plant-insect herbivore interactions on the inevitable brink of re-synthesis. *ESAB* 73:28–34.

Stamp, N. E. 1996. Developing a theory of plant-insect herbivore interactions: Are we there yet? *ESAB* 77:51–61.

Stamp, N. E. 2004. Misconceptions about plant-herbivore interactions, especially plant defenses. *ESAB* 85:201–205.

Stamp, N. E. 2005. The problem with the messages of plant-herbivore interactions in ecology textbooks. *ESAB* 86:27–31.

Stearns, F. 1961. Resolution of respect: John T. Curtis, 1913–1961. *ESAB* 42:167–170.
Stearns, F. 1985. Distinguished Service Citation: Arthur Wells Cooper. *ESAB* 66:21–22.
Steenstrup, J. J. S. 1842. Geonostic-geologisk unterögelse af Skovmoserne Vidnesdam og Lillemose I det nordlige Sjaelland, ledsaget af sammenlignde Bemaekninger. *Kongelige Danske Videnskaberner Selskabs Skriften* 9:17–120.
Steere, W. C., ed. 1958. *Fifty years of botany: Golden jubilee volume of the Botanical Society of America.* McGraw–Hill, New York.
Stegner, W., ed. 1955. *This is Dinosaur: Echo Park country and its magic rivers.* Knopf, New York.
Steiger, T. L. 1930. Structure of prairie vegetation. *Ecology* 11:170–217.
Steiner, F., M. Simmons, M. Gallagher, J. Ranganthan & C. Robertson. 2013. The ecological imperative for environmental design and planning. *Frontiers* 11:355–361.
Stejneger, L. 1936. *Georg Wilhelm Steller: The pioneer of Alaskan Natural History.* Harvard University Press, Cambridge, MA.
Sterling, K. B. 1977. *Last of the naturalists: The career of C. Hart Merriam.* Edition 2. Arno Press, New York.
Sterling, K. B. 1997a. Warder Clyde Allee, 1885–1955, ecologist. *BDAC* 17–18.
Sterling, K. B. 1997b. (Charles) William Beebe, 1877–1962, naturalist, oceanographer, zoological society executive. Pp. 71–73 in *BDAC* 1997.
Sterling, K. B. 1999. William Beebe (1877–1962), oceanographer, and zoological society executive. *ANB* 2:461–462.
Sterling, K. B., R. P. Harmond, G. A. Cevasco & L. F. Hammond, eds. 1997. *Biographical Dictionary of American and Canadian Naturalists and Environmentalists.* Greenwood Press, Westport, CT.
Stern, P. C. 1993. A second environmental science: Human-environment interactions. *Science* 260:1897–1899.
Stiling, P. 1994. What do ecologists do? *ESAB* 75:116–121.
Stoddard, H. L. 1931. *The bobwhite quail: Its habits, preservation and increase.* Scribners, New York.
Stoddard, H. L. 1969. *Memoirs of a naturalist.* University of Oklahoma Press, Norman, OK.
Stolzenburg, W. 2008. *Where the wild things were: Life, death, and ecological wreckage in a land of vanishing predators.* Bloomsbury, New York.
Storer, J. H. 1953. *The web of life.* Devin-Adair, New York.
Strain, B. R. 1997. Resolution of respect: William Dwight Billings, 1910–1997. *ESAB* 78:115–117.
Strong, D. R., Jr. 1980. Null hypotheses in ecology. *Synthese* 43:271–285.
Strong, D. R., Jr. 2008a. Ecologists and environmentalism. *Frontiers* 6:347.
Strong, D. R., Jr. 2008b. The author responds. *Frontiers* 6:468.
Strong, D. R., Jr. 2009. Labels and values: A reply to Burke and Lauenroth. *Frontiers* 7:240.
Strong, D. R., Jr. & A. Hastings. 2001. Resolution of respect: Gary Allen Polis, 28 August 1946–29 March 2000. *ESAB* 82:110–111.
Strong, D. R., Jr., D. Simberloff, L. G. Abele & A. B. Thistle, eds. 1984. *Ecological communities: Conceptual issues and the evidence.* Princeton University Press, Princeton, NJ.
Strain, B. R. 1997. William Dwight Billings, 1910–1997. *ESAB* 78:115–117.
Stuckey, R. L. 1973. E. Lucy Braun (1889–1971), outstanding botanist and conservationist: A biographical sketch, with bibliography. *Michigan Botanist* 12:83–106. Reprinted in Stuckey 2001:4–28.
Stuckey, R. L. 1981. Contributions to the development of the prairie peninsula concept: Emma Lucy Braun. *Ohio Biological Survey Biological Notes* 15:24–36.
Stuckey, R. L., ed. 1994. *E. Lucy Braun, Ohio's foremost woman botanist: A collection of biographical accounts, maps and photos.* Columbus, OH.

Stuckey, R. L. 1997a. (Emma) Lucy Braun, 1889–1971, botanist, ecologist, conservationist, educator. *BDAC* 103–106.

Stuckey, R. L. 1997b. Edgar Nelson Transeau, 1875–1960, botanist, ecologist, educator. *BDAC* 783–785.

Stuckey, R. L., ed. & compiler. 2001. *E. Lucy Braun (1889–1971), Ohio's foremost woman botanist: An anthology of papers*. Ohio State University, Columbus, OH.

Stuckey, R. L. & K. B. Sterling. 1997. Paul Bigelow Sears, 1891–1990, botanist, ecologist, conservationist. *BDAC* 717–719.

Sumner, F. B. 1945. *The life history of an American naturalist.* Jacques Cattell Press, Lancaster PA.

Sumner, F. B., G. B. Louderback & W. L. Schmitt. 1914. A report upon the physical conditions in San Francisco Bay, based upon the operations of the United States steamer *Albatross. U. CA Publications in Zool.* 14:1–198 + 13 pls.

Sumner, F. B., R. G. Osburn, L. J. Cole & B. M. Davis. 1913 (for 1911). *A biological survey of the waters of Woods Hole and vicinity.* 2 vols. *U.S. Bureau of Fisheries Bull.* 31.

Suter, G. W. 1981. Ecosystem theory and NEPA assessment. *ESAB* 62:186–192.

Svenson, H. K. 1944. C. Stuart Gager (1873–1943). *Ecology* 25: 1–2 + photo.

Swank, W. T., J. L. Meyer & D. A. Crossley, Jr. 2001. Long-term ecological research: Coweeta history and perspectives. Pp. 143–163 in Barrett & Barrett 2001.

Swihart, R. K., J. B. Dunning Jr. & P. W. Wieser. 2002. Gray matter in ecology: Dynamics of pattern, process, and scientific progress. *ESAB* 83:149–155.

Sylvester, K. M. & E. S. A. Rupley. 2012. Revising the Dust Bowl: High above the Kansas grasslands. *Environ. Hist.* 17:603–633.

Tallis, H. R. et al. 2009. Integrating conservation and development in the field: Implementing ecosystem service projects. *Frontiers* 7:12–20.

Tamarin, R. H. 1994. A critique for ecology continued. *ESAB* 75:101.

Tanner, J. T. 1950. Edwin Booth Powers, 1880–1949. *Ecology* 31:478–479.

Tansley, A. G. 1913–1914. International Phytogeographic Excursion (I. P. E.) in America. *New Phytologist* 12:322–336, 13:30–41, 82–92, 263–275, 325–333.

Tansley, A. G. 1935. The use and abuse of vegetational concepts and terms. *Ecology* 16:284–307. Reprinted in Real & Brown 1991:318–341.

Tansley, A. G. 1945. Frederic Edward Clements, 1874–1945. *J. Ecology* 34:194–196.

Taub, D. R. 1998. Ecology and the geotome: Calling a spade a spade. *ESAB* 76:54.

Taylor, N. 1911. Review: Harshberger 1911. *Torreya* 11:190–199.

Taylor, N. 1923. The vegetation of Montauk: A study of grassland and forest. *BBGM* 2:3–107.

Taylor, N. 1938. Beginnings of ecology. *Ecology* 19:352.

Taylor P. J. 1998. Technological optimism: H. T. Odum and the partial transformation of ecological metaphor after World War II. *JHB* 21:213–244.

Taylor, W. P. 1927. Ecology or bio-ecology. *Ecology* 8:280–281.

Taylor, W. P. & J. V. G. Leftfield. 1922. Damage to range grasses by the Zuni prairie dog. *USDA Bull.* 1227:1–16.

Thienemann, A. F. 1926. Der Nahrungskreislauf im Wasser. *Verhandlungen der Deutschen Zoologischen Gesellschaft* 31:29–79.

Thoreau, H. D. 1988. *The natural history essays.* R. Sattelmeyer, introd. & notes. Gibbs Smith, Salt Lake City, UT.

Tilman, G. D. 1988. *Plant strategies and the dynamics and structure of plant communities.* Princeton University Press, Princeton, NJ.

Tilman, G. D. 2000. What *Issues in Ecology* is and isn't. *ESAB* 81:240.

Tilman, G. D. et al. 2004. *NEON: Addressing the nation's environmental challenges.* National Academy Press, Washington, DC.

Tinbergen, N. 1953a. *Social behaviour in animals, with special reference to vertebrates.* Menthuen, London.

Tinbergen, N. 1953b. *The herring gull's world.* Collins, London.

Tino, C. 1997. Marston Bates, 1906–1974, zoologist, cultural biologist, university professor, author. *BDAC* 68–69.

Tobey, R. C. 1977. American grassland ecology, 1895–1955; The life of a professional research community. Pp. 1–56 in Egerton 1977a.

Tobey, R. C. 1981. *Saving the prairies: The life cycle of the founding school of American ecology, 1895–1955.* University of California Press, Berkeley, CA.

Twombly, S. 1999. Yvette H. Edmondson receives ASLO's Distinguished Service Award. *ASLO Bulletin* 8:20–21.

Ulanowicz, R. E. 1997. *Ecology: The ascendant perspective.* Columbia University Press, New York.

Van Dobben, W. H. & R. H. Lowe-McConnell, eds. 1975. *Unifying concepts in ecology.* W. Junk, The Hague.

Van Dyne, G. M., ed. 1969. *The ecosystem concept in natural resource management.* Academic Press, New York.

Van Dyne, G. M. 1978. Foreword: Perspectives on the ELM Model and modeling efforts. Pp. v–xxii in Innis 1978.

Van Name, W. G. 1921. Threatened extinction of the bald eagle. *Ecology* 2:76–78.

Verduin, J. 1956. Review: Langlois 1954. *Ecology* 37:400.

Vernadsky, V. I. 1998. *The biosphere.* D. B. Langmuir, transl. M. A. S. McMenamin, ed. Copernicus/Springer-Verlag, New York.

Vestal, A. G. 1931. Review: Weaver & Clements 1929. *Ecology* 12:232–239.

Vetter, J. 2011. Rocky Mountain high science: Teaching, research, and nature at field stations. Pp. 108–134 in *Knowing global environments: New historical perspectives on the field sciences.* J. Vetter, ed. Rutgers University Press, New Brunswick, NJ.

Vetter, J. 2012. Labs in the field: Rocky Mountain biological stations in the early twentieth century. *JHB* 45:587–611.

Visher, S. S. 1948. Memoir to Ellsworth Huntington, 1876–1947. *Assoc. Amer. Geogr. Ann.* 38:38–50.

Volterra, V. 1926. Fluctuations in the abundance of a species considered mathematically. *Nature* 118:558–560. Reprint. Pp. 283–285 in Real & Brown 1991.

Volterra, V. 1931. Variations and fluctuations of the number of individuals in animal species living together. M. E. Wells, transl. Pp. 409–448 in Chapman 1931.

Waksman, S. A. 1927. *Principles of soil microbiology.* Williams & Wilkins, Baltimore, MD.

Waksman, S. A. 1954. *My life with the microbes.* Simon & Schuster, New York.

Waller, A. E. 1921. Review: Weaver 1919, 1920. *Ecology* 2:145–147.

Wang, D. 1979. Letter on professional ethics. *ESAB* 60:2–3.

Ward, H. B. 1934. Obituary: Robert Henry Wolcott. *Science* 79:422–423.

Wardle, D. L., M. A. Huston et al. 2000. Biodiversity and ecosystem function: An issue in ecology. *ESAB* 81:235–239.

Waring, R. H. 1981. Review: Bormann & Likens 1970. *Ecology* 62:871–872.

Warming, J. E. B. 1895. *Plantesamfund: Grundträk af den ökologiske Plantegaeografi.* P.G. Philipsen, Copenhagen, Denmark.

Waterman, W. G. 1926. Report of Committees on the preservation of natural conditions. *Ecology* 7:235–237.

Watt, K. E. F. 1966. *Systems analysis in ecology.* Academic Press, New York.

Way, A. G. 2011. *Conserving southern longleaf: Herbert Stoddard and the rise of ecological land management.* University of Georgia Press, Athens, GA.

Weart, S. R. 2008. *The discovery of global warming.* Edition 2. Harvard University Press, Cambridge, MA.

Weaver, J. E. 1919. *The ecological relations of roots.* Carnegie Institution, Washington, DC.

Weaver, J. E. 1920. *Root development in the grassland formation.* Carnegie Institution, Washington, DC.
Weaver, J. E. 1930. Underground plant development in its relation to grazing. *Ecology* 11:543–557.
Weaver, J. E. 1954. *The North American prairie.* Johnson, Lincoln, NE.
Weaver, J. E. & F. E. Clements. 1929. *Plant ecology.* McGraw-Hill, New York.
Webb, G. E. 1983. *Tree rings and telescopes: The scientific career of A. E. Douglas.* University of Arizona Press, Tucson, AZ.
Webb, G. E. 1999. Andrew Ellicott Douglass (1867–1962), astronomer and dendrochronologist. *ANB* 6:814–815.
Wegener, A. L. 1924. *The origin of continents and oceans.* J. G. A. Skerl, transl., Methuen, London.
Weinberg, A. 1961. Impact of large-scale science on the United States. *Science* 131:161–164.
Welker, R. H. 1975. *Natural man: The life of William Beebe.* Indiana University Press, Bloomington, IN.
Welsh, B. L. 1972, December. ESA does not need a code of ethics and a certification program. *ESAB* 53(4):4–5.
Werner, E. 1991. Changing perspectives on some long-standing problems in ecology. *ESAB* 72:43–47.
West, D. C., H. H. Shugart & D. B. Botkin, eds. 1981. *Forest succession.* Springer-Verlag, New York.
Westbroek, P. 1991. *Life as a geological force: Dynamics of the earth.* Norton, New York.
Westman, W. E. 1982. Eminent Ecologist: Robert Harding Whittaker. *ESAB* 63:11–12.
Westman, W. E. & R. K. Peet. 1982. Robert H. Whittaker (1920–1980): The man and his work. *Vegetatio* 48:97–122.
Wetzel, R. G. 1975. *Limnology.* W. B. Saunders, Philadelphia, PA.
Wherry, E. T. 1920a. Plant distribution around salt marshes in relation to soil acidity. *Ecology* 1:42–48.
Wherry, E. T. 1920b. Soil acidity and a field method for its measurement. *Ecology* 1:160–173.
Whittaker, R. H. 1958. Recent evolution of ecological concepts in relation to the eastern forests of North America. Pp. 340–358 in Steere 1958.
Whittaker, R. H. 1962. Classification of natural communities. *Bot. Rev.* 28:1–239. 1977. Arno Press, New York.
Whittaker, R. H. 1967. Gradient analysis of vegetation. *Biol. Reviews* 42:207–264.
Whittaker, R. H. 1973. *Handbook of vegetation science*: vol. 5: *Ordination and classification of communities.* W. Junk, The Hague, the Netherlands.
Whittaker, R. H. 1994. Autobiographical letter, 20 Dec. 1957. Pp. 617–621 in Burgess 1994.
Wiegert, R. G., J. J. Gilbert, M. A. R. Koehl, P. J. Morin, R. B. Root & S. J. Turner. 1997. Robert H. MacArthur Award: David Tilman. *ESAB* 78:23–24.
Wilbur, H. M. 1998. 1996 Sewall Wright Award: Robert T. Paine. *Amer. Nat.* 15(1):i–ii.
Wilbur, H. M. & D. E. Gill. 2008. Resolution of respect: Nelson George Hairston, Sr., 1917–2008. *ESAB* 89:266–277.
Williamson, C. S. et al. 2008. Lakes and streams as sentinels of environmental change in terrestrial and atmospheric processes. *Frontiers* 6:247–254.
Wilson, E. O. 1971. *The insect societies.* Harvard University Press, Cambridge, MA.
Wilson, E. O. 1973. Eminent Ecologist, 1973, Robert Helmer MacArthur. *ESAB* 54, no. 3:11–12.
Wilson, E. O. 1975. *Sociobiology: The new synthesis.* Harvard University Press, Cambridge, MA.
Wilson, E. O. 1976. The central problems of sociobiology. Pp. 205–217 in May 1976.
Wilson, E. O. 1992. *The diversity of life.* Harvard University Press, Cambridge, MA.
Wilson, E. O. 1994. *Naturalist.* Island Press/Shearwater Books, Washington, DC.

Wilson, E. O. & G. E. Hutchinson. 1989. Robert Helmer MacArthur, April 7, 1930–November 1, 1972. *Nat. Acad. Sci. Biog. Mem.* 58:318–327.
Woodbury, A. M. 1954. *Principles of general ecology*. Blakiston, NY.
Woods, L. P. & R. R. Miller. 1948. Review: Hubbs & Lagler 1947. *Ecology* 29:222.
Worster, D. 1979. *Dust Bowl: The Southern Plains in the 1930s.* Oxford University Press, New York.
Worster, D. 2008. *A passion for nature: The life of John Muir.* Oxford University Press, New York.
Wright, R. D. 1971. Review: Reichle 1970. *Ecology* 52:725.
Wright, S. 1960. Thomas Park, president-elect. *Science* 131:502–503.
Wu, J. & O. L. Loucks. 1995. From the balance of nature to hierarchical patch dynamics: A paradigm shift in ecology. *QRB* 70:439–466.
Yaich, S. C. 2005. The ethics of hunting. *Frontiers* 3:397.
Yong, E. 2013. Dynasty: Bob Paine fathered an idea—and an academic family—that changed ecology. *Nature* 493:286–289.
Zamon, J. 1992. The literature glut: Some general advice to ecologists (a response to Hairston). *ESAB* 73:45–46.
Zedler, J. 2014. Patterns in nature and their causes. *ESAB* 95:27–28.
Zeide, B. 1998a. Affirmative action or parasitism? *ESAB* 79:130–131.
Zeide, B. 1998b. Biodiversity: A mixed blessing. *ESAB* 79:215–216.
Zeide, B. 1998c, January. Another look at Leopold's land ethic. *J. Forestry* 96(1):13–19.
Zeide, B. 2004. Nature's surprises. *ESAB* 85:120–124.

Index

F

G

J

K

L

M

O

P

For Product Safety Concerns and Information please contact our EU representative GPSR@taylorandfrancis.com Taylor & Francis Verlag GmbH, Kaufingerstraße 24, 80331 München, Germany

Batch number: 10726532

Printed by Printforce, the Netherlands